Biofilm Infections

Thomas Bjarnsholt · Claus Moser ·
Peter Østrup Jensen · Niels Høiby
Editors

Biofilm Infections

Springer

Editors
Thomas Bjarnsholt
Department of International Health
Immunology and Microbiology
University of Copenhagen
Faculty of Health Sciences
Blegdamsvej 3B
DK-2200 Copenhagen N, Denmark

H:S Rigshospitalet
Department for Clinical Microbiology
Afsnit 9301
Juliane Maries Vej 22
DK-2100 Copenhagen Ø, Denmark
tbjarnsholt@sund.ku.dk

Claus Moser
H:S Rigshospitalet
Department of Clinical Microbiology
Afsnit 9301
Juliane Maries Vej 22
DK-2100 Copenhagen Ø, Denmark
moser@dadlnet.dk

Peter Østrup Jensen
H:S Rigshospitalet
Department for Clinical Microbiology
Afsnit 7602
Juliane Maries Vej 22
DK-2100 Copenhagen Ø, Denmark
peter.oestrup.jensen@rh.regionh.dk

Niels Høiby
Department of International Health
Immunology and Microbiology
University of Copenhagen
Faculty of Health Sciences
Blegdamsvej 3B
DK-2200 Copenhagen N, Denmark

H:S Rigshospitalet
Department for Clinical Microbiology
Afsnit 9301, Juliane Maries Vej 22
DK-2100 Copenhagen Ø, Denmark
hoiby@hoibyniels.dk

ISBN 978-1-4419-6083-2 e-ISBN 978-1-4419-6084-9
DOI 10.1007/978-1-4419-6084-9
Springer New York Dordrecht Heidelberg London

Library of Congress Control Number: 2010937636

Front cover pictures: Background and first picture to the left show in vitro interaction between *Pseudomonas aeruginosa* and human leukocytes depicted by scanning electron microscopy. Middle picture show a *Staphylococcus aureus* biofilm in a chronic wound and picture to the right show a *P. aeruginosa* biofilm in a cystic fibrosis lung. The bacteria in both the middle and right picture are visualized by PNA FISH (green and red) and host cells by DAPI (blue).

Printed on acid-free paper

Springer is part of Springer Science+Business Media (www.springer.com)

Preface

The aim of this book is to give an overview of important medical biofilm infections, the pathogenesis of these infections, and the current concepts of how biofilm infections can be prevented, diagnosed, and treated. The current definition of bacterial biofilms is *A coherent cluster of bacterial cells (one or several species) imbedded in a matrix – which are more tolerant to: most antimicrobials and the host defence, than planktonic bacterial cells*. The clinical feature of biofilms infections is the increased tolerance to the defense mechanism of the body and to antibiotics and disincentives. The consequence of these features is chronic infections which are defined as infections which persists in spite of the body's innate and adapted immune response and in spite of antibiotic therapy and – in contrast to the normal colonization of the body's surfaces (skin, mucosa) – induce an immune response and gives rise to pathology in the neighboring tissues and therefore clinical signs of disease. A schematic view of the features of medical biofilms is given in the Table 1. With respect to diagnosing biofilms, adequate biopsies, pus, swabs or removed foreign bodies is used for culture, Gram-stain and FISH stain for microscopy, and to release the biofilm material on, e.g., foreign bodies, sonication is recommended. Adequate culture technique for at least one week may be necessary, and in case of no growth, PCR amplification of 16S rRNA genes should be performed to detect and identify non-culturable but viable bacteria.

The chapters of this book are written by well-known international experts in medical biofilms and the result is scientifically frontline information, which hopefully can be used both in the clinical and basic science work. Since several aspects of biofilms are not fully understood, certain opinions might deviate slightly between the chapters, for the benefit of the doubt.

The focus of the book is on the medical biofilms. It is the hope of the editors that inspirations from the chapters of the book may lead to further research in other areas of clinical biofilms which at the present time are not so well studied. To this purpose we have included chapters on technical methods to study biofilms.

Biofilm infections were introduced into human medicine by professor J.W. Costerson (at that time at the University of Calgary) about 30 years ago, but during the first decades it remained a research area for a few dedicated scientist and their results did not have much influence in human medicine. This has changed

Table 1 Some general features of biofilm infections in humans compared to acute planktonic infections and superficial colonozation/normal flora on skin and mucosal membranes. The bold fonts indicate biofilm specific features. (N. Høiby 18-11-2009)

Features of biofilm infections	Necessary condition for biofilm infections	Sufficient condition for biofilm infecions	Also found in acute planktonic infections	Also found in colonization/ normal flora on skin and mucosal membranes
Aggregates of bacteria embedded in a self-produced polymer matrix	Yes	Yes	No	No/Yes
Tolerant to clinical relevant PK/PD dosing of antibiotics in spite of susceptibility of planktonic cells	Yes	Yes	No	No/Yes
Tolerant to both innate and adaptive immune reponse	Yes	Yes	No	No/Yes – unknown (s-IgA)
Inflammation	Yes	No	Yes	No
Biofilm-specific antigens	No and Yes – seldom – e.g. Pseudomonas aeruginosa alginate	No and Yes – seldom – e.g. Pseudomonas aeruginosa alginate	No	No
Antibody response	Yes – after some weeks	No	Yes – after some weeks	No
Chronic infections	Yes	Yes	No	No
Foreign body associated infections	No	Yes	No but yes the first day of infection	No
Located on surfaces	No	No	Yes	Yes
Localized infetion	Yes	No	Yes	Yes
Focus for spreading or local exacerbation	Yes	No	Yes	Yes

dramatically during the last decade probably due to the organization of regular scientific biofilm meetings in USA, Europe, and Japan which included both basic scientists and clinicians. Cystic fibrosis *P. aeruginosa* lung infection, foreign body infections, chronic osteomyelitis, and dental infections were soon recognized as important biofilm infections and have ever since been in frontline of the interests of clinicians. It is our hope that this book will further promote both basic and clinical relevant biofilm research in these and other areas to the benefit of patients suffering from chronic biofilm infections.

Copenhagen, Denmark Thomas Bjarnsholt

Copenhagen, Denmark Claus Moser

Copenhagen, Denmark Peter Østrup Jensen

Copenhagen, Denmark Niels Høiby

Contents

Contributors

Morten Alhede Department of Systems Biology, Technical University of Denmark, DK-2800, Lyngby, Denmark

Leif Percival Andersen Department of Infection Control 9101, Copenhagen University Hospital, Rigshospitalet, Juliane Maries Vej 18, DK-2100 Copenhagen Ø, Denmark, lpa@rh.dk

Randall Basaraba Department of Microbiology, Immunology and Pathology, Colorado State University, Fort Collins, CO 80524, USA

Thomas Bjarnsholt University of Copenhagen, Faculty of Health Sciences, Department of International Health, Immunology and Microbiology, Blegdamsvej 3B, DK-2200 Copenhagen N, Denmark; H:S Rigshospitalet, Department for Clinical Microbiology, afsnit 9301, Juliane Maries Vej 22, DK-2100 Copenhagen Ø, Denmark, tbjarnsholt@sund.ku.dk

Bradley R. Borlee Department of Microbiology, University of Washington, Seattle, WA, USA

Rebecca A. Brady CBER/FDA, Building 29, Room 418, 8800 Rockville Pike, Bethesda, MD 20892, USA

Jason H. Calhoun Department of Orthopedics, Ohio State University Medical Center, Suite 1503 of Tower Building, 1492 E Broad St, Columbus, OH, USA

Nancy L. Carty Department of Biological Sciences, Texas Tech University, MS43131, Lubbock, TX 794, USA, nancy.carty@ttu.edu

Oana Ciofu Faculty of Health Sciences, Department of International Health, Immunology and Microbiology, Faculty of Health Sciences, University of Copenhagen, Blegdamsvej 3B, DK-2200 Copenhagen, Denmark

Noam Cohen Department of Otorhinolaryngology-Head and Neck Surgery, University of Pennsylvania Medical Center, Philadelphia, PA 19104, USA

Kelly K. Colvin Department of Microbiology, University of Washington, Seattle, WA, USA

Emily K. Cope Center for Microbial Genetics and Genomics, Northern Arizona University, Flagstaff, AZ 86011, USA

Richard Darveau Department of Periodontics, University of Washington, Seattle, WA, USA

Scot Dowd Medical Biofilm Research Institute, Lubbock, TX 79407, USA

Michael Givskov Department of International Health, Immunology and Microbiology, Faculty of Health Sciences, University of Copenhagen, Blegdamsvej 3B, DK-2200 Copenhagen N, Denmark, mgivskov@sund.ku.dk

Luanne Hall-Stoodley Wellcome Trust Clinical Research Facility, U Southampton Faculty of Medicine, Southampton NIHR Respiratory BRU, Division of Infection, Inflammation and Immunity, Southampton University Hospitals NHS Trust, University of Southampton, Tremona Road, Southampton, SO16 6YD, l.hall-stoodley@soton.ac.uk

Joe J. Harrison Department of Microbiology, University of Washington, Seattle, WA, USA

Preben Homøe Department of Otolaryngology, Head, and Neck Surgery F 2071, Rigshospitalet, Copenhagen, DK-2100, Denmark, phom@rh.regionh.dk

Darrell Hunsaker United States Naval Medical Center, San Diego, CA 92134, USA

Niels Høiby University of Copenhagen, Faculty of Health Sciences, Department of International Health, Immunology and Microbiology, Blegdamsvej 3B, DK-2200 Copenhagen, Denmark; H:S Rigshospitalet, Department for Clinical Microbiology, afsnit 9301, Juliane Maries Vej 22, DK-2100 Copenhagen Ø, Denmark, hoiby@hoibyniels.dk

Yasuhiko Irie Department of Microbiology, University of Washington, Seattle, WA, USA

Tim Holm Jakobsen Department of International Health, Immunology and Microbiology, University of Copenhagen, DK-2200, Copenhagen, Denmark

Garth James Center for Biofilm Engineering, Montana State University, Bozeman, MT 59717-3809, USA, gjames@erc.montana.edu

Peter Østrup Jensen H:S Rigshospitalet, Department for Clinical Microbiology, afsnit 7602, Juliane Maries Vej 22, DK-2100 Copenhagen Ø, Denmark, peter.oestrup.jensen@rh.regionh.dk

Helle Krogh Johansen Department of Clinical Microbiology, Rigshospitalet, Juliane Maries Vej 22, 2100, Copenhagen, Denmark

Klaus Kirketerp-Møller Koege University Hospital, DK-4600 Koege, Denmark, kkm@dadlnet.dk

Mette Kolpen Department of Clinical Microbiology and Danish Cystic Fibrosis Center, Rigshospitalet, Copenhagen, Denmark

Richard J. Lamont Department of Oral Biology, University of Florida College of Dentistry, Gainesville, FL, USA

Jeff G. Leid Center for Microbial Genetics and Genomic, Northern Arizona University, Flagstaff, AZ 86011, USA, jeff.leid@nau.edu

Lotte Mandsberg Institute of International Health, Immunology and Microbiology, University of Copenhagen, Copenhagen, Denmark

Søren Molin Department of Systems Biology, Technical University of Denmark, Building 301, 2800 Kgs Lyngby, sm@bio.dtu.dk

Claus Moser H:S Rigshospitalet, Department for Clinical Microbiology, afsnit 9301, Juliane Maries Vej 22, DK-2100 Copenhagen Ø, Denmark, moser@dadlnet.dk

Keiji Murakami Department of Microbiology, University of Washington, Seattle, WA, USA

Bente Nyvad School of Dentistry, Faculty of Health Sciences, University of Aarhus, Aarhus, Denmark

Graeme A. O'May Department of Microbial Pathogenesis, Dental School, University of Maryland, Baltimore, 650 West Baltimore Street, Baltimore, MD 21201, USA

James Palmer Department of Otorhinolaryngology - Head and Neck Surgery, University of Pennsylvania Medical Center, Philadelphia, PA 19104, USA

Robert J. Palmer Jr. Oral Infection and Immunity Branch, National Institute of Dental and Craniofacial Research, National Institutes of Health, Bethesda, MD, USA, rjpalmer@dir.nidcr.nih.gov

Stacy Parmenter Center for Microbial Genetics and Genomics, Northern Arizona University, Flagstaff, AZ 86011, USA

Matthew R. Parsek Department of Microbiology - Box 357242, School of Medicine, University of Washington, 1959 NE Pacific Street, Seattle, WA 98195-7242, USA, parsem@u.washington.edu

S. Brook Peterson Department of Microbiology, University of Washington, Seattle, WA, USA

Ranjani Prabhakara Department of Microbial Pathogenesis, Dental School, University of Maryland – Baltimore, 650 West Baltimore Street, Baltimore, MD 21201, USA

Lone Rasmussen Infection Control 9101, Copenhagen University Hospital, Rigshospitalet, Juliane Maries Vej 18, DK-2100 Copenhagen, Denmark

Kendra P. Rumbaugh Department of Surgery, Texas Tech University Health Sciences Center, 3601 4th Street, Lubbock, TX 79430, USA, kendra.rumbaugh@ttuhsc.edu

Mark E. Shirtliff Department of Microbial Pathogenesis, Dental School, University of Maryland – Baltimore, 650 West Baltimore Street, Baltimore, MD 21201, USA, mshirtliff@umaryland.edu

Paul Stoodley National Centre for Advanced Tribology at Southampton (nCATS), School of Engineering Sciences, Room 4081 Lanchester Building (7), University of Southampton, Highfield, Southampton, SO17 1BJ, UK, pstoodley@gmail.com

Ricardo P. Teles Department of Periodontology, The Forsyth Institute, Boston, MA, USA

Trine Rolighed Thomsen Department of Biotechnology, Chemistry, and Environmental Engineering, Aalborg University, Sohngaardsholmsvej 49, DK-9000, Aalborg, Denmark; Life Science division, The Danish Technological Institute, Kongsvang Allé, DK-8000 Århus C, Denmark, trt@bio.aau.dk

Tim Tolker-Nielsen Department of International Health, Immunology and Microbiology, Faculty of Health Sciences, University of Copenhagen, Blegdamsvej 3B, DK-2200 Copenhagen, Denmark, ttn@sund.ku.dk

Andrej Trampuz Division of Infectious Disease, Department of Internal Medicine, University Hospital Lausanne, 1011 Lausanne, Switzerland

Randall Wolcott Medical Biofilm Research Institute, Lubbock, TX 79407, USA

Werner Zimmerli Department of Infectious Diseases and Internal Medicine, Medical University Clinic, Kantonsspital, Rheinstrasse 26, CH-4410 Liestal, Switzerland, werner.zimmerli@unibas.ch

Karen Zulkowski College of Nursing, Montana State University, Bozeman, MT, 59717-3809, USA, karenz@montana.edu

Chapter 1
Introduction to Biofilms

Thomas Bjarnsholt

1.1 Introduction

Bacterial growth is characterized by two life forms, one being as single cells (planktonic) and the other being in sessile aggregates. The later is commonly referred to as the biofilm mode of growth. Many more or less divergent definitions of bacterial biofilm exist; all agreeing that biofilms are multiple bacteria in an aggregate. The different definitions present in the literature differ mainly in whether the cells have to be attached to a surface or whether the bacteria exist in a structured community. As for medical microbiology and biofilm infection, the topic of this book, biofilm is defined as:

> A coherent cluster of bacterial cells imbedded in a matrix – which are more tolerant to most antimicrobials and the host defence, than planktonic bacterial cells.

1.2 The Development of the Biofilm Era

Mainstream microbiology, starting in 1880 up until the middle of the twentieth century is popularly referred to as "the pure culture period" (Atlas and Bartha 1997). All this time bacteria were viewed as being purely free floating single cells, also referred to as the planktonic state. Most work on characterizing bacteria was done by propagating the bacteria in a liquid media in test tubes, or on agar plate. This is very peculiar with the present day's knowledge, since we estimate that less than 0.1% of the total microbial biomass lives in the planktonic mode of growth (Costerton et al. 1995, Potera 1998).

The first observation of surface associated bacteria was made by Anthony van Leeuwenhoek (published in 1684). He observed "animals" in the scurf of the teeth.

T. Bjarnsholt (✉)
University of Copenhagen, Faculty of Health Sciences, Department of International Health, Immunology and Microbiology, Blegdamsvej 3B, DK-2200 Copenhagen N, Denmark; H:S Rigshospitalet, Department for Clinical Microbiology, afsnit 9301, Juliane Maries Vej 22, DK-2100 Copenhagen Ø, Denmark
e-mail: tbjarnsholt@sund.ku.dk

T. Bjarnsholt et al. (eds.), *Biofilm Infections*, DOI 10.1007/978-1-4419-6084-9_1,

Photomicrographs of aggregating bacteria was presented in 1933 by Henrici (1933). He describes an observation: "It is quite evident that for the most part water bacteria are not free floating organisms, but grow upon submerged surfaces." Microbiology may be divided into two fields: the environmental and the medical. Environmental microbiology accepted the aggregation of bacteria almost 20 years before medical microbiology even thought about it. Aggregates or flocks of bacteria have long been used in waste water treatment plants, and the first articles using the term biofilm was by Rogovska et al. and published in Microbiology-USSR (MIKROBIOLOGIYA) in 1961.

The medical field, at least publication wise, started to describe clumps or heeps of bacteria in 1977 (Høiby 1977); here Høiby described aggregating (heeps) *Pseudomonas aeruginosa* in the lungs of chronic infected patients. In 1978 Costerton et al. (Costerton et al. 1978) defined the term biofilm for the first time, and the phenomenon was reviewed and described in 1987 by Costerton et al. (1987), as a matrix-enclosed mode of growth.

In 1993 the American Society for Microbiology (ASM) recognized that the biofilm mode of growth was relevant to microbiology (Costerton et al. 1994). As a result the biofilm mode of growth became increasingly accepted as an important bacterial trait. In 1999 Costerton et al. (1999) defined a biofilm as "a structured community of bacterial cells enclosed in a self produced polymeric matrix, adherent to a surface." During the last 15 years, the literature of biofilms have experienced a dramatic increase in numbers of publications each year (more than 10,000) and numerous books on the subject as well.

1.3 What Is a Biofilm?

Biofilms have probably been present on earth since the first bacteria evolved. In medical microbiology biofilms are typically involved in chronic persistent infections. Before the antibiotic era common bacterial infections were very serious; people died of pneumonia and other acute infections which are easily cured today, at least in the developed part of the world. On the other hand the chronic infections, such as tuberculosis, leprosy, etc. slowly degraded the patients eventually leading to death. With the development of antibiotics an increase in "low-grade" infections were observed. These infections occur in all age groups and the patients experience discomfort, fever, and other clinical signs of an infection. However, often bacteria are not detected and the effects of antibiotics are often disappointing. This is in contrast to the acute infections which are fairly easy to treat with antibiotics if diagnosed in time. The breakthroughs of identifying the origin of these persistent infections were a series of both in vitro and in vivo observation in the 1980s (Costerton 2007). Peculiarly biofilms and their extreme tolerance towards antimicrobial agents were discovered 300 years earlier by Leeuwenhoek (1684). He observed that the animals (bacteria) inside the previously mentioned scurfs were protected against vinegar in contrast to the animals outside the scurfs which were killed. This is today one of the major hallmarks of biofilms – the extreme tolerance against antimicrobial agents.

It is to be noted that the biofilm antibiotic tolerance is not to be mistaken with antibiotic resistance, since the bacteria when being in biofilm survive antibiotic treatment but if the biofilm is disrupted the bacteria become susceptible to the treatment (Bayles 2007) (Chapter 13).

Today we know from thousands of in vitro and in vivo biofilm-related observations that the origin of most persistent infections is aggregates of bacteria. We know that the bacteria in these aggregates are physically joined together and they have an extracellular matrix consisting of all kinds of extracellular-produced substances (EPS), such as proteins, DNA, and polysaccharides. We also know that these aggregates withstand very high doses of antibiotics which will kill planktonic cells. Also their tolerance towards the host defense is dramatically increased. These characteristics are combined in the biofilm definition presented in the beginning of this chapter. This definition differs in one way from most other biofilm definitions since we no longer necessitate a surface biotic or abiotic as a hallmark for biofilms. Many chronic infections involve a surface, such as infections on implants, catheters, artificial heart valves, teeth and, contact lenses. On the other hand many observations from non-surface related infections, such as CF, otits media, chronic wounds osteomylitis, etc. revealed the same patterns except the missing surface (Fig. 1.1).

Based on this chronic infections are defined as *an infection which (A) persists in spite of antibiotic therapy, and the host's innate and adaptive immune response, and (B) is characterized by persisting pathology* (Høiby 2008).

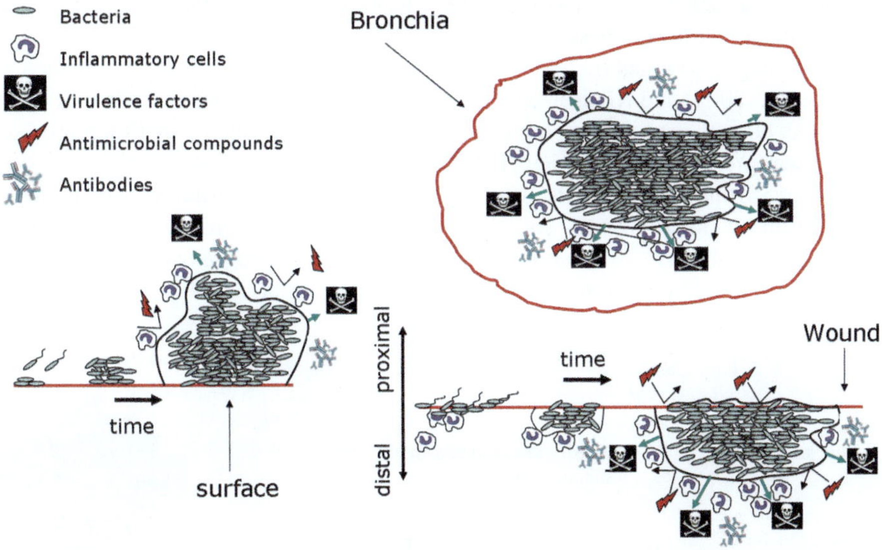

Fig. 1.1 The different scenarios of biofilms in chronic infections, either on a surface (implants), in the lumen of an organ (the CF lung) or imbedded under the surface (chronic wound). The biofilm mode of growth enable density dependent gene regulation (Quorum sensing) and protects against inflammatory cells, antibodies, and antimicrobial agents

1.4 How Biofilms Are Formed

The developmental processes of biofilms have been thoroughly studied in surface-based in vitro systems. The most studied bacterium in this context is *P. aeruginosa*.

The ability of *P. aeruginosa* to form biofilms is thought to be one of its main survival strategies in an infectious process and is considered an important pathogenicity trait.

The in vitro formed biofilm consists of microcolonies encapsulated by EPS produced by the bacteria itself, though most of the biofilms are made up of water channels which are thought to operate as a distribution system of nutrients and oxygen. An oxygen gradient is present from the surface decreasing downwards to the substratum (Walters III et al. 2003, Worlitzsch et al. 2002). *P. aeruginosa* forms biofilms on almost any surface and in any condition both nutritional and environmental.

From time to time it is stated that biofilm formation, from planktonic to sessile mode of growth, is a complex and highly regulated process (O'Toole et al. 2000). This slight variation in biofilm structure by the same strain of *P. aeruginosa* indicates that biofilm formation, and the successive growth and development, is a complex but somewhat arbitrary process. It has also been suggested that biofilm formation is dependent on the expression of a specific biofilm program (O'Toole et al. 2000, Sauer 2003). However, based on all the in vivo observations present today it is more likely that biofilm formation proceeds through a series of temporal events that probably reflect adaptation to nutritional and environmental conditions (Hentzer et al. 2004, Purevdorj et al. 2002, Reimmann et al. 2002).

The in vitro, surface-based, biofilm developmental process can be divided into different stages (i) attachment, (ii) maturation, and (iii) dispersion, as indicated by Sauer et al. (2002) and Klausen et al. (2003). The model for biofilm in vitro development has been the subject to changes over the years. One such experimental-based model was presented by Klausen et al. (2003).

As seen from Fig. 1.2, cells attach to the surface and form the microcolonies of the biofilm by clonal growth. The surface is covered with motile bacteria by means of twitching motility. Twitching motility is used by *P. aeruginosa* and many other bacteria for moving over moist surfaces (Semmler et al. 1999). The mushrooms

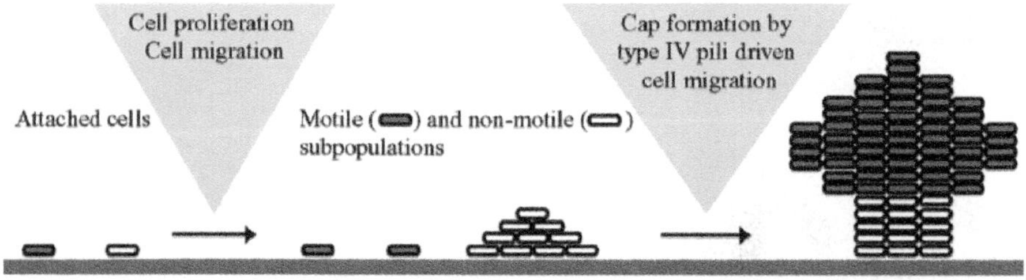

Fig. 1.2 Formation of differentiated *P. aeruginosa* biofilm. After attachment a subpopulation of non-motile bacteria forms the stalks of the microcolonies by clonal growth, while another migrating subpopulation of motile bacteria spread out on the surface, a type IV-pili dependent process. Migrating bacteria using type IV-pili climbs the stalks forming the hat on the mushroom structures (Figure from Klausen et al. (2003)). Reproduced with permission from Blackwell Publishing

are formed by these motile bacteria which eventually climb up on the stalks using type IV pili, over time forming the caps of the mushrooms. Throughout the years great emphasis has been put on the development of these biofilm structures. Klausen et al. (2003) showed that the structure formation is influenced by nutritional and environmental conditions.

Apparently the nutrient availability is one of the most important factors in the microcolony formation of *P. aeruginosa* biofilm in vitro. Klausen et al. (2003) showed that *P. aeruginosa* forms a highly differentiated biofilm composed of mushroom structures on glucose as the carbon source; the same strain forms a flat and undifferentiated biofilm on citrate as the carbon source (see Fig. 1.3). The reason for this is *P. aeruginosa* cells grown on glucose are less motile. The sub-population forming the stalks of the mushrooms with glucose as the carbon source are not present when grown on citrate.

Klausen et al. also showed that the initial development of the biofilm grown on citrate is the same as for biofilm grown on glucose. The microcolony formation occurs by clonal growth, and then the bacteria spread on the surface by means of twitching motility and form the flat biofilm with no towers and mushrooms.

Biofilms are thought to maintain equilibrium due to growth and dispersal. The dispersal occurs either as single cells or as small microcolonies ripped off from the biofilm as seen on Fig. 1.4 (Costerton et al. 1999, Stapper et al. 2004, Stoodley et al. 2001). The mechanism of dispersion is not fully understood, but Costerton et al. (1999) suggest the planktonic dispersion to be a somehow programmed process whereas the clusters are ripped off by shear forces (Fig. 1.4).

Dispersal has severe implications in medical biofilms since it is a mechanism by which biofilm bacteria might spread throughout the infected organ and to the whole body.

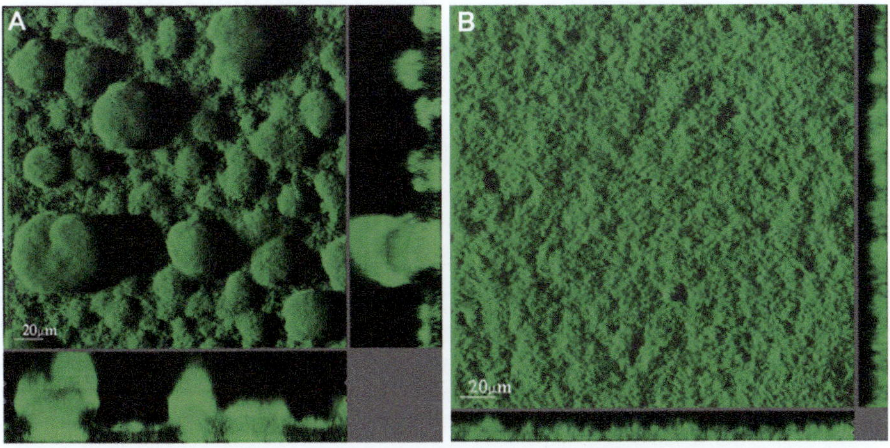

Fig. 1.3 The same strain of *P. aeruginosa* form different in vitro biofilms in terms of architecture, depending on the carbon source. The biofilm shown in Frame (**a**) is grown using glucose as the carbon source and the biofilm shown in frame (**b**) is grown using citrate as the carbon source. Both biofilms are 4 days old

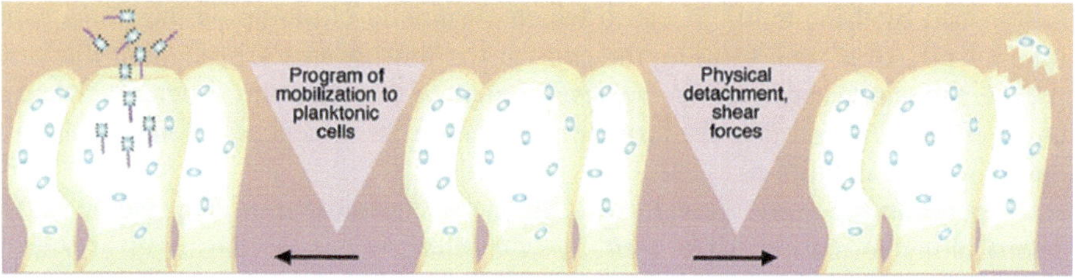

Fig. 1.4 Dispersal of bacteria from a biofilm. Bacteria detach from the biofilms, a process called dispersion. Dispersion takes place either as single cells or in clusters. (Figure from Costerton et al. (1999) and reproduced with permission from AAAS)

The question is what drives the bacteria to produce or form a biofilm? As for *P. aeruginosa,* biofilm formation is possibly just what it does the best, and all four driving forces listed by Jefferson (2004) probably apply (see Fig. 1.5).

Fig. 1.5 Artistic interpretation of the four possible driving forces behind bacterial biofilm formation by Taverna (Jefferson 2004). Reproduced with permission from Blackwell Publishing

When comparing these in vitro observations to the in vivo observations in the literature it raises the possibility that the highly organized structures on in vitro surfaces could be of artificial nature since all in vivo observations so far show unstructured aggregates of bacteria.

1.5 Who Forms Biofilm and Where Are They Found?

As stated earlier close to the entire biomass of bacteria exists as biofilms. All bacteria could theoretically be relevant in medical biofilms. Biofilms have so far been identified in several chronic infections as will be described in the subsequent chapters. Also many different bacteria have been identified in the investigated chronic infections both by standard culturing and different molecular methods, but only a few bacteria have visually been observed and identified in these infections. An example is the chronic wounds; here the presence of a great bacterial diversity has been established (Dowd et al. 2008, Gjodsbol et al. 2006, James et al. 2008, Kirketerp-Moller et al. 2008). This has resulted in a large number of microbes that have been identified in wounds. The drawback is that these techniques are qualitative which means that they do not reveal on the proportions between the different bacteria and how they are organized and distributed in the wounds. Another just as important drawback is that these techniques cannot be used to identify which bacteria plays a key role in the impairment of the wound healing process. Only *S. aureus* and *P. aeruginosa* have directly been visually identified as being present in biofilms in the wounds.

The key challenge for colonizing bacteria is first of all whether they can survive the encounter with the defense system second whether they can survive the administered antibiotics.

1.6 What Is the Implication of Biofilms?

One of the major implications of biofilms in the human body is the identification or diagnosis of the biofilm, subsequently the treatment of it. The scope of this book is to assemble the known knowledge about biofilms in chronic infections to suggest general guideline for diagnosis and treatment of these infections. This collection of knowledge and subsequent guidelines is thought to be the basis for improved treatment of all known biofilm infections and for those to be identified and developed in the future.

References

Atlas RM, Bartha R (1997) Microbial Ecology 4th edition, Benjamin/Cummings Science Publishing, Menlo Park, CA, USA

Bayles KW (2007) The biological role of death and lysis in biofilm development. Nat Rev Microbiol 5(9):721–726

Costerton JW (2007) Replacement of acute planctonic by chronic biofilm diseases. In: Eckey C (ed) The biofilm primer. Springer, New York
Costerton JW, Cheng KJ, Geesey GG, Ladd TI, Nickel JC, Dasgupta M, Marrie TJ (1987) Bacterial biofilms in nature and disease. Annu Rev Microbiol 41:435–464
Costerton JW, Geesey GG, Cheng KJ (1978) How bacteria stick. Sci Am 238(1):86–95
Costerton JW, Lewandowski Z, Caldwell DE, Korber DR, Lappin-Scott HM (1995) Microbial biofilms. Annu Rev Microbiol 49:711–745
Costerton JW, Lewandowski Z, DeBeer D, Caldwell D, Korber D, James G (1994) Biofilms, the customized microniche. J Bacteriol 176(8):2137–2142
Costerton JW, Stewart PS, Greenberg EP (1999) Bacterial biofilms: a common cause of persistent infections. Science 284(5418):1318–1322
Dowd SE, Sun Y, Secor PR, Rhoads DD, Wolcott BM, James GA, Wolcott RD (2008) Survey of bacterial diversity in chronic wounds using pyrosequencing, DGGE, and full ribosome shotgun sequencing. BMC Microbiol 8:43
Gjodsbol K, Christensen JJ, Karlsmark T, Jorgensen B, Klein BM, Krogfelt KA (2006) Multiple bacterial species reside in chronic wounds: a longitudinal study. Int Wound J 3(3): 225–231
Henrici AT (1933) Studies of freshwater bacteria. J Bacteriol 25:2132
Hentzer M, Eberl L, Givskov M (2005) Transcriptome analysis of Pseudomonasaeruginosa biofilm development: anaerobic respiration and iron limitation. Biofilms 2(01):37–61
Høiby N (1977) *Pseudomonas aeruginosa I* Infection in Cystic fibrosis. APMIS, 262(Suppl):1–96
James GA, Swogger E, Wolcott R, Pulcini ED, Secor P, Sestrich J, Costerton JW, Stewart PS (2008) Biofilms in chronic wounds. Wound Repair Regen 16(1):37–44
Jefferson KK (2004) What drives bacteria to produce a biofilm? FEMS Microbiol Lett 236(2): 163–173
Kirketerp-Moller K, Jensen PO, Fazli M, Madsen KG, Pedersen J, Moser C, Tolker-Nielsen T, Høiby N, Givskov M, Bjarnsholt T (2008) Distribution, organization, and ecology of bacteria in chronic wounds. J Clin Microbiol 46(8):2717–2722
Klausen M, Aes-Jorgensen A, Molin S, Tolker-Nielsen T (2003) Involvement of bacterial migration in the development of complex multicellular structures in Pseudomonas aeruginosa biofilms. Mol Microbiol 50(1):61–68
Klausen M, Heydorn A, Ragas P, Lambertsen L, Aes-Jorgensen A, Molin S, Tolker-Nielsen T (2003) Biofilm formation by Pseudomonas aeruginosa wild type, flagella and type IV pili mutants. Mol Microbiol 48(6):1511–1524
O'Toole G, Kaplan HB, Kolter R (2000) Biofilm formation as microbial development. Annu Rev Microbiol 54:49–79
Potera, C. (1998) Studying slime. Environ Health Perspect 106(12):A604–A606
Purevdorj B, Costerton JW, Stoodley P (2002) Influence of hydrodynamics and cell signaling on the structure and behavior of Pseudomonas aeruginosa biofilms. Appl Environ Microbiol 68(9):4457–4464
Reimmann C, Ginet N, Michel L, Keel C, Michaux P, Krishnapillai V, Zala M, Heurlier K, Triandafillu K, Harms H, Defago G, Haas D (2002) Genetically programmed autoinducer destruction reduces virulence gene expression and swarming motility in Pseudomonas aeruginosa PAO1. Microbiology 148(Pt 4):923–932
Sauer K (2003) The genomics and proteomics of biofilm formation. Genome Biol 4(6):219
Sauer K, Camper AK, Ehrlich GD, Costerton JW, Davies DG (2002) Pseudomonas aeruginosa displays multiple phenotypes during development as a biofilm. J Bacteriol 184(4): 1140–1154
Semmler AB, Whitchurch CB, Mattick JS (1999) A re-examination of twitching motility in Pseudomonas aeruginosa. Microbiology 145(Pt 10):2863–2873
Stapper AP, Narasimhan G, Ohman DE, Barakat J, Hentzer M, Molin S, Kharazmi A, Hoiby N, Mathee K (2004) Alginate production affects Pseudomonas aeruginosa biofilm development and architecture, but is not essential for biofilm formation. J Med Microbiol 53(Pt 7):679–690

Stoodley P, Wilson S, Hall-Stoodley L, Boyle JD, Lappin-Scott HM, Costerton JW (2001) Growth and detachment of cell clusters from mature mixed-species biofilms. Appl Environ Microbiol 67(12):5608–5613
Walters MC III, Roe F, Bugnicourt A, Franklin MJ, Stewart PS (2003) Contributions of antibiotic penetration, oxygen limitation, and low metabolic activity to tolerance of Pseudomonas aeruginosa biofilms to ciprofloxacin and tobramycin, Antimicrob. Agents Chemother 47(1):317–323
Worlitzsch D, Tarran R, Ulrich M, Schwab U, Cekici A, Meyer KC, Birrer P, Bellon G, Berger J, Weiss T, Botzenhart K, Yankaskas JR, Randell S, Boucher RC, Doring G (2002) Effects of reduced mucus oxygen concentration in airway Pseudomonas infections of cystic fibrosis patients. J Clin Invest 109(3):317–325

Chapter 2
Chronic Wound Colonization, Infection, and Biofilms

Klaus Kirketerp-Møller, Karen Zulkowski, and Garth James

2.1 Introduction

The term "chronic wound" is generally accepted, but yet no simple definition has been agreed upon. A mechanistic definition such as "those not following normal wound healing trajectory" have been proposed but the most common definition have been "ulcers (wounds) older than 3 months of age". Indeed, some ulcers have been present for years. One to two percent of the population in the developed countries will experience a chronic wound in their lifetime (Gottrup 2004). It is expected that the number of chronic wounds will increase worldwide due to the increase of lifestyle diseases, such as diabetes, obesity, and cardiovascular diseases. It is estimated that 246 million people had diabetes worldwide in 2007 and this is expected to increase to 380 million people by the year 2025 (Diabetes Atlas).

Chronic wounds cause a significant burden to healthcare systems as well as morbidity and mortality to mankind. In Denmark, it has been estimated that the prevalence of non-healing wounds is about 1% of the population. The prevalence and incidence of ulcers are similar to elsewhere in the industrialized world. The total expenses for treatment of wounds are estimated to be approximately 2–3% of the total budget of the health care system in Denmark (Ayello et al. 2004, Gottrup 2004) which matches the percentages from the European Community and probably United States of America as well. Notably in the United States there has been an increase of hospital stays for pressure ulcers by approximately 80% from 1993 to 2006. The 2007 Medicare data indicated that the development of a pressure ulcer increased a person's hospital costs by $43,180. This adds up to 11 billion dollars a year in increased health care cost in the United States. According to the American Diabetes Association (ADA), 17.9 million Americans are diagnosed with either type I or type II diabetes, which is a major cause of premature mortality and disability. Additionally, it has been estimated that there are 6.2 million undiagnosed cases of diabetes in the United States. Often a chronic wound is the first manifestation of diabetes complications. Roughly 15% of people living with diabetes are estimated to

K. Kirketerp-Møller (✉)
Koege University Hospital, DK-4600 Koege, Denmark
e-mail: kkm@dadlnet.dk

T. Bjarnsholt et al. (eds.), *Biofilm Infections*, DOI 10.1007/978-1-4419-6084-9_2,

develop lower extremity ulcers, and 14–24% of people with a lower extremity ulcer eventually suffer an amputation (ADA). The annual cost for treating diabetic foot ulcers has been predicted to be around $20,000 per patient. This amounts to billions of dollars spent on diabetic foot ulcers annually in USA alone. Diabetes and, therefore, diabetic foot ulcer complications are growing at double digit rates and have the potential of becoming even a more devastating epidemic. Despite efforts directed at prevention, the rate of amputation in patients with DM continues to rise. Novel treatment strategies are required to reverse the rising trend in the rate of amputation. Focusing on healing these wounds will improve human health as well as reducing healthcare cost.

Interest in the potential association of biofilms in the pathology of chronic wounds has been demonstrated by the recent influx of publications referencing the key words "biofilm" and "chronic wound" (Fig. 2.1). Increasing evidence from these studies suggests that biofilms may play a role in wound chronicity. Microscopic evaluation of specimens from chronic wounds often indicates the presence of biofilms (Fig. 2.2). However, the role of bacteria and biofilms in wound healing is poorly understood. The following chapter will describe chronic wounds, and current treatments as well as review evidence that biofilms may be a major barrier to wound healing.

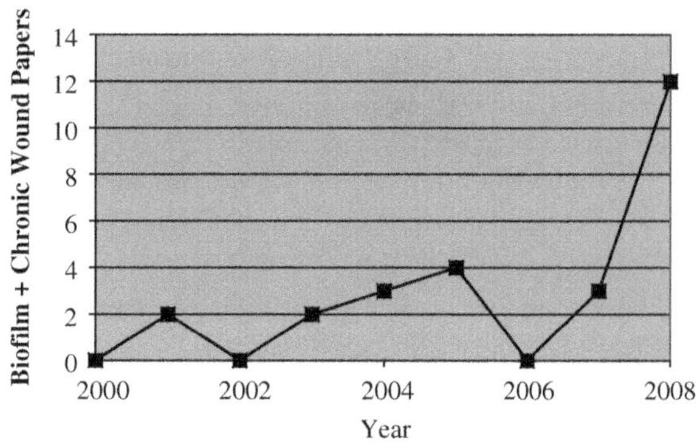

Fig. 2.1 Number of papers, by year, in the Institute for Scientific Information Web of Knowledge database (www.isiwebofknowledge.com) indexing to the combination of keywords "biofilm" and "chronic wound"

2.2 Types of Chronic Wounds

Chronic wounds are traditionally divided into three major groups: venous leg ulcers, diabetic foot ulcers, and pressure ulcers. Additionally there are other subgroups such as ischemic ulcers, cancer ulcers, and inflammatory ulcers. Common for all these groups are that all have underlying predisposing factors for origin and persistence.

2.2.1 Venous Leg Ulcers

Venous leg ulcers are the most common ulcers of the lower leg. Persons with venous ulcers often have lower leg edema and a history of previous phlebitis, tired aching legs, and a minor traumatic event (e.g., bumping the leg). Most venous

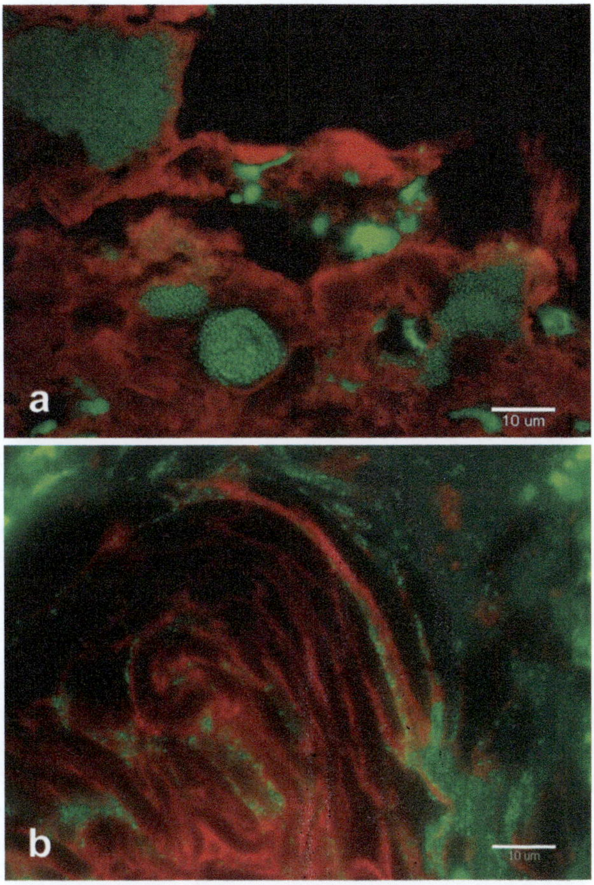

Fig. 2.2 Confocal scanning laser micrographs showing clusters of cocci that formed biofilms in a specimen from a venous leg ulcer specimen (**a**) and rod-shaped bacteria that formed biofilm in a specimen from a diabetic foot ulcer (**b**). Wound tissue was formalin fixed, processed, and embedded in paraffin. Sections (3–5 um) were then deparaffinized and stained with Molecular Probes' ViaGram™ Red+ Bacterial Gram Stain and Viability Kit following their fluorescent staining protocol. The kit includes SYTOX® Green nucleic acid stain, which stains DNA (Gram+ and Gram– DNA and host DNA) and Texas Red®-X dye-labeled wheat germ agglutinin (WGA), which stains host extracellular tissue and also selectively binds to the surface of gram-positive bacteria . Thus, small areas of scattered green staining indicate bacterial cells while the larger green regions are host cells. The host tissue was stained red. Microscopy by Dr. Kelly Kirker

ulcers are located on the medial malleolus and the skin surrounding the wound is hyper-pigmented.

2.2.2 Pressure Ulcers

Pressure ulcers are a localized injury to the skin and underlying tissue, usually over a bony prominence as a result of pressure or pressure in combination with friction and shear. They usually occur in patients who are immobilized either temporarily or permanent especially with concomitant impaired sensation. Pressure ulcers are also called decubitus ulcers, bedsores, or pressure sores.

2.2.3 Diabetic Ulcers

This term is frequently used to describe ulcers that occur at the bottom of the feet in persons with diabetes mellitus. The direct cause of these ulcers is pressure. Peripheral neuropathy, structural foot deformities, or limited range of motion in the foot may increase pressure and contribute to the development of these ulcers. Diabetics often have ischemia as well, and the term "neuro-ischemic foot" is often used. The neuropathic foot is characterized by dry atrophic skin, deformity, and limited range of motion. The sensory function is often impaired and leaves the foot susceptible to injury.

2.2.4 Neuropathic Ulcers

The most common cause of neuropathy is diabetes. However, other causes include alcohol or drug abuse, hereditary defects, and some metabolic diseases or malnutrition. The frequency of neuropathy increases with age and is not uncommon in the age group of 80+ years.

2.2.5 Traumatic Ulcers

Although almost all other ulcers have a history of minor trauma, the term traumatic ulcer is used for ulcers that occur in patients who do not have a prior history of predisposing factors such as diabetes, venous or arterial vascular disorder, or immune-incompetence. Normally traumatic ulcers heal uneventfully but the size, necrotic tissue, contamination, or local edema may disturb healing. This may also include skin tears which are common in elderly persons. Skin tears may provide an ideal site for biofilm formation.

2.2.6 Arterial Ulcers

Peripheral vascular occlusive disease, which often presents as intermittent claudication, can also lead to slow-to-heal wounds. Many of the risk factors are similar to those for coronary artery disease. As with venous ulcers, the direct cause of an arterial ulcer is often a minor bump or bruise. Arterial ulcers often have a "punched out" appearance and there are signs of impaired tissue perfusion, such as pale skin and diminished or absent pedal pulses. However, arterial wounds often heal after successful revascularization or bypass of the arterial blockage.

2.3 Wound Treatment

Ideally the wounds are treated according to their origin, i.e., pressure ulcers are treated with off-loading and ambulation, venous ulcers with compression therapy, and diabetic ulcers with off-loading and management of ischemia. However, in our experience, even referrals to tertiary centers have not received this treatment and have often been on several courses of antibiotic treatment, and numerous different dressings have been applied. The lack of recognition of underlying predisposing factors leads to insufficient treatment and delays or even prevents healing. This includes restoration of the patient's health status in general if at all possible.

2.3.1 Local Wound Treatment

Removal of necrotic or devitalized tissue is perhaps the most important part of local treatment. Several wound treatment regimens include this in their concept as in TIME (Tissue, Infection or Inflammation, Moisture balance and Edge effect) (Falanga et al. 2008, Rhoads. et al. 2008, Wolcott et al. 2009) and others (Werthen et al. 2004).

Removal of devitalized tissue minimizes the bioburden of the wounds by decreasing the presence of bacteria, reducing the hypoxic part of the wound and diminishing the local inflammatory reaction.

Moisture balance is an important aspect of wound healing and this is the main purpose of modern dressings. It has been demonstrated using occlusive and semi-occlusive dressings that a moist wound environment speeds epithelialization (Eaglstein and 1978) and collagen synthesis (Alvarez et al. 1983). However, occlusive dressings can also promote the growth of pathogens (Mertz and Eaglstein 1984), particularly anaerobic bacteria (Marshall et al. 1990). In contrast to wounds where moisture retention is necessary, highly exudative wounds require moisture removal. This has been achieved using absorbent dressings and more recently Negative Pressure Wound Therapy (NPWT).

A large variety of wound dressings are available; some of the most common types are shown in Table 2.1. Selection of a wound dressing should be based on several factors including (a) exudate amount, (b) presence of necrotic tissue, (c) bacterial burden, (d) presence of undermining or tunneling, and (e) providing protection from the environment. Dressings may also help with wound pain. Dressing choice can change as the wound progresses or deteriorates. Recently a number of antimicrobial wound dressings have become available. These dressings contain various antimicrobial agents including PHB (Wolcott and Rhoads 2008a) and silver (Percival et al. 2008). A number of non-dressing wound therapies are also currently used. Negative Pressure Wound Therapy (NPWT) is the application of a controlled level of sub-atmospheric pressure to a wound at 50–175 mm Hg (intermittent or continuous). The pressure is generated by a portable programmable pump. The suction effect is applied to the entire interior surface of a clean wound through open-cell polyurethane or polyvinyl alcohol (PVA) foam or most recently gauze, either impregnated with antimicrobial agents or plain. The fluid from chronic wounds may have a detrimental effect on the wound and its surroundings. Trengove et al. (1996a) demonstrated increased levels of cytokines and metalloproteases in chronic wounds compared to acute wounds. These and other factors likely contribute a negative

Table 2.1 Common wound dressings

Dressing type	Description
Foam	Inert material which is hydrophilic and non-adherent, modified polyurethane foam
Transparent film	Polyurethane and polyethylene membrane film coated with a layer of acrylic hypoallergenic adhesive. Moisture vapor transmission rates (MVTR) vary
Hydrocolloid	Gelatin, pectin, carboxymethylcellulose in a polyisobutylene adhesive base with polyurethane or film backing. Hydrophilic colloid particles bound to polyurethane foam
Hydrogel	Water or glycerin-based, non-adherent, cross-linked polymer. May or may not be supported by a fabric net, high water content, and varying amounts of gel-forming material (glycerin, co-polymer, water, propylene glycol, and humectant)
Calcium alginate	Calcium sodium salts of alginic acid (naturally occurring polymer in seaweed). Nonwoven composite of fibers from a cellulose-like polysaccharide which acts via an ion exchange mechanism, absorbing serous fluid or exudate and forming a hydrophilic gel which conforms to the shape of the wound
Silver	Topical silver has broad spectrum antimicrobial activity that has been shown to be effective against many antibiotic-resistant wound pathogens and can be added to a variety of composite dressings
Honey	Use of or addition of certain honeys as a part of a variety of composite dressings. Honey provides a hypertonic environment thought to assist with microbial control

effect on healing and excess amounts of wound fluid should be removed by the dressing.

Other non-dressing therapies include growth factors and skin grafting. Skin grafts may be of the persons own skin or tissue engineered skin. Hyperbaric oxygen is also considered a useful wound therapy.

2.3.2 Debridement

The removal of necrotic tissue and wound exudates is recognized as an important aspect of wound care. Debridement strategies range from occlusive dressings, which provide a moist environment for digestion by endogenous enzymes, to surgical removal of tissue. Wolcott and Rhoads (2008b) have stressed the importance of frequent (weekly) debridement as an important aspect of their biofilm-based wound care regimen. They used surgical debridement for non-healing ischemic leg ulcers and ultrasonic debridement as an adjunct therapy. Maggot larvae have also been used for wound debridement. Not only do the maggots digest and consume necrotic tissue and bacteria, but they also produce antimicrobial secretions (Bexfield et al. 2008, Huberman. et al. 2007). However experimental research on maggot excretions has shown different effects on biofilm formation in *Staphylococcus aureus* and *P. aeruginosa* (van der Plas et al. 2008). At least in the beginning of biofilm formation the maggot excretions seem to enhance the biofilm formation in *P. aeruginosa*, but not in *S. aureus* (Leid et al. 2002a, van der Plas et al. 2008). Indeed the experience in Copenhagen Wound Healing Center is that maggot debridement in heavily *P. aeruginosa* colonized wounds was ineffective. Experimental research (Andersen et al. 2010) have shown that a Quorum Sensing dependent virulence factor from *P. aeruginosa* is capable of killing maggots. Debridement of chronic wounds likely plays an important role in biofilm control by not only removing the biofilm but also necrotic tissue which can harbor subsequent biofilm re-growth. Additionally, debridement can increase blood and oxygen supply to the wound bed.

2.4 Bacteria and Wounds

The role of bacteria in wound healing has been debated over the years. Some have suggested that bacteria may play a beneficial role in normal wound healing and wounds will heal despite the presence of large numbers of microorganisms (Edwards and Harding 2004). Nonetheless, the detrimental effects of specific pathogens, such as *Clostridium perfringes* and *Streptococcus pyogenes,* have been well recognized. These are typically invasive bacteria that are not normal members of the human skin microflora. In contrast, *S. aureus*, which is part of the microflora of many humans, commonly also causes wound infections. In cases of acute wound infection, wound cleansing, dressings, and systemic antibiotic therapy often provide an effective cure by killing or inhibiting the growth of bacteria thereby allowing the immune system to clear the infection.

The clinical definition of wound infection includes pain, erythema, edema, heat, and purulent exudates (Robson et al. 1990). Attempts have also been made to correlate the quantity of bacteria in a wound to infection. Robson et al. have suggested that quantitative bacterial counts from tissue biopsy samples of $<10^5$ CFU/gram indicate wound infection. Lower tolerance for specific organisms, such as haemolytic *Streptococci* is accepted due to higher virulence.

There has been some controversy over the method used to quantify wound bacteria, with some arguing for a simple swab culture (Bill et al. 2001) and others advocating tissue biopsy specimens (Robson et al. 1990). Biopsy specimens likely provide the most complete specimen through the depth of the wound, including bacteria invading the tissue. However, superficial specimens are less invasive and less difficult to collect, handle, and analyze, and may provide useful information regarding the quantity and types of microorganisms within a wound (Bowler et al. 2001). Furthermore, a swab culture may provide a specimen from a larger surface area of the wound. In our research, we have analyzed specimens obtained during sharp debridement of chronic wounds (Dowd et al. 2008, James et al. 2008) which provided a tissue specimen without involving a non-routine procedure. However, processing these specimens required specialized handling and analysis, and this approach is likely unpractical for routine clinical microbiology. In another work, Kirketerp-Moller et al. (2008) found discrepancy between clinical culturing and detection of bacteria with PNA-FISH and confocal light microscopy. *P. aeruginosa* was detected in several cases where culturing failed, despite the fact that *P. aeruginosa* is considered easy to culture. In Fazli et al. (2009), it was shown that *S. aureus* is more likely to be at the surface of the wounds and that *P. aeruginosa* resides deeper in the tissue. This again adds fuel to the debate whether to use swabs or biopsies in detection of presence of bacteria.

Traditionally, wound microbiology has been described in three phases: contamination, colonization, and infection. Contamination refers to the presence of bacteria that are not multiplying, whereas colonization refers to bacteria which are growing within the wound but not causing tissue damage. Bacteria causing tissue damage and clinical signs of infection, discussed above, indicate an infected wound. The concept of "critical colonization" has also been used to describe bacteria that are growing within the wound and, while not causing classical clinical symptoms of infection, are adversely affecting wound healing (Edwards and Harding 2004).

The polymicrobial nature of wound communities has long been recognized. Various bacteria can be introduced into a wound from exogenous (soil and water) and endogenous (skin, saliva, urine, and feces) sources. As discussed above, most of these organisms are contaminants that do not multiply within the wound and even bacteria that do multiply are not considered "infective" unless they cause detrimental effects. This is particularly the case for bacteria considered commensal skin organisms, such as Corynebacteria and coagulase-negative Staphylococcus. However, the interactions of multiple microbial populations in chronic infections is poorly understood. In a study of 52 patients by Trengove et al., bacterial diversity was correlated with wound chronicity (Trengove et al. 1996b). Selective media were used for culture of six groups of bacteria; no particular group was associated with delayed healing but wounds that yielded four or more groups had a significantly higher proportion (42%) that failed to heal.

Over the past several years molecular-based methods have been increasingly applied in skin and wound microbiology research. Molecular studies of wound microbiology have revealed very diverse bacterial communities. These studies involved polymerase chain reaction (PCR) amplification of the bacterial gene encoding the small ribosomal subunit RNA (16S). This gene contains both highly conserved and variable regions of DNA, which enables bacterial speciation and forms the basis of bacterial phylogeny. Pioneering studies by Davies et al. (2001, 2004) and Hill et al. (2003) used PCR of 16S DNA along with denaturing gradient gel electrophoresis and clone libraries for culture-independent analysis of the microflora of chronic venous leg ulcers. More recent studies have used additional molecular methods including metagenomics for the evaluation of chronic wound microflora (Dowd et al. 2008, Frank et al. 2009, Price et al. 2009). The results of these studies indicate that chronic wounds contain diverse polymicrobial communities and similar community features, such as the presence of strictly anaerobic bacteria, even though the studies were from diverse geographic regions. However, Kirketerp-Moller et al. (2008) found only a few polymicrobial colonies in a study of venous leg ulcers. Thus, the prevalence and importance of polymicrobial biofilms in chronic wounds remains unclear.

2.5 Immune Response to Biofilms

In-vitro studies have demonstrated the ability of human leukocytes to penetrate *S. aureus* biofilms (Leid et al. 2005) (see also Chapter 15). In a murine model of acute wound infection, Akiyama et al. found that antimicrobial efficacy against *S. aureus* biofilms was significantly better in normal mice than those depleted of leukocytes (Akiyama et al. 2002) (see also Chapter 16). They suggested that the primary mechanism of antimicrobial action in normal mice was invasion of PMNs into the biofilm. For *P. aeruginosa* biofilms, production of the extracellular polysaccharide, alginate, protected them from IFN-γ mediated phagocytosis by human leukocytes (primarily monocytes) (Leid et al. 2005, Vuong et al. 2004). Similarly, polysaccharide intercellular adhesin (PIA) protected *Staphylococcus epidermidis* against phagocytosis and killing by polymorphonuclear leukocytes (PMN) (Gardner et al. 2001, Vuong et al. 2004). Thus, extracellular polysaccharides seem to be an important factor in biofilm resistance to phagocytosis. Furthermore, PIA also protected *S. epidermidis* against the antibacterial peptides, cathelicidin/hCAP18, human b-defensin 3 and dermcidin (Vuong et al. 2004). In addition to limiting the effectiveness of innate immune factors, *P. aeruginosa* biofilms cause killing of PMN through the production of rhamnolipids (Jensen et al. 2007). *S. aureus* is also capable of producing leukotoxins, including the Panton-Valentine leukocidin associated with severe cutaneous infections.

Chronic wounds often show increased levels of proinflammatory cytokines including IFN-α, IFN-γ, TNF-α, and interleukin 1. These cytokines are commonly produced in response to bacterial virulence determinants, such as lipopolysaccharides, peptidoglycans, and DNA. Chronic wounds also have high levels of matrix metalloproteinases (MMPs) including collagenases and gelatinases that are produced by PMN (Tarnuzzer and Schultz 1996) and correspondingly low levels of tissue inhibitors of matrix metalloproteinases (Yager and Nwomeh 1999). It has been suggested that these chronic wound characteristics are due to the presence of biofilms (Bjarnsholt et al. 2008).

In addition to the primarily innate immune mechanisms discussed above, *S. aureus* biofilms also can elicit adaptive immune responses, which may result in the development of biofilm-specific diagnostics and possibly vaccines (Leid et al. 2002b).

2.6 Biofilm Formation in Acute Wounds: In Vivo Models

Biofilm formation in wounds has been demonstrated in vivo using both murine and porcine models (see also Chapter 16). Akiyama et al. used neutropenic mice to evaluate *S. aureus* and *Streptococcus* biofilm formation within incisional wounds (Akiyama et al. 1996, Akiyama et al. 2002, 2003). The mice were treated with cyclophosphamide to inhibit leukocytes because normal mice were found to quickly clear inoculated bacteria with a strong PMN response. Biofilms were identified using light microscopy (LM) and electron microscopy (SEM) as well as confocal scanning laser microscopy (CSLM). Aggregated clusters of bacteria were apparent. Polysaccharide extracellular polymeric substance (EPS) surrounding the bacteria was imaged after staining with ruthenium red (EM) and FITC-labeled concanavalin A (ConA, CSLM). Schaber et al. also used a murine model to evaluated biofilm formation by *P. aeruginosa* in burn wounds. They demonstrated that *P. aeruginosa* rapidly colonized burn wounds and formed biofilms primarily around blood vessels. Again the criteria used for biofilm classification was the presence of cell clusters and EPS. In addition to the general polysaccharide stain, ruthenium red, the presence of alginate was specifically detected using an anti-alginate antibody. Of all in vivo models developed to date, porcine skin structure and wound healing is the most similar to humans. Davis et al. demonstrated the formation of *S. aureus* biofilms and EPS on excisional wounds in pigs and further showed that planktonic bacteria recovered from the wounds were more susceptible to topical antimicrobial agents than the biofilm bacteria recovered from the same wounds (Davies 2003). All of the in vivo models described above utilized acute wound models. Further development is required to establish an in vivo model of chronic wounds. Nonetheless, these studies clearly demonstrate that bacteria readily form biofilms in animal models of wound infection. It should also be noted that, with the exception of anti-alginate antibody, they used general carbohydrate stains to identify EPS. Similar methods have been used in studies of human specimens. Due to the wide variety of carbohydrate residues associated with mammalian cells and tissues, it is impossible to determine whether the EPS is of bacterial or mammalian origin. As this material forms the matrix encapsulating the biofilm cells in vivo this should be considered an important part of the biofilm regardless of origin. Furthermore, incorporating host components into the EPS could help biofilms evade the immune system.

2.7 Evidence of Biofilms in Human Wounds

It had been speculated as early as 2001 that bacteria colonizing human chronic wounds exist as biofilm communities. Akiyama et al. used safranin, ConA, and immunofluorescent staining with CSLM to study and demonstrate the presence of *S. aureus* biofilms in specimens collected from patients with the skin diseases bullous impetigo, atopic dermatitis, and pemphigus foliaceus (Akiyama et al. 2003). Kirketerp-Møller et al. evaluated specimen wounds of 22 patients suspected of *P. aeruginosa* colonization (Kirketerp-Moller et al. 2008). They used PNA FISH and anti-alginate antibodies and found that *P. aeruginosa* existed as biofilms rather than single cells in these wounds. James et al. microscopically evaluated specimens from 50 chronic wounds and 16 acute wounds for the presence of biofilms and found biofilms in 60% of the chronic wounds and only 6% of the acute wounds (James et al. 2008).

2.8 Clinical Implications of Chronic Wound Biofilm

The recognition of bacterial biofilm in chronic wounds may give us the opportunity to explain many of the characteristics of the chronic wound. It may explain why chronic wounds does not heal despite adequate treatment of underlying condition

(Bjarnsholt et al. 2008) and it gives us new paths of research that may lead to new treatments.

The bacterial biofilm communities interfere with the human immune system in numerous ways. This interference facilitates establishment of further bacterial communities and inflammation of the chronic wound, and prevents healing. It is tempting for the clinician to start antibiotic treatment, but in case of established, mature biofilm this treatment often has only temporary effect on both inflammation and healing. In addition the clinician has to rely on the results from a swab or biopsy, which rarely reflects all specimens present in the wound. The bacteria in biofilm are up to 1000 times less susceptible to antibiotics (Bjarnsholt et al. 2007), and MIC is not reached in the chronic wound fluid. Even silver treatment, as incorporated in several wound dressings, has limited effect in biofilm in vitro (Falanga 2000, Wolcott and Rhoads 2008b). With this in mind the clinician should exercise restraint in admission of antibiotics. Administering antibiotics favors biofilm capable bacteria and promotes resistance to the administered antibiotic. Mechanical removal of wound debris and even granulation tissue is an effective way of diminishing the bacterial load and is an important part of treatment protocols as TIME and others.

Even in the case of extensive surgical debridement in combination with split skin transplant the presence of *P. aeruginosa* prior to surgery seams to influence the healing (Hoegsberg et al., unpublished results). This indicates that the bacteria reside deep in what is thought to be normal tissue, probably protected in biofilm.

From our point of view the recognition and acceptance of bacterial biofilm in chronic wounds already have changed wound care. "Biofilm managing strategies" have been made, but none have yet proved to be more effective than others or even better than "Best Practice." But the rapid change reflects the need for new and more efficient treatment regimens and the research in biofilm may provide wound care specialists with new ways to heal the wounds.

2.9 Future Aspects

Understanding bacterial biofilm communities will provide us with knowledge to design new treatments for chronic wound patients. These treatments might work alone or act together in helping the host immune defense system to fight pathogens in the chronic wound. There could be several ways to do this. See Table 2.2.

Table 2.2 Future treatment options

Biofilm formation	Turn off biofilm production
Quorum sensing	Manipulating QS
Antibiotic resistance	QS Manipulation New biofilm penetrating drugs
Biofilm disrupting agents	New agents have to disrupt the biofilm in order to reach bacteria residing deep in the wound
Mechanical debridement	Surgery Ultra sound assisted surgery

Acknowlegment Chronic wound research in Dr. James' laboratory is supported by grant number 5 P20 GM078445 from the National Institute of General Medical Sciences (NIGMS). The contents of this publication are solely the responsibility of the authors and do not necessarily represent the official views of the NIGMS.

References

Akiyama H, Huh WK, Fujii K, Yamasaki O, Oono T, Iwatsuki K (2002) Confocal laser microscopic observation of glycocalyx production by Staphylococcus aureus in vitro. J Dermatol Sci 29(1):54–61

Akiyama H, Kanzaki H, Tada J, Arata J (1996) Staphylococcus aureus infection on cut wounds in the mouse skin: experimental staphylococcal botryomycosis. J Dermatol Sci 11(3):234–238

Akiyama H, Morizane S, Yamasaki O, Oono T, Iwatsuki K (2003) Assessment of Streptococcus pyogenes microcolony formation in infected skin by confocal laser scanning microscopy. J Dermatol Sci 32(3):193–199

Alvarez OM, Mertz PM, Eaglstein WH (1983) The effect of occlusive dressings on collagen synthesis and re-epithelialization in superficial wounds. J Surg Res 35(2):142–148

Andersen AS, Joergensen B, Bjarnsholt T et al (2010) Quorum-sensing-regulated virulence factors in Pseudomonas aeruginosa are toxic to Lucilia sericata maggots. Microbiology 156:400–407

Ayello EA, Dowsett C, Schultz GS, Sibbald RG, Falanga V, Harding K, Romanelli M, Stacey M, TEot L, Vanscheidt W (2004) TIME heals all wounds. Nursing 34(4):36–41

Bexfield A, Bond AE, Roberts EC, Dudley E, Nigam Y, Thomas S, Newton RP, Ratcliffe NA (2008) The antibacterial activity against MRSA strains and other bacteria of a <500 Da fraction from maggot excretions/secretions of Lucilia sericata (Diptera: Calliphoridae). Microbes Infect 10(4):325–333

Bill TJ, Ratliff CR, Donovan AM, Knox LK, Morgan RF, Rodeheaver GT (2001) Quantitative swab culture versus tissue biopsy: a comparison in chronic wounds. Ostomy Wound Manage 47(1):34–37

Bjarnsholt T, Kirketerp-Moller K, Jensen PO, Madsen KG, Phipps R, Krogfelt K, Hoiby N, Givskov M (2008) Why chronic wounds will not heal: a novel hypothesis. Wound Repair Regen 16(1):2–10

Bjarnsholt T, Kirketerp-Moller K, Kristiansen S, Phipps R, Nielsen AK, Jensen PO, Hoiby N, Givskov M (2007) Silver against Pseudomonas aeruginosa biofilms. APMIS 115(8):921–928

Bowler PG, Duerden BI, Armstrong DG (2001) Wound microbiology and associated approaches to wound management. Clin Microbiol Rev 14(2):244–269

Davies D (2003) Understanding biofilm resistance to antibacterial agents. Nat Rev Drug Discov 2(2):114–122

Davies CE, Hill KE, Wilson MJ et al (2004) Use of 16S ribosomal DNA PCR and denaturing gradient gel electrophoresis for analysis of the microfloras of healing and nonhealing chronic venous leg ulcers. J Clin Microbiol 42:3549–3557

Davies CE, Wilson MJ, Hill KE et al (2001) Use of molecular techniques to study microbial diversity in the skin: chronic wounds reevaluated. Wound Repair Regen 9:332–340

Dowd SE, Sun Y, Secor PR, Rhoads DD, Wolcott BM, James GA, Wolcott RD (2008) Survey of bacterial diversity in chronic wounds using pyrosequencing, DGGE, and full ribosome shotgun sequencing. BMC Microbiol 8, 43

Eaglstein WH Mertz PM (1978) New methods for assessing epidermal wound healing: the effects of triamcinolone acetonide and polyethelene film occlusion. J Invest Dermatol 71(6): 382–384

Edwards R, Harding KG (2004) Bacteria and wound healing. Curr Opin Infect Dis 17(2):91–96

Falanga V (2000) Classifications for wound bed preparation and stimulation of chronic wounds. Wound Repair Regen 8(5):347–352

Falanga V, Brem H, Ennis WJ, Wolcott R, Gould LJ, Ayello EA (2008) Maintenance debridement in the treatment of difficult-to-heal chronic wounds. Recommendations of an expert panel. Ostomy Wound Manage Suppl:2–13

Fazli M, Bjarnsholt T, Kirketerp-Moller K et al (2009) Nonrandom distribution of Pseudomonas aeruginosa and Staphylococcus aureus in chronic wounds. J Clin Microbiol 47:4084–4089

Frank DN, Wysocki A, Specht-Glick DD, Rooney A, Feldman RA, St Amand AL, Pace NR, Trent JD (2009) Microbial diversity in chronic open wounds. Wound Repair Regen 17(2):163–172

Gardner SE, Frantz RA, Troia C, Eastman S, MacDonald M, Buresh K, Healy D (2001) A tool to assess clinical signs and symptoms of localized infection in chronic wounds: development and reliability. Ostomy. Wound Manage 47(1):40–47

Gottrup F (2004) A specialized wound-healing center concept: importance of a multidisciplinary department structure and surgical treatment facilities in the treatment of chronic wounds. Am J Surg 187(5A):38S–43S

Hill KE, Davies CE, Wilson MJ, Stephens P, Harding KG, Thomas DW (2003) Molecular analysis of the microflora in chronic venous leg ulceration. J Med Microbiol 52:365–369

Huberman L, Gollop N, Mumcuoglu KY, Breuer E, Bhusare SR, Shai Y, Galun R (2007) Antibacterial substances of low molecular weight isolated from the blowfly, Lucilia sericata. Med Vet Entomol 21(2):127–131

James GA, Swogger E, Wolcott R, Pulcini E, Secor P, Sestrich J, Costerton JW, Stewart PS (2008) Biofilms in chronic wounds. Wound Repair Regen 16(1):37–44

Jensen PO, Bjarnsholt T, Phipps R, Rasmussen TB, Calum H, Christoffersen L, Moser C, Williams P, Pressler T, Givskov M, Hoiby N (2007) Rapid necrotic killing of polymorphonuclear leukocytes is caused by quorum-sensing-controlled production of rhamnolipid by Pseudomonas aeruginosa. Microbiology 153(Pt 5):1329–1338

Kirketerp-Moller K, Jensen PO, Fazli M, Madsen KG, Pedersen J, Moser C, Tolker-Nielsen T, Hoiby N, Givskov M, Bjarnsholt T (2008) Distribution, organization, and ecology of bacteria in chronic wounds. J Clin Microbiol 46(8):2717–2722

Leid JG, Costerton JW, Shirtliff ME, Gilmore MS, Engelbert M (2002b) Immunology of Staphylococcal biofilm infections in the eye: new tools to study biofilm endophthalmitis. DNA Cell Biol 21(5–6):405–413

Leid JG, Shirtliff ME, Costerton JW, Stoodley AP (2002a) Human leukocytes adhere to, penetrate, and respond to Staphylococcus aureus biofilms. Infect Immun 70(11):6339–6345

Leid JG, Willson CJ, Shirtliff ME, Hassett DJ, Parsek MR, Jeffers AK (2005) The exopolysaccharide alginate protects Pseudomonas aeruginosa biofilm bacteria from IFN-gamma-mediated macrophage killing. J Immunol 175(11):7512–7518

Marshall DA, Mertz PM, Eaglstein WH (1990) Occlusive dressings. Does dressing type influence the growth of common bacterial pathogens? Arch Surg 125(9):1136–1139

Mertz PM, Eaglstein WH (1984) The effect of a semiocclusive dressing on the microbial population in superficial wounds. Arch Surg 119(3):287–289

Percival SL, Bowler P, Woods EJ (2008) Assessing the effect of an antimicrobial wound dressing on biofilms. Wound Repair Regen 16(1):52–57

Price LB, Liu CM, Melendez JH, Frankel YM, Engelthaler D, Aziz M, Bowers J, Rattray R, Ravel J, Kingsley C, Keim PS, Lazarus GS, Zenilman JM (2009) Community analysis of chronic wound bacteria using 16S rRNA gene-based pyrosequencing: impact of diabetes and antibiotics on chronic wound microbiota. PLoS One 4(7):e6462

Rhoads DD, Wolcott RD, Percival SL (2008) Biofilms in wounds: management strategies. J Wound Care 17(11):502–508

Robson MC, Stenberg BD, Heggers JP (1990) Wound healing alterations caused by infection. Clin Plast Surg 17(3):485–492

Tarnuzzer RW, Schultz GS (1996) Biochemical analysis of acute and chronic wound environments. Wound Repair Regen 4(3):321–325

Trengove NJ, Langton SR, Stacey MC (1996a) Biochemical analysis of wound fluid from nonhealing and healing chronic leg ulcers. Wound Repair Regen 4(2):234–239

Trengove NJ, Stacey MC, McGechie DF, Mata S (1996b) Qualitative bacteriology and leg ulcer healing. J Wound Care 5(6):277–280

van der Plas MJ, Jukema GN, Wai SW, Dogterom-Ballering HC, Lagendijk EL, van, Gulpen C, van Dissel JT, Bloemberg GV, Nibbering PH (2008) Maggot excretions/secretions are differentially effective against biofilms of Staphylococcus aureus and Pseudomonas aeruginosa. J Antimicrob Chemother 61(1):117–122

Vuong C, Voyich JM, Fischer ER, Braughton KR, Whitney AR, Deleo FR, Otto M (2004) Polysaccharide intercellular adhesin (PIA) protects Staphylococcus epidermidis against major components of the human innate immune system. Cell Microbiol 6(3):269–275

Werthen M, Davoudi M, Sonesson A, Nitsche DP, Morgelin M, Blom K, Schmidtchen A (2004) Pseudomonas aeruginosa-induced infection and degradation of human wound fluid and skin proteins ex vivo are eradicated by a synthetic cationic polymer. J Antimicrob Chemother 54(4):772–779

Wolcott RD, Rhoads DD (2008a) A study of biofilm-based wound management in subjects with critical limb ischaemia. J Wound Care 17(4):145–2, 154

Wolcott RD, Rhoads DD (2008b) A study of biofilm-based wound management in subjects with critical limb ischaemia. J Wound Care 17(4):145–2, 154

Wolcott RD, Kennedy JP, Dowd SE (2009) Regular debridement is the main tool for maintaining a healthy wound bed in most chronic wounds. J Wound Care 18(2):54–56

Yager DR, Nwomeh BC (1999) The proteolytic environment of chronic wounds. Wound Repair Regen 7(6):433–441

Chapter 3
The Relation of Biofilms to Chronic Otitis Media and Other Ear-Related Chronic Infections

Preben Homøe and Helle Krogh Johansen

3.1 Introduction

Ear, nose and throat infections (rhinitis, rhinosinusitis, otitis media, adenoiditis and tonsillitis), also termed, upper respiratory tract infections (URTI) are extremely frequent diseases worldwide, especially in childhood. The diseases are caused by viruses or bacteria and occur in both acute and chronic forms. The chronic forms are often progressions of the acute infections and have a tendency to recur even when relevant antibiotic therapy has been instigated. In some diseases, such as chronic otitis media with effusion (COME), chronic rhinosinusitis and chronic tonsillitis, it has been difficult to culture pathogenic bacteria. It has also been a problem to explain the frequent recurrences of acute exacerbations, for example, chronic suppurative otitis media (CSOM). Therefore, local inflammatory reactions have been suggested as causes for the maintenance of these diseases. Recently, evidence of biofilm involvement in URTI has been suggested and presented (see Rayner et al. 1998, Post 2001, Cryer 2004, Homøe et al. 2009).

Biofilms can often be found on inert medical devices. In the head and neck region, they have been found after insertion of voice prostheses between the trachea and the oesophagus after laryngectomy to improve voice rehabilitation, and on ventilation tubes inserted in the tympanic membrane to resolve COME and recurrent acute otitis media (see Biedlingmaier et al. 1998, Everaert et al. 1999). Recently, Pawlowski et al. (2005) found biofilm formation on a human cochlear implant after removal due to infection.

The mucosa of the middle ear is usually sterile while this is not the case for tonsils, adenoids and some of the sinuses. URTIs are frequent and often progress to chronic states with repeated and recalcitrant infections in the throat, adenoids, tonsils, sinuses and in the middle ear. Bacteria in free-floating forms usually associated with acute URTI are predominantly *Streptococcus pneumoniae*, non-typeable

P. Homøe (✉)
Department of Otolaryngology, Head, and Neck Surgery F 2071, Rigshospitalet, Copenhagen, DK-2100, Denmark
e-mail: phom@rh.regionh.dk

T. Bjarnsholt et al. (eds.), *Biofilm Infections*, DOI 10.1007/978-1-4419-6084-9_3,

Haemophilus influenzae, Moraxella catarrhalis, group A and G streptococci and *Staphylococcus aureus* (see Rovers et al. 2004, Bakaletz 2007) while free-floating bacteria associated with chronic URTI most often are *Escherichia coli, Pseudomonas aeruginosa* and *S. aureus* (see Bendouah et al. 2006). All these bacteria are known to be capable of producing biofilms (see Kirketerp et al. 2008). Anaerobic bacteria may also be involved in these infections.

The anatomy of the upper respiratory tract is characterized by rather small, pneumatized compartments with narrow openings, such as the Eustachian tube and the nasal sinus ostiums, and the adenoid and tonsillar regions are characterized by deep tissue grooves.

Microbiological culturing of potential pathogenic aerobic URTI bacteria after sampling may not show growth, and in COME the bacteria are only cultured from middle ear effusions in approximately 25–40% of the cases, indicating sterile effusions in 60–75% (see Giebink 1989, Post et al. 1995). By use of the polymerase chain reaction (PCR) for potential pathogenic bacterial DNA it has been demonstrated that DNA products from these bacteria are present in the effusions in approximately 80% of cases despite the fact that the bacteria cannot be cultured (see Post et al. 1995, Ueyama et al. 1995, Dingman et al. 1998, Aul et al. 1998). The importance of these findings was not readily understood and was not taken as proof of viable bacteria in the effusions. Experimental animal studies have documented that only DNA from viable bacteria can be demonstrated after 4 weeks in otherwise culture sterile effusions while inoculation of purified DNA or DNA from heat-killed bacteria cannot (see Post et al. 1996). Rayner et al. (1998) performed a study in which reverse transcription (RT)-PCR for bacterial mRNA was applied on human COME specimens. The study revealed presence of the *H. influenzae* glyceraldehyde-3-phosphate dehydrogenase gene (GAPDH) mRNA, implying bacterial metabolic activity and thereby presence of viable bacteria despite the lack of culturability. These findings, together with the recurrent and chronic clinical nature of COME and other URTIs, have given rise to the theory that bacterial biofilm may be involved in the pathogenesis of these diseases. Several experimental animal and human clinical studies have been performed during recent years to explore this biofilm hypothesis.

3.2 Otitis Media

3.2.1 Experimental Studies

The basis for the first experimental studies concerning biofilm in COME comes from a clinical study by Rayner et al. (1998) who suggested the biofilm hypothesis concerning the pathogenesis of COME. They proposed that culture negative chronic middle ear effusions as seen in COME harboured viable and metabolically active *H. influenzae* bacteria that could be detected by PCR and RT-PCR. The study involved 93 middle ear effusions from children with a median age of 17 months who underwent myringotomy or had ventilation tubes inserted for COME lasting longer than 3 months. They found that 31% of the effusions were PCR positive for *H.*

influenzae but negative by culture for *H. influenzae*. All PCR positive samples were also positive for *H. influenzae* GAPDH-specific mRNA by RT-PCR. This suggested recent presence of metabolically active *H. influenzae* or presence of *H. influenzae*. Unfortunately, biofilm was not directly visualized. Post (2001) demonstrated direct visual evidence of non-typable *H. influenzae* bacterial biofilms on middle ear mucosa by scanning electron microscopy (SEM) in an animal model of chinchillas with experimentally induced non-typeable *H. influenzae* otitis media. The animals were killed 6 days after the inoculation. Post (2001) also studied the temporal development of non-typeable *H. influenzae* biofilm by SEM and found the first evidence of extracellular matrix formation surrounding the bacteria 48 h after inoculation, and biofilm growth of the bacteria after 96 h. Animals treated with antibiotics 96 h after the inoculation did not show evidence of biofilm in the middle ear mucosa specimens. In the same series of studies, SEM and confocal microscopy documented the presence of biofilm on ventilation tubes from children with posttympanostomy suppuration. Ehrlich et al. (2002) confirmed the presence of biofilm in chinchillas after inoculation of non-typeable *H. influenzae* by SEM and confocal laser scanning microscopy, employing differential fluorescence live/dead dye staining of bacteria and cells.

Dohar et al. (2005) studied biofilm in middle ear mucosa specimens by SEM in a primate model with chronic suppurative otitis media (CSOM) caused by *P. aeruginosa*. All ears infected by inoculation of *P. aeruginosa* (strain PITT27853) in the cynomolgus monkeys revealed direct histological evidence of *P. aeruginosa* biofilm formation whereas the control ears did not.

3.2.2 Clinical Studies

Hall-Stoodley et al. (2006) were the first to show the presence of biofilm in human clinical samples of middle ear biopsies. These were from 26 children who underwent tympanostomy tube insertion for COME and recurrent acute otitis media (rAOM). Gram staining, aerobic culturing, PCR detection of *H. influenzae*, *S. pneumoniae* and *M. catarrhalis* were carried out on the middle ear specimens. For examination of middle ear mucosa biopsies, confocal laser scanning microscopy (CLSM) was applied, and for detection and visualization of bacteria in biofilms they used bacterial species-specific fluorescence in situ hybridization (FISH) with probes against *H. influenzae*, *S. pneumoniae*, *M. catarrhalis* and eubacterial probe in a smaller subset of specimens. Mucosal biofilm was found in 92% of the ears by CLSM. Only six of the 27 middle ear effusions were culture positive for a bacterial pathogen while 24 of 24 (100%) tested with PCR were positive for at least one bacterial pathogen. Bacteria in biofilm were visualized with FISH in 18 of 20 samples tested and in four samples more than one bacterium was found indicating multiple bacteria in the biofilms. The study was quite extensive but the study design was hampered by small amounts of effusion and middle ear mucosa material. The conclusion of this study was that chronic middle ear disorders may be biofilm-related.

Chole and Faddis (2002) found evidence of biofilm in humans suffering from the chronic ear disease, cholesteatoma. This was reported along with biofilm findings in experimentally induced cholesteatoma in gerbils. Cholesteatoma is a destructive ear disease. It consists of a keratin matrix and is often combined with infection, resulting in chronic recurrent suppuration from the middle ear. Histological sections of surgically removed cholesteatomas were examined by light microscopy and transmission electron microscopy. Sixteen of 24 (66%) human cholesteatomas showed morphological evidence of biofilm. Gram-staining showed evidence of both Gram-positive and Gram-negative bacteria indicating multiple bacterial species in the biofilms. In another study, the same authors (2002) used the experimental cholesteatoma model in gerbils and found microscopical evidence of biofilm formation in 21 of 22 keratin accumulations. This was done by light and transmission electron microscopy.

Wang et al. (2005) demonstrated biofilm formation in vitro using a crystal violet assay. They used 10 otopathogenic *P. aeruginosa* strains obtained from human patients with cholesteatoma. *P. aeruginosa* were allowed to adhere to keratinocytes in a human keratinocyte culture, and expression of intracellular signalling quorum-sensing genes (*lasR, lasI* and *rhlR, rhlI*) was demonstrated using primers for RT-PCR, twitching motility, as wells as alginate production as a measure of the ability of the bacteria to produce extracellular matrix. The otopathogenic *P. aeruginosa* strains were all biofilm producers and all isolates had increased adherence capability to keratinocytes compared to the laboratory strain. The *las* and *rhl* genes were expressed in 11 of 12 of the otopathogenic *P. aeruginosa* strains but in none of the laboratory strains. Also, the twitching ability was found to be significantly lower in all but one of the otopathogenic *P. aeruginosa* strains compared to the laboratory strains. This is consistent with findings from *P. aeruginosa* strains obtained from the lungs of cystic fibrosis patients (see Li et al. 2005 and Chapter 10). Finally, alginate production was also significantly higher in 11 of the 12 otopathogenic strains compared to the laboratory strain.

In a study by Pinar et al. (2008), *S. aureus* and *P. aeruginosa* from human patients with CSOM were examined in vitro for presence of intercellular adhesion genes for *S. aureus* (*ica* A and *ica* D) and for extracellular matrix formation by culturing on Congo red agar plates (Sigma Chemical Company, St Louis, MO, USA). Four of the five *S. aureus* strains tested were positive for *ica* but only two of these were positive for extracellular matrix formation on the plates, while seven of the 11 *P. aeruginosa* strains were positive for extracellular matrix formation. These studies indicate that biofilm forming bacteria can be cultured from different specimens obtained from different chronic middle ear diseases.

Recently, Moriyama et al. (2009) reported that 84% of non-typeable *H. influenzae* cultured from children with intractable acute otitis media using crystal violet assay were biofilm formers and that the level of biofilm formation was higher in ampicillin-resistant strains.

Homøe et al. (2009) studied the presence of biofilm in patients from Greenland where chronic otitis media is widespread. Conventional culturing of bacteria, Gram-staining and peptide nucleic acid (PNA) FISH of smears of aspirated otorrhea from children with CSOM and of aspirated middle ear effusions without mucosa biopsies from children with COME were performed along with middle ear mucosa biopsies from adult patients operated for CSOM. The specimens were examined by light immersion microscopy and CLSM. Morphological evidence of biofilm was found in five of six specimens of otorrhea from children with CSOM and in eight of ten middle ear mucosa biopsies from adults with CSOM (see Fig. 3.1). There was no evidence of biofilm in seven smears of middle ear effusions from the four COME children. As CSOM is a frequent problem in children and adults in Greenland and is often recalcitrant to the treatment applied, it was postulated that biofilm may be involved in CSOM. The species-specific PNA-FISH analyses and eubacterial PNA-FISH analyses indicated the presence of multibacterial biofilms in the specimens obtained.

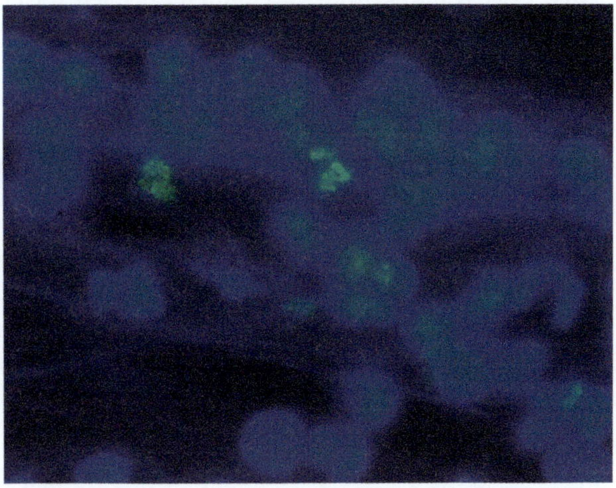

Fig. 3.1 *S. aureus* biofilms in smear from a patient with chronic suppurative otitis media (CSOM)

Biofilm in chronic otitis media being COME, rAOM, CSOM and cholesteatoma may therefore be of pathogenetic importance. Biofilm has been found in specimens from these diseases and biofilm forming bacteria have been cultured and detected in viable forms in the biofilms. The exact role of these findings still awaits resolution. Furthermore, treatment modalities directed against biofilm characteristics have not yet been developed. Until now, treatment therefore still relies on local and systemic antiobiotics, local mechanical irrigation and cleaning, sometimes using alcohol or acetic acid solutions, and surgical eradication of the mucosa in the middle ear (see Verhoeff et al. 2006).

3.3 Tonsils and Adenoids

3.3.1 Experimental Studies

Studies concerning biofilm in tonsillar and adenoid diseases are all clinical, which is probably related to the easy access to these tissues and there are no published experimental studies on tonsils and adenoids.

3.3.2 Clinical Studies

Chole and Faddis (2003) were the first to document morphological evidence of biofilm as seen in transmission electron microscopy on histological sections from the tonsils of 15 patients operated for recurrent tonsillitis and 4 patients operated for hyperplastic tonsils resulting in sleep apnoea. Evidence of biofilm was found in 11 of 15 infected tonsils and in 3 of 4 hypertrophic tonsils. The biofilms were mostly seen in the crypts of the tonsils and consisted of both Gram-positive and Gram-negative bacteria.

Lately, Kania et al. (2008) reported almost similar results using scanning electron microscopy and CLSM of tonsils double stained with propidium iodide and FITC concavalin A from 24 children operated for chronic or recurrent tonsillitis. Morphological evidence of biofilm was seen in 17 of the 24 patients using CLSM but only in 5 of 14 suitable specimens using scanning electron microscopy. Zuliani et al. (2006) and Galli et al. (2007) examined by scanning electron microscopy adenoid tissue in patients with URTI and found dense biofilm formation. Al-Mazrou and Al-Khattaf (2008) have reported almost identical results.

Biofilm has been found morphologically in both tonsillar and adenoid tissue from patients with chronic tonsillitis or tonsillar hyperplasia and in patients with chronic URTI in need for tonsillectomy or adenoidectomy. Also, bacteria have been detected in viable forms in the biofilms. The cause and relationship with disease in these organs is still to be proven and specific biofilm treatment is therefore not an issue at the moment. Treatment of these disease entities still consists of prolonged systemic antiobiotics and surgical eradication with tonsillectomy and adenoidectomy. Biofilm in tonsils and adenoids also in individuals without local disease may serve as reservoir for repeated infections, especially in the lungs in patients with cystic fibrosis or with chronic obstructive lung disease. This needs to be further examined, however.

3.4 Medical Devices in Otorhinolaryngology

Implants are widely used in otorhinolaryngology, especially ventilation tubes for treatment of COME. Contamination on implants (see also Chapter 5) are believed to cause device failure and recalcitrant infections. The implants may be extruded or must be surgically removed. Biofilms on implants are most often examined by conventional culture and scanning electron microscopy of the devices.

Biedlingmaier et al. (1998), Saidi et al. (1999), Berry et al. (2000), Post (2001), and Bothwell et al. (2003) have all demonstrated bacterial biofilm formation using scanning electron microscopy or CLSM experimentally and in clinical studies in animals and patients with posttympanostomy otorrhea on tympanostomy tubes of different materials ranging from silicone to titanium and with different coatings (e.g. silver-oxide) of the surfaces. It is hypothesized that the biofilm colonization on the tubes causes recurrent suppuration. One study by Berry et al. (2000) reported that, in

S. aureus and *P. aeruginosa*, in vitro biofilm formation was not seen on ventilation tubes with phosphorylcholine coated fluoroplastic polytetrafluoroethylene surfaces. It is the usual procedure to remove the ventilation tubes if suppuration continues (so-called posttympanostomy otorrhea).

Everaert et al. (1999) and Leunisse et al. (2001) have, among others, demonstrated biofilm formation using scanning electron microscopy of voice prostheses (tracheo-oesophageal prostheses). It is hypothesized that the biofilm coating causes impaired airflow through the valve, thereby affecting speech quality and causing a leak of the valve in the opposite direction. Experiments have been done with different materials ranging from silicone- to titanium-coated implants and other coatings (e.g. perfluoro-alkylsiloxane). It is well recognized that *Candida albicans* and other candida species are often found in the biofilms along with different bacteria like streptococcus and staphylococcus species. Some recommend probiotics or local application of chemotherapeutics, such as miconazole. Often, the prostheses will have to be replaced.

Emery et al. (2003) studied in vitro bacterial biofilm formation on different reconstruction plates for maxillofacial surgery and found that a strain of *S. aureus* did not adhere to and did not show biofilm formation on titanium plates and resorbable implant plates (polylactide plates), but did so on silicone ventilation tubes. This is in accordance with the relatively low incidence of infections in reconstructive maxillofacial surgery.

Cochlear implants are used for hearing rehabilitation in patients with severe hearing loss. Sometimes, an infection demands removal of the implant. Antonelli et al. (2004) and Pawlowski et al. (2005) detected biofilm formation on such a removed human cochlear implant after intractable infection. The latter case was caused by *S. aureus*.

It is important that implantable devices are tested for resistance to bacterial adherence and biofilm production and to develop resistant materials or coatings for these devices even when they are implanted in otherwise sterile regions. As a rule, all medical implants may be contaminated before implantation or during implantation.

3.5 Other Biofilm Findings in Otolaryngologic Diseases

Nason and Chole (2007) recently described evidence of biofilm in recalcitrant osteoradionecrosis of the temporal bone in a patient formerly undergoing treatment with irradiation for a nasopharyngeal carcinoma. The patient suffered from chronic otorrhea, pain and hearing loss and the bone was exposed in several areas. Biopsies obtained during eradication surgery showed biofilm formation in several spots with extracellular matrix without penetration of inflammatory cells. After surgery, the patient was treated with antibiotics intravenously and hyperbaric oxygen and improved on this treatment. These findings are in accordance with findings of bacteria seen in transmission and scanning electron microscopy in osteoradionecrosis of the mandible (see Store and Olsen 2005). It is not yet clear whether biofilm infection in osteoradionecrosis is a result of the osteoradionecrosis or whether the bacterial infection may play a role in the process leading to osteoradionecrosis. Only experimental studies can reveal this and these are not yet available.

3.6 Future Implications

Till now, there are no studies or documentation of effective treatment of biofilm in ear, nose and throat infections, except for surgical removal of the affected mucosal sites in the middle ear and the sinuses and in adenotonsillectomies. Probiotics are used in voice prosthesis but hitherto this has not been successfully applied in tissue infections. Wang et al. (2008) recently published in vitro results of the efficacy of gentian violet and ferric ammonium citrate on biofilm volume created by two different strains of otopathogenic *P. aeruginosa* using fluorescence and confocal laser scanning microscopy. It was found that gentian violet reduced the biofilm volume and the rate of growth in both strains while ferric ammonium citrate only reduced biofilm volume in one of the strains. It is suggested from the study that different strains of biofilm forming bacteria may show difference in their response to treatment.

Understanding the mechanisms of biofilm development in chronic otitis media is important when trying to develop strategies for its prevention. One of the most promising tools could be vaccination against *S. pneumoniae* and *H. influenzae* since they are the predominant bacterial pathogens that cause acute otitis media. Bacterial vaccination may be able to prevent colonisation, thereby reducing the burden of bacteria in the nasopharynx of children and the retrograde ascension via the Eustachian tube of bacteria from the nasopharynx to the middle ear. In addition, viral vaccines have been shown to be able to prevent episodes of acute otitis media, thereby preventing development into COME and eventually a biofilm disease. Furthermore, understanding the unique properties of bacteria residing in biofilm would be an attractive approach for developing biofilm-specific target therapies for preventing otitis media (see Rovers et al. 2004).

References

Al-Mazrou KA, Al-Khattaf AS (2008) Adherent biofilms in adenotonsillar diseases in children. Arch Otolaryngol Head Neck Surg 134(1):20–23

Antonelli PJ, Lee JC, Burne RA (2004) Bacterial biofilms may contribute to persistent cochlear implant infection. Otol Neurotol 25(6):953–957

Aul JJ, Anderson KW, Wadowsky RM et al (1998) Comparative evaluation of culture and PCR for the detection and determination of persistence of bacterial strains and DNAs in the Chinchilla laniger model of otitis media. Ann Otol Rhinol Laryngol 107(6):508–513

Bakaletz LO (2007) Bacterial biofilms in otitis media evidence and relevance. Ped Infect Dis J 26:S17–S19

Bendouah Z, Barbeau J, Hamad WA et al (2006) Biofilm formation by *Staphylococcus aureus* and *Pseudomonas aeruginosa* is associated with unfavourable evolution after surgery for chronic sinusitis and nasal polyposis. Otolaryngol Head Neck Surg 134:991–996

Berry JA, Biedlingmaier JF, Whelan PJ (2000) In vitro resistance to bacterial biofilm formation on coated fluoroplastic tympanostomy tubes. Otolaryngol Head Neck Surg 123(3):246–251

Biedlingmaier JF, Samaranayake R, Whelan P (1998) Resistance to biofilm formation on otologic implant materials. Otolaryngol Head Neck Surg 118(4):444–451

Bothwell MR, Smith AL, Phillips T (2003) Recalcitrant otorrhea due to *Pseudomonas* biofilm. Otolaryngol Head Neck Surg 129:599–601

Chole RA, Faddis BT (2002) Evidence for microbial biofilms in cholesteatomas. Arch Otolaryngol Head Neck Surg 128:1129–1133

Chole RA, Faddis BT (2003) Anatomical evidence of microbial biofilms in tonsillar tissues: a possible mechanism to explain chronicity. Arch Otolaryngol Head Neck Surg 129:634–636

Cryer J, Schipor I, Perloff JR et al (2004) Evidence of bacterial biofilms in human chronic sinusitis. ORL J Otorhinolaryngol Relat Spec 66:155–158

Dingman JR, Rayner MG, Mishra S et al (1998) Correlation between presence of viable bacteria and presence of endotoxin in middle-ear effusions. J Clin Microbiol 36(11):3417–3419

Dohar JE, Hebda PA, Veeh R et al (2005) Mucosal biofilm formation on middle-ear mucosa in a nonhuman primate model of chronic suppurative otitis media. Laryngoscope 115:1469–1472

Ehrlich GD, Veeh R, Wang X et al (2002) Mucosal biofilm formation on middle-ear mucosa in the chinchilla model of otitis media. JAMA 287(13):1710–1715

Emery BE, Dixit R, Formby CC et al (2003) The resistance of maxillofacial reconstruction plates to biofilm formation in vitro. Laryngoscope 113(11):1977–1982

Everaert EP, Mahieu HF, van de Belt-Gritter B et al (1999) Biofilm formation in vivo on perfluoro-alkylsiloxane-modified voice prostheses. Arch Otolaryngol Head Neck Surg 125(12): 1329–1332

Galli J, Ardito F, Calo L et al (2007) Recurrent upper airway infections and bacterial biofilms. J Laryngol Otol 121:341–344

Giebink GS (1989) The microbiology of otitis media. Pediatr Infect Dis J 8(1 Suppl):S18–20

Hall-Stoodley L, Hu FZ, Gieseke A et al (2006) Direct detection of bacterial biofilms on the middle-ear mucosa of children with chronic otitis media. JAMA 296:202–211

Homøe P, Bjarnsholt T, Wessman M et al (2009) Morphological evidence of biofilm formation in Greenlanders with chronic suppurative otitis media. Eur Arch Otorhinolaryngol 266(10): 1533–1538

Kania RE, Lamers GE, Vonk MJ et al (2007) Demonstration of bacterial cells and glycocalyx in biofilms on human tonsils. Arch Otolaryngol Head Neck Surg 133(2):115–121

Kirketerp-Møller K, Jensen PØ, Fazli M et al (2008) Distribution, organisation and ecology of bacteria in chronic wounds. J Clin Microbiol 46:2717–2722

Leunisse C, van Weissenbruch R, Busscher HJ et al (2001) Biofilm formation and design features of indwelling silicone rubber tracheoesophageal voice prostheses—an electron microscopical study. J Biomed Mater Res 58(5):556–563

Li Z, Kosorok MR, Farrell PM et al (2005) Longitudinal development of mucoid *Pseudomonas aeruginosa* infection and lung disease progression in children with cystic fibrosis. JAMA 293:581–588

Moriyama S, Hotomi M, Shimada J et al (2009) Formation of biofilm by Haemophilus influenzae isolated from pediatric intractable otitis media. Auris Nasus Larynx 161(2):162–170

Nason R, Chole RA (2007). Bacterial biofilms may explain chronicity in osteoradionecrosis of the temporal bone. Otol Neurotol 28:1026–1028

Pawlowski KS, Wawro D, Roland PS (2005) Bacterial biofilm formation on a human cochlear implant. Otol Neurotol 26:972–975

Pinar E, Öncel S, Karagöz Ü et al (2008) Demonstration of bacterial biofilms in chronic otitis media. Mediterr J Otol;4:64–68

Post JC (2001) Direct evidence of bacterial biofilms in otitis media. Laryngoscope 111:2083–2094

Post JC, Aul JJ, White GJ et al (1996) PCR-based detection of bacterial DNA after antimicrobial treatment is indicative of persistent, viable bacteria in the chinchilla model of otitis media. Am J Otolaryngol 17(2):106–111

Post JC, Preston RA, Aul JJ et al (1995) Molecular analysis of bacterial pathogens in otitis media with effusion. JAMA 273(20):1598–1604

Rayner MG, Zhang Y, Gorry MC et al (1998) Evidence of bacterial metabolic activity in culture-negative otitis media with effusion. JAMA 279:296–299

Rovers MM, Schilder AGM, Zielhuis GA et al (2004) Otitis media. Lancet 363:465–473

Saidi IS, Biedlingmaier JF, Whelan P (1999) In vivo resistance to bacterial biofilm formation on tympanostomy tubes as a function of tube material. Otolaryngol Head Neck Surg 120:621–627

Store G, Olsen I (2005) Scanning and transmission electron microscopy demonstrates bacteria in osteoradionecrosis. Int J Oral Maxillofac Surg 34:777–781

Ueyama T, Kurono Y, Shirabe K et al (1995) High incidence of Haemophilus influenzae in nasopharyngeal secretions and middle ear effusions as detected by PCR. J Clin Microbiol 33(7):1835–1838

Verhoeff M, van der Veen EL, Rovers MM et al (2006) Chronic suppurative otitis media: a review. Int J Pediatr Otorhinolaryngol 70:1–12

Wang EW, Agostini G, Olomu O et al (2008) Gentian violet and ferric ammonium citrate disrupt Pseudomonas aeruginosa biofilms. Laryngoscope 118(11):2050–2056

Wang EW, Jung JY, Pashia ME et al (2005) Otopathogenic Pseudomonas aeruginosa strains as competent biofilm formers. Arch Otolaryngol Head Neck Surg 131(11):983–989

Zuliani G, Carron M, Gurrola J et al (2006) Identification of adenoid biofilms in chronic rhinosinusitis. Int J Pediatr Otorhinolaryngol 70(9):1613–1617

Chapter 4
Human Oral Bacterial Biofilms: Composition, Dynamics, and Pathogenesis

Robert J. Palmer Jr., Richard Darveau, Richard J. Lamont, Bente Nyvad, and Ricardo P. Teles

4.1 Introduction

The oral cavity is home to *ca.* 700 bacterial species, many of which engage in biofilm formation through a sequential and ordered accumulation on tooth surfaces. Supra- and subgingival oral biofilms are complex and dynamic multispecies communities even at very early stages when only a few cells are present. The arrangement of cells is influenced by specific cell–cell recognition which establishes particular combinations of bacteria. Inter-bacterial communication is likely to play a significant role in metabolism within these communities. In its early stages of development, plaque is dominated by streptococci and actinomyces, and it exists in commensal harmony with the host. However, in the absence of adequate oral hygiene, ecological shifts occur within the microbial community which initiate two complex oral diseases: caries and periodontal diseases. These diseases do not arise from sudden infection by, or even chronic presence of, a single pathogen – instead they result from over-representation of several pathogenic bacteria that are typically found in low numbers in the healthy commensal plaque microflora and the concomitant under-representation of the normally more abundant commensal organisms. Virulence factors have been described for several pathogenic bacteria, but little is understood about the manner in which commensal bacteria promote health. Furthermore, oral bacteria that are normally commensal may play a role in certain systemic diseases.

Dental caries (tooth decay that ultimately leads to cavitation) result from ecologically controlled metabolic processes in the plaque biofilm that disturb mineral balance at the biofilm–tooth interface. Caries present a range of clinical manifestations, varying from so-called white spot lesions with intact surfaces to overt cavities and tooth destruction. Comprehensive epidemiological studies show that caries and subsequent periapical infections remain the single most important reason for

R.J. Palmer Jr. (✉)
Oral Infection and Immunity Branch, National Institute of Dental and Craniofacial Research, National Institutes of Health, Bethesda, MD, USA
e-mail: rjpalmer@dir.nidcr.nih.gov

T. Bjarnsholt et al. (eds.), *Biofilm Infections*, DOI 10.1007/978-1-4419-6084-9_4,

tooth extractions worldwide, even greater than the number for periodontal disease (Baelum et al. 2008).

Periodontal diseases are complex polymicrobial biofilm infections that involve interactions between a large subset of oral bacteria and the host. Accumulation of plaque at the gingival margin triggers an inflammatory reaction (gingivitis). In susceptible subjects, the inflammatory process expands deeper into the tissues resulting in the loss of collagen fibers that support the tooth (attachment loss), as well as the resorption of alveolar bone: the two hallmarks of periodontitis. The interface between the oral microbial biofilm and the host immune system is unique; a non-shedding highly mineralized hydroxylapatite surface protrudes from normally sterile host tissue. Dental plaque grows at this interface, and the balance between coexistence with commensal bacteria and attack by pathogens places special burdens on the immune protective mechanisms of the host tissue. Pathogens are rarely eliminated by the inflammatory immune responses they cause. Instead, inflammation leads to increased flow of gingival crevicular fluid (GCF), a transudate that leaves the gingival capillaries and emerges at the gingival sulcus, which is an abundant source of bacterial nutrients (Socransky and Haffajee 2005). Thus, interplay between the host and the pathogenic bacterial community results in chronic inflammation that leads not only to tissue destruction but also fosters further proliferation of the subgingival biofilm. Disease manifestation depends not only on the composition of the subgingival microbiota but also on the magnitude and quality of the host response to the oral bacteria, as well as on predisposing genetic and environmental factors (e.g., smoking) in the host (Page and Kornman 1997).

Here we provide an introduction to microbial composition and physiology of oral biofilms, to virulence mechanisms of oral pathogens, to immunobiology of host/microbe interactions, and to the clinical indications and treatment of caries and of periodontal diseases. The broad scope limits detail, yet we hope this contribution makes clear that the human oral microbial ecosystem is one of the best characterized bacterial ecosystems anywhere.

4.2 Composition of Oral Biofilms

According to the latest molecular census data, the oral cavity is home to roughly 700 bacterial phylotypes, about 50% of which are known from previous bacteriological studies (Aas et al. 2005). Molecular and bioinformatics techniques that create 16S rRNA catalogues reveal a few unusual taxa that were not recovered by classical approaches (Paster et al. 2006), at least one of which may play a role in oral disease: TM7 (Brinig et al. 2003). Furthermore, these data show that the inventory of bacteria from a single healthy individual reaches about 100 organisms; the much higher diversity of 700 phylotypes is obtained by combining lists across many individuals and includes organisms found in extreme disease states such as gangrenous stomatitis (Paster et al. 2006). While these data are seminal to our understanding of bacterial diversity, they do not describe the proportions and activities of the various bacteria, i.e., the structure and physiology of the biofilm community.

The oral cavity presents an assortment of surfaces for colonization. An obvious distinction can be made between hard surfaces (enamel and dentin), epithelial surfaces (tongue, cheek, and gum), and saliva. Comprehensive data on the proportion of various species at these sites are few. One study (Mager et al. 2003) catalogued and compared species composition at various locations by enumerating major oral isolates via a panel of 40 DNA-based probes. The choice of probes mirrored the prevalence of streptococci and actinomyces obtained in non-selective isolation studies. Seven *Streptococcus* spp. and five *Actinomyces* spp. were included, along with probes representative of lesser but important species, such as fusobacteria and capnocytophagae, as well as probes covering putative periodontal pathogens, such as *Porphyromonas gingivalis*, *Bacteroides forsythis* (now *Tannerella forsythia*), and *Actinobacillus* (now *Aggregatibacter*) *actinomycetemcomitans*. Tooth-surface biofilm compositions clustered together and were distinct from those of other sites, regardless of whether the sample was supragingival (from above the gum line) or subgingival (from below the gum line); *Actinomyces* spp. made up a significantly higher proportion of the flora from tooth surfaces than that from epithelial surfaces or from saliva. Flora of the lateral and dorsal surfaces of the tongue clustered together with salivary flora; these environments were dominated by *Veillonella* spp. and *Prevotella* spp. Composition of biofilms on all other epithelial surfaces clustered together and differed from those on the tongue through enrichment in three *Streptococcus* species (*oralis*, *mitis*, and *constellatus*), in *Gemella morbillorum* (formerly a *Streptococcus* sp.), and in *Capnocytophaga ochracea*. Such studies emphasize habitat diversity within the oral cavity while providing baseline data for studies of changes within these different biomes.

A more comprehensive approach to quantitative community analysis is reverse-checkerboard hybridization: digoxygenin-labeled 16S rDNA PCR amplicons are generated from the total nucleic acid of the community, captured via hybridization to an array bearing spots of probes complementary to known rDNA sequences, and then detected using chemifluorescence (Paster et al. 1998). The resulting fluorescence intensity map yields the cell number of the target species (currently 272 phylotypes) within the sample; the standardized hybridization/analysis procedure is offered commercially (http://mim.forsyth.org) (Dewhirst et al. 2010). The ability to quantify hundreds of phylotypes simultaneously within a single sample has highlighted the role of bacterial communities in pathogenesis; the ecological complexity of the oral cavity dictates that disease studies must take into account not only the overabundance of "pathogenic" phylotypes, but also the underabundance of phylotypes associated with health. Studies on caries (Becker et al. 2002, Aas et al. 2008) show that, in addition to bacteria historically recognized as caries pathogens (e.g., *S. mutans* and lactobacilli), other bacteria including *Veillonella* spp., *Bifidobacteria* spp., and *Actinomyces gerencseriae* are statistically tied to caries lesions. In turn, *Eubacteria* spp., *Leptotrichia* spp., and certain streptococci are associated with healthy enamel. Similar studies are underway to define periodontal disease from the standpoint of the healthy subgingival microflora (Cotton et al. 2009). In addition to the widely recognized pathogens *P. gingivalis* and *T. forsythia*, other bacteria associated with periodontitis include *Campylobacter* spp. and *Parvimonas micra*.

Periodontal health is characterized by streptococci, actinomyces, and *Rothia* spp. Thus an exchange of saccharolytic organisms for proteolytic organisms characterizes the transition from health to disease. Methanogens from the domain Archaea represent a minor component of the oral microbial population and therefore are not typically included in molecular surveys. However, molecular evidence exists to link them with a subset of periodontal disease cases (Lepp et al. 2004)— an interesting syntrophic relationship can be postulated by which these organisms utilize hydrogen and thereby promote growth of hydrogen-sensitive organisms. Regardless of the taxonomic and physiological details of the microflora in periodontally diseased patients, treatment of the disease must result in restoration and maintenance of the healthy flora during subsequent recolonization of treated sites (Quirynen et al. 2005). This strategy requires manipulation of the community not only away from pathogens but also towards commensal bacteria.

4.3 Spatiotemporal Development of Oral Biofilms

Developmental biology of natural biofilms, as for that of multicellular organisms, has temporal and spatial components. Biofilm studies generally approach only one aspect—spatial studies often consist of a snapshot that provides information obtained at a single timepoint, and temporal studies typically track species composition without regard to the positioning of bacteria.

Spatiotemporal changes in community composition are assessed primarily through model systems in which substrata placed in the oral cavity are retrieved at different times. Studies on supragingival plaque dominate because of sampling simplicity—the substratum is easily placed and retrieved with minimal patient discomfort. One exemplary early study coupled traditional bacteriological taxonomy with electron microscopy to examine spatial and temporal aspects of smooth-surface plaque development (Nyvad and Fejerskov 1987a, b, Nyvad and Kilian 1987, Nyvad 1993). While the power of bacterial taxonomy was lower than it is today, the study's basic conclusions remain seminal to our understanding of oral biofilm development. The number of cultivable bacteria colonizing a clean enamel surface increases exponentially during the first 8 h and reaches a plateau of roughly 10^6 cells per 0.025 cm^2 by 12 h. Streptococci make up at least 60% of the flora and can reach 90%. Gram positive rods (predominantly *Actinomyces* spp.) and Gram negative cocci (predominantly *Veillonella* spp.) made up the remaining biomass. A shift within the streptococcal community occurs as the percentage of *S. oralis* increases at the expense of other streptococci. After 4 h of colonization, scanning electron microscopy shows widely separated single cells and pairs of cells which develop into a monolayer across large areas by 12 h. Although morphological diversity is limited in oral bacteria, some rod-shaped organisms can be seen among the largely coccoid cells of the community. Transmission electron microscopy reveals Gram negative and Gram positive cell walls abutting one another, which suggests that biofilm development does not occur exclusively through expansion of clonal colonies.

A temporal molecular study of early smooth-surface plaque (Diaz et al. 2006) supported conclusions of earlier bacteriological studies while clearly demonstrating that individual variation in community composition can be significant—of the 97 recovered phylotypes, only ten (eight streptococci, one *Neisseria* sp., and one *Gemella* sp.) were common to all three of the study's subjects. Streptococci made up 60–80% of the phylotypes from any one subject and *S. oralis* was the only phylotype that clearly increased with time. In contrast, 68 of the phylotypes were unique to a particular subject: one subject bore 44 phylotypes not seen in the other two subjects, whereas the other two subjects bore 13 and 11 unique phylotypes. The predominant Gram negative phylotypes were *Veillonella* spp. and *Prevotella* spp. While the multispecies nature of small clusters of cells in this early plaque was demonstrated using fluorescence in situ hybridization (FISH), those data were from a single timepoint. Clear evidence of spatiotemporal development is provided through an immunofluoresence approach (Palmer et al. 2003). The data support those of Nyvad et al. (Nyvad and Fejerskov 1987a, b, Nyvad and Kilian 1987, Nyvad 1993), while further showing that phenotypically different bacteria can be located within the same cluster of morphologically indistinguishable cells. These multispecies communities develop further by growth and attachment/integration of additional phenotypically different cells. Cell–cell recognition plays a role in the developmental process. Specific recognition and binding of oral isolates to one another, termed coaggregation (Gibbons and Nygaard 1970, Kolenbrander 1988), is a characteristic of oral bacteria known from in vitro studies. In the immunofluorescence study cited above, adjacent cells were identified as phenotypically different by sets of antibodies specific for coaggregating organisms. The most convincing example was that of cells reactive with antibody raised against type-2 fimbriae (an adhesin on certain *Actinomyces naeslundii* strains) juxtaposed with cells reactive with antibody against receptor polysaccharide (RPS), a cell-surface carbohydrate on certain streptococci (Cisar et al. 1995) that recognizes complementary adhesins such as type-2 fimbriae. Other cell–cell contacts also suggestive of adhesin–receptor interaction were identified using less-specific antibodies. The diversity within an initial cell cluster, and therefore the cellular arrangement and interaction within the developing biofilm, seems to depend to a certain degree on cell–cell recognition as a driving force in community formation. This concept was further underscored by the capture of a small cell cluster from early plaque and the subsequent isolation/culture of the cluster's members to yield coaggregating streptococci (Chalmers et al. 2008). Cell–cell recognition seems critical to establishment of functioning communities, especially in early plaque formation. Figures 4.1 and 4.2 document the development of supragingival plaque and illustrate the concepts outlined above.

Spatial arrangement in late (thick) supragingival plaque has been studied with electron microscopy in a removable-crown model (Listgarten 1994). The strikingly complex architecture is characterized by abutted columns, termed palisades, in which a particular cell morphology (or mixture of morphologies) predominates within the adjoining columns. In some cases, columns are demarcated by differences in spacing between cells within the column. Figure 4.3a,b illustrates spatial arrangement in late (thick) supragingival plaque. Subgingival plaque biofilms retrieved

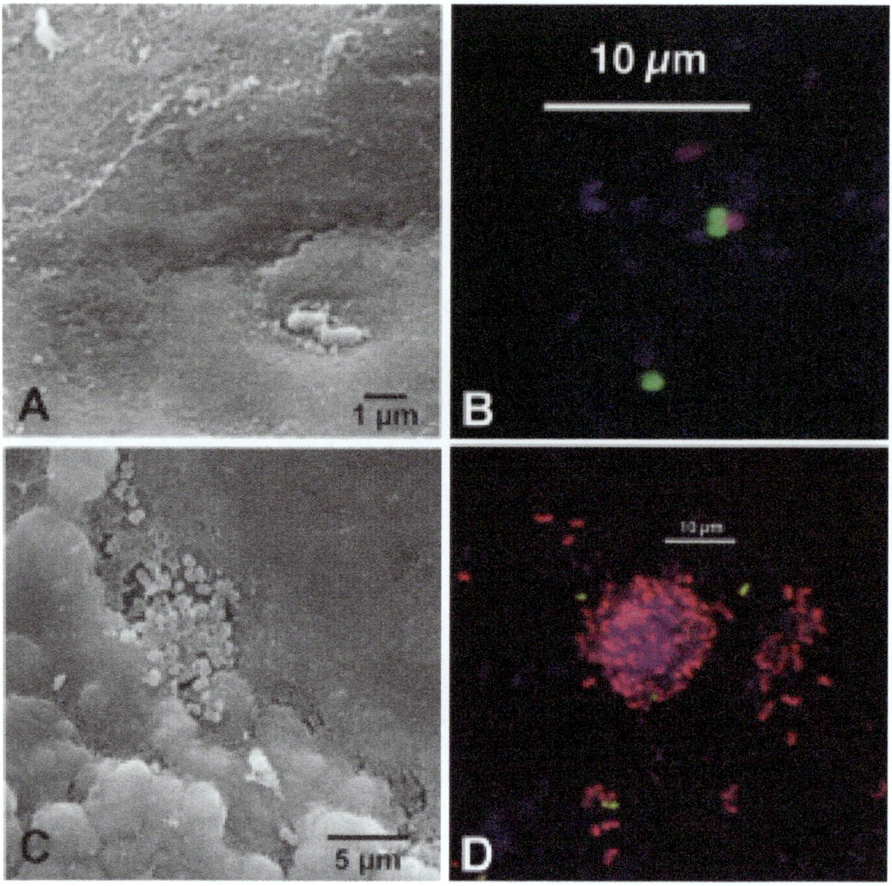

Fig. 4.1 Comparison of spatiotemporal development of smooth-surface plaque as documented by scanning electron microscopy (SEM) and by confocal microscopy. (**a**) SEM; initial colonization (4 h of growth). Sparse adjacent cells. (**b**) Confocal micrograph; 4 h. Cells reactive with an antibody against *Streptococcus gordonii* DL1 (anti-DL1) are *green* and cells reactive with antibody against receptor polysaccharide (anti-RPS) are *red*. An initial mixed-species community is present that contains two streptococcal cells reactive with anti-DL1 and one streptococcal cell that bears RPS. Other unidentified cocci (including pairs of cells) reactive with the nucleic acid stain Syto 59 are *blue*. (**c**) SEM; 8 h of growth. Cluster of cells containing primarily cocci but also scattered rods. (**d**) Confocal micrograph, 8 h. Mixed-species cluster of coccoid cells containing streptococci that bear RPS (*red*), streptococci reactive with anti-DL1 (*green*), and unidentified cells (*blue*). Clusters can be communities of phenotypically different cells. SEM images reprinted from (Nyvad and Fejerskov 1987b); confocal micrographs reprinted from Palmer Jr. et al. (2003)

in the crown model had a mosaic architecture also known from FISH studies of subgingival biofilms retrieved from periodontally diseased patients (Wecke et al. 2000). Figure 4.3c–e document cellular arrangement in subgingival plaque. However, much of our knowledge of cellular arrangement in subgingival biofilms comes from immunoelectron microscopy and immunohistochemical studies on extracted teeth, particularly teeth excised with the surrounding tissue intact. For healthy teeth, these studies describe a small sparsely colonized "plaque-free" zone

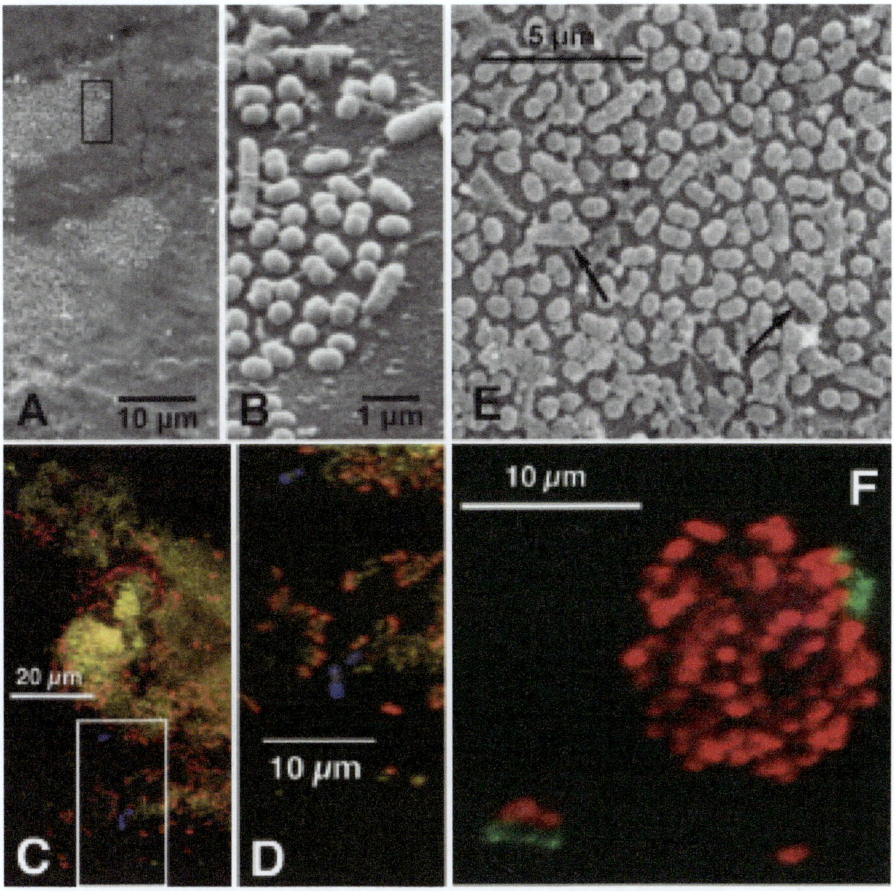

Fig. 4.2 Comparison of spatiotemporal development of smooth-surface plaque as documented by scanning electron microscopy (SEM) and by confocal microscopy. (**a**) SEM; spreading sheet of biofilm at 12 h of growth. Advancing edge of biofilm is boxed. (**b**) SEM; magnification of box from panel a. At the edges of the primarily coccoid cell mass, some rod-shaped cells are present. (**c**) Confocal micrograph; sheet-like 8-h-old biofilm. Acridine orange (*green*) stains DNA within cells. Anti-RPS (*red*) stains the periphery of the subset of streptococci that bear receptor polysaccharide. Anti-type-1 (*blue*) stains type 1 fimbriae of genospecies 2 *Actinomyces naeslundii* (now *A. oris*) strains. Advancing edge of biofilm is boxed. (**d**) Magnification of box from panel C. Rod-shaped *A. naeslundii* cells (*blue*) are present at the edges of the coccoid cell mass. The cocci are a mixture of RPS-bearing streptococci (*orange/yellow*) and cocci that lack RPS (*green*). (**e**) SEM; cell mass containing rod-shaped cells (*arrows*) in close proximity to cocci. (**f**) Confocal micrograph of *A. naeslundii* cells reactive with antibody against type-2 fimbriae (adhesin, *green*) in contact with streptococci reactive with anti-RPS (receptor, *red*). The spatial relationship between these two cell types is established by the adhesin–receptor pair identified by the antibodies. SEM images reprinted from Nyvad and Fejerskov (1987b); confocal micrographs reprinted from Palmer et al. (2003)

just below the gingival margin, but above the tissue attachment site. A small amount of subgingival plaque is found above this region but is difficult to clearly demarcate from supragingival plaque lying close to the gingiva. In stark contrast, periodontally diseased teeth have lost much of their attachment to surrounding tissue, and the sulcus is enlarged. The plaque-free zone is close to the root and contains small clusters of periopathogenic bacteria such as *P. gingivalis*. The disease-associated subgingival plaque mass fills the enlarged sulcus and, within the mass, *P. gingivalis* is found in the deeper region while *Prevotella* spp. are located away from the

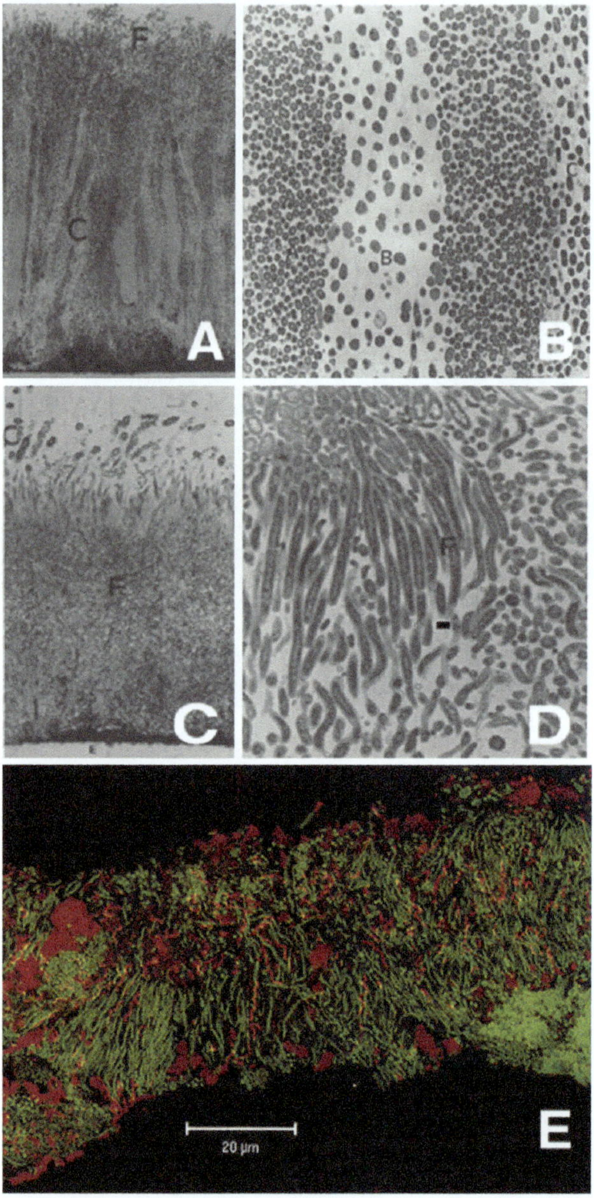

Fig. 4.3 (continued)

tooth surface and nearer to the gingival tissue. *Actinomyces naeslundii*, a species not regarded as disease-associated, is located in the upper region of the plaque mass at the tooth surface. Thus, species composition changes three dimensionally, presumably as a consequence of host inflammatory response and entry of serum exudate in the deeper gingival regions, and entry of saliva and oxygen at the top of the sulcus. Figure 4.4 shows the locations of differently colonized zones on healthy and diseased teeth.

4.4 Signaling Between Bacteria Within Biofilms

Bacteria in biofilms are phenotypically distinct from their planktonic counterparts. Density- or contact-dependant signaling and biofilm-specific gene regulation occur in oral biofilms. Communication mediated by secreted diffusible molecules, such as autoinducer (AI)-2 (a group of molecules produced by the enzyme LuxS) is important for oral biofilm formation. Aberrant monospecies biofilm formation is seen in *LuxS*-deficient mutants of the cariogenic bacterium *S. mutans* and of the non-cariogenic early colonizer *S. gordonii* (Blehert et al. 2003, Merritt et al. 2003, Lemos and Burne 2008). Moreover, LuxS-dependent signaling is required for biofilm development in *A. actinomycetemcomitans* (Shao et al. 2007), as well as for co-culture biofilms of *P. gingivalis* and *S. gordonii* (McNab et al. 2003).

Contact-dependant signaling occurs between various oral streptococci and *P. gingivalis*. Arginine deiminase on the surface of *S. cristatus* initiates a signal transduction cascade in *P. gingivalis* that results in down-regulation of *fimA*, the gene which encodes the subunit of long fimbriae. Because the long fimbriae are required for initial attachment, *P. gingivalis* cells that do not synthesize FimA also fail to accumulate into heterotypic biofilms with *S. cristatus* (Xie et al. 2000, Xie et al. 2007). In contrast, contact with *S. gordonii* provides a biofilm-compatible signal to *P. gingivalis* thereby encouraging formation of heterotypic *P. gingivalis*–*S. gordonii* communities (Simionato et al. 2006). A gene upregulated in *P. gingivalis* following contact with *S. gordonii* is *ltp1* which encodes an eukaryote-like low-molecular-weight tyrosine phosphatase (Maeda et al. 2008). Activity of this enzyme in *P. gingivalis* constrains not only monospecies biofilm development but also dual-species

Fig. 4.3 Structural features of thick supragingival plaque and of subgingival plaque. (**a**) Transmission electron micrograph of 1-week-old supragingival plaque. Low magnification image shows adjacent palisades. (**b**) Transmission electron micrograph of 1-week-old supragingival plaque. High-magnification image shows detailed structure within adjacent palisades. Bar = 1 μm. (**c**) Transmission electron micrograph of 2-month-old subgingival plaque. Low magnification image shows mosaic architecture rather than palisade architecture. (**d**) Transmission electron micrograph of 2-month-old subgingival plaque. High-magnification image shows spirochete and filamentous morphologies. Bar = 1 μm. (**e**) Confocal micrograph of FISH-stained subgingival plaque from a periodontally diseased sulcus. Spirochetes are stained *red* and other bacteria are stained *green*. Mosaic organization of cell types; cf., panels c and d. Electron microscopic images reprinted from Listgarten (1994). Confocal micrograph reprinted from Wecke et al. (2000)

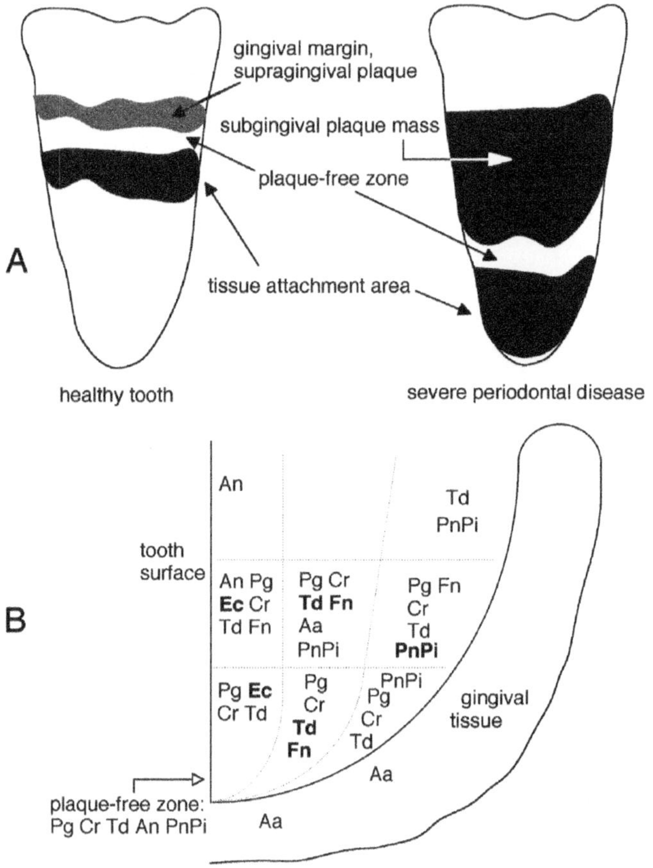

Fig. 4.4 Diagrammatic representation of bacterial colonization patterns on teeth in health and in severe periodontal disease. (**a**) Macroscopic comparison of colonization on healthy tooth with that on periodontally diseased tooth as revealed by osmium staining. Tissue attachment area is not part of the plaque biomass. Plaque-free zone is immediately above the tissue attachment site. After Friedmann et al. (1992). (**b**) Diagram of colonization within periodontal pocket cross section as revealed by immunolabeling. Pocket is divided into nine regions (*dotted lines*). Organism abbreviations: An, *Actinomyces naeslundii*; Pg, *Porphyromonas gingivalis*; Cr, *Campylobacter rectus*; Ec, *Eikenella corrodens*; Td, *Treponema denticola*; Fn, *Fusobacterium nucleatum*; PnPi, *Prevotella nigrescens/Prevotella intermedia*; Aa, *Actinobacillus (Aggregatibacter) actinomycetemcomitans*. Boldface indicates heavy colonization relative to other sites. Drawing after Noiri et al. (2001). Data compiled from Christersson et al. (1987), Kigure et al. (1995), Noiri et al. (1997), Noiri and Ebisu (2000), and Noiri et al. (2001). Figure reprinted from Kolenbrander and Palmer (2004)

biofilm development with *Streptococcus gordonii*. Limitation of biofilm development thus appears to be important for *P. gingivalis*, and may derive from a need to reduce exposure to oxygen, or to facilitate influx of nutrients and efflux of waste. Mechanistically, Ltp1 restricts biofilm development by transcriptional down-regulation of multiple genes involved in exopolysaccharide biosynthesis and transport. Furthermore, Ltp1 regulates transcriptional activity of *luxS* and thus impacts AI-2-dependent signaling in biofilm communities. Adaptation of *S. mutans* to the biofilm lifestyle also occurs at multiple levels (Chong et al. 2008, Lemos and Burne 2008). Within biofilms, proteins of *S. mutans* associated with carbon uptake and cell division are down-regulated. In contrast, *S. mutans* upregulates components of the transformation machinery, two of which are competence-stimulating peptide (CSP) and the two-component system ComED. Competence in *S. mutans* increases in biofilms (Senadheera and Cvitkovitch 2008). Moreover, the development of competence is coordinated with bacteriocin production (Kreth et al. 2007). This facilitates acquisition of DNA from elsewhere in the biofilm, and thereby contributes to genetic diversity.

4.5 Dental Caries

4.5.1 Current Concepts of Caries

Caries is a cumulative disease, the prevalence of which increases with age (Baelum et al. 2008). Moreover, the prevalence and severity of caries differ between individuals from different parts of the world and over time (Petersen 2003). Over the past 25 years in developed countries, a trend can be seen away from the cavitated stages of caries among 12-year-old children. This observation results primarily from school dental health care programs and systematic use of fluoride toothpastes, although general improvement in living conditions plays some role. In contrast, in underdeveloped countries, a trend toward increased caries incidence is associated with growth in sugar consumption.

Central to development and progression of caries is bacterial metabolism of sugars. The result of sugar metabolism is local alteration of pH that can be measured directly using pH-microelectrodes within the biofilm (Fig. 4.5). When

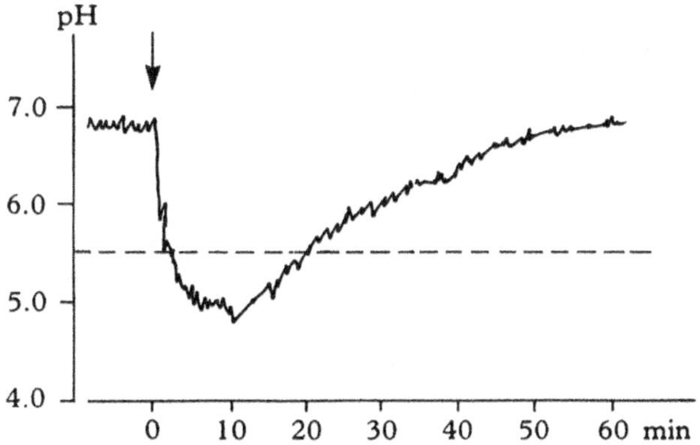

Fig. 4.5 Classical Stephan curve produced in dental biofilm by a 1-min mouth rise with 10% sucrose solution (*arrow*). The dotted line indicates "critical pH" below which tooth minerals may dissolve

dental biofilms are at rest (i.e., not in active metabolism, several hours following sugar consumption), the pH oscillations in the biofilm due to endogenous bacterial metabolism are small in amplitude: less than half a pH unit (Newman et al. 1979). However, bacterial metabolism increases significantly following consumption of dietary carbohydrates. Within the first two minutes after sucrose exposure, bacterial production of organic acids results in a marked decrease in pH. This low pH is maintained over 10–30 min by continued sugar metabolism. Thereafter a gradual increase in pH to its previous resting value occurs, partly through slow diffusion of sugars and organic acids out of the biofilm into saliva and partly through neutralization of organic acids by buffer systems in saliva as well as within the biofilm itself. These pH changes in the biofilm are graphed in the Stephan curve (Fig. 4.5), named after the researcher who first described the process (Stephan 1940). The Stephan curve follows the same general pattern every time carbohydrate is consumed by the host. However, the type and concentration of the carbohydrate can influence the shape of the curve. Most important to caries is the age of the biofilm. Only microbial communities older than 2 days are mature enough to reduce the pH below 5.5 (Imfeld and Lutz 1980). This is an important clinical observation because dissolution of tooth mineral (hydroxyapatite) begins at about this pH, the so-called "critical pH" (Larsen 1990). At pH above 5.5, mineralogical ion concentrations within the biofilm fluid are typically supersaturated, and thus favor redeposition of minerals in dental hard tissues. Hence, tooth surfaces covered by a metabolically active biofilm experience minute mineral losses and gains dependent on the metabolic activity within the biofilm. Also, because the critical pH depends on a number of biogeochemical factors (Dawes 2003), it is a generally accepted simplification to set the value at 5.5. Therefore, it is important to understand caries lesion development and progression as a cumulative effect of de- and remineralization in which net mineral loss (Manji et al. 1991) exceeds remineralization. When de- and remineralization processes result in a net mineral gain, as facilitated by regular exposure to fluoride toothpaste, deposition of minerals in the tooth surface and arrest of lesion progression occur.

These observations form the basis of the caries model promoted by Fejerskov and colleagues (Fejerskov and Manji 1990) (Fig. 4.6). According to this concept, de- and remineralization processes occur unremittingly in the metabolically active biofilm under the influence of multiple biological factors that may either promote, reduce, or arrest lesion development (Fig. 4.6, inner circle). Because these factors can be present simultaneously and in varying proportions over time, it is impossible to truly prevent the disease (Fejerskov 1997). Nevertheless, this model can help the clinician identify the most relevant biological factors that drive caries processes towards net mineral gain. A clear benefit of this model is that it offers an explanation for why biofilm-covered tooth surfaces do not always develop clinically visible caries lesions: over time, mineral loss is balanced by mineral gain. Importantly, this model emphasizes that the biofilm is a prerequisite for caries. Also, in the real world, successful caries control is equally dependent on the behavioral and/or societal factors (Fig. 4.6, outer circle) that influence biological processes.

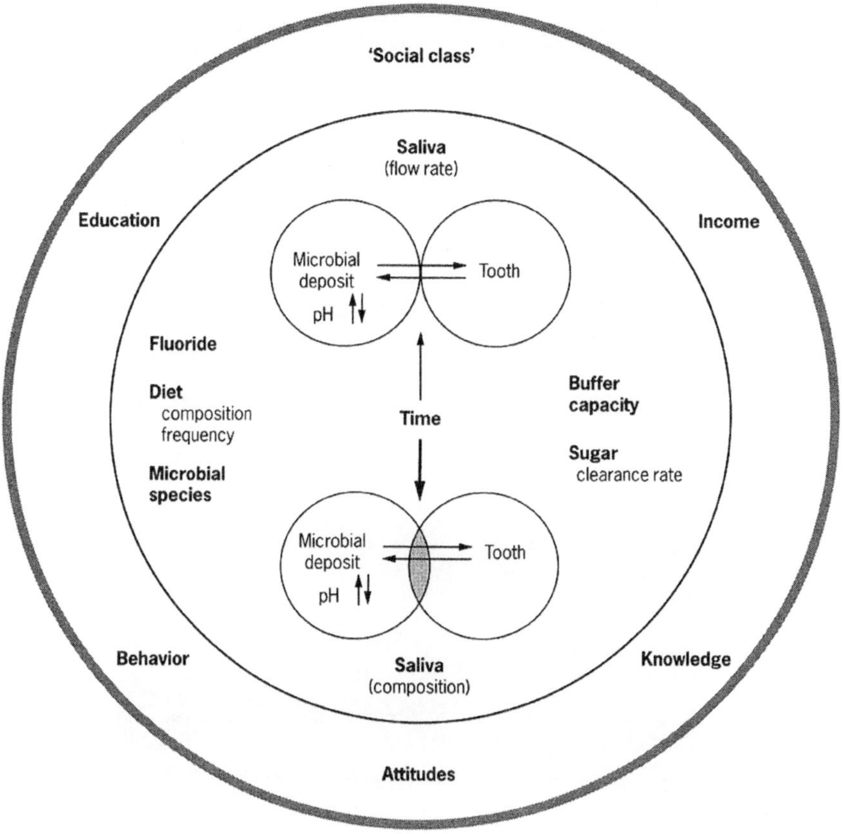

Fig. 4.6 The Fejerskov and Manji caries concept. For description, see text

4.5.2 Caries Lesion Dynamics

A generation ago it was commonly held that once caries lesion progression had started, cavity formation would occur within a limited period of time. Clinical dentistry was based on filling holes in teeth with little attempt to look for pre-cavitation lesions. The current use of fluoride supplements in drinking water as well as in toothpastes has resulted in a more dynamic diagnosis of caries that includes the presence of non-cavitated lesions as well as overt cavities (Fig. 4.7a,b). Precise monitoring of lesion behavior has convincingly shown that lesion progression can be arrested at any stage (Dirks 1966, Nyvad et al. 2003), even at the cavitation stage (Lo et al. 1998), when local environmental conditions are made favorable through biofilm control and exposure to topical fluorides (Nyvad and Fejerskov 1997). Analyses of lesion progression rate have demonstrated that fluoride affects the outcome of ongoing caries activity more than it does the development of new lesions (Baelum et al. 2003). These observations confirm existing hypotheses about

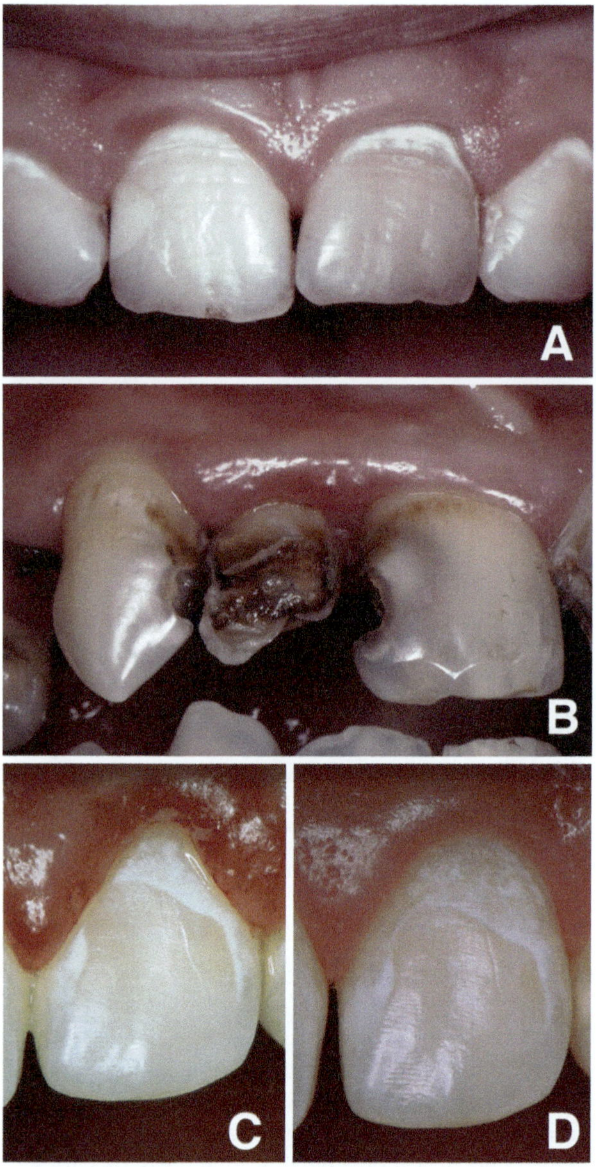

Fig. 4.7 (**a**) Whitish non-cavitated caries lesions along gingival margins of upper anterior teeth. (**b**) Upper anterior teeth with cavitated caries lesions. (**c, d**) Active enamel caries lesion (**c**) converted into inactive enamel caries (**d**). c and d courtesy of A. Thylstrup (deceased)

the mechanism of fluoride in caries control (Fejerskov et al. 1981) and may explain why, in epidemiological studies, fluoride results in a lower number of cavities with minimal change in the total number of lesions (Groeneveld 1985). When fluoridated toothpaste is used in conjunction with biofilm removal (Nyvad 2008), many non-cavitated lesions can be arrested and will never become cavitated.

The dynamic nature of caries has been confirmed in several clinical studies that adopted this non-operative approach (Nyvad and Fejerskov 1986, Holmen et al. 1987, Årtun and Thylstrup 1989). These studies report that, as a result of lesion arrest, significant changes occur in surface texture and in appearance of non-cavitated caries lesions (Fig. 4.7c,d). Accordingly, progressive (active) lesions in the enamel have been described as matt and rough, while inactive/arrested lesions are shiny and smooth (Holmen et al. 1987, Nyvad et al. 1999). Arrest of active lesions in the root/dentin may change the surface characteristics from soft to hard (Nyvad and Fejerskov 1986). These surface changes are believed to reflect metabolic adaptation in the biofilm that favors net mineral gain (Takahashi and Nyvad 2008) and to result from mild abrasion during toothbrushing (Thylstrup et al. 1994). The clinical characteristics of non-cavitated caries lesions are important from a prognostic point of view because it has been estimated that untreated active lesions run a considerably higher risk of cavity formation than do arrested/inactive lesions or clinically sound surfaces (Nyvad et al. 2003). Therefore, to obtain the best possible outcome, treatment should be based on a caries classification that reflects the activity status of lesions (Baelum et al. 2006).

4.5.3 From Specific Infection to Ecological Imbalance

For many years dental caries has been considered a transmissible infectious disease: concepts that grew out of rodent studies (Keyes 1960). Keyes observed that rodents got caries only when they were caged with (or ate fecal pellets of) caries-active rodents. Further proof of the transmissibility of caries emerged when it was observed that only certain streptococci isolated from carious lesions in hamsters caused uncontrolled decay in previously caries-inactive animals (Fitzgerald and Keyes 1960). These bacteria were later identified as *Streptococcus mutans*, and this organism was hypothesized to be the single-pathogen cause of caries. *S. mutans* converts hexose sugars through the intermediate pyruvate to lactic acid, one of the stronger organic acids. *S. mutans* is highly acidogenic and converts about 90% of pyruvate to lactic acid in a reaction catalyzed by the enzyme lactate dehydrogenase. In the absence of sufficient salivary dilution or neutralization, homofermentative lactic acid formation can disrupt the balance between enamel demineralization and remineralization thereby resulting in a carious lesion (Banas 2004). *S. mutans* has other attributes that contribute to caries virulence, one of which (aciduranee) is the ability to resist toxic effects of low pH. Mechanisms of acid resistance include: ATPase-dependent pumping of protons out of the cytoplasm; induction of efficient stress–response mechanisms; increased proportion of long-chained, mono-unsaturated fatty acids in the cell membrane; and elevation of pH through ammonia-generating systems (Quivey et al. 2000, Quivey et al. 2001). Hence, areas of the oral biofilm that are rich in fermentable carbohydrate eventually become dominated by *S. mutans* and a few other aciduric organisms such as lactobacilli (Takahashi and Nyvad 2008). Another cariogenic property of *S. mutans* is the ability to convert extracellular sucrose into polymers of glucose (glucans) and fructose (fructans) through the action of cell-surface glucosyl- and fructosyltransferases (Banas 2004). These polysaccharides can act as carbohydrate storage products and thereby extend the duration of biofilm acidification. In addition, as the degree of α1,3 linkages between glucose monomers increases, solubility of glucan decreases. Insoluble glucans contribute to biofilm structure and adherence. Moreover, because *S. mutans* possesses cell-surface glucan-binding proteins, the insoluble glucans provide a substratum that is specifically recognized and promotes further colonization.

While it is clear that *S. mutans* is highly cariogenic, in recent years the concept of caries as a specific infection has been largely replaced by the theory that caries results from an ecological shift (Marsh 2003). In this "ecological plaque hypothesis," dental caries is a consequence of changes in the natural balance of the resident microflora brought about by an alteration in local environmental conditions. Intra-oral factors such as sucrose consumption reduce the biofilm pH, which in turn stimulates a shift in biofilm bacterial composition in favor of acidogenic and aciduric species. The outcome is a change in biofilm metabolism towards enhanced acidity and thus towards demineralization. An ecological hypothesis for caries is attractive because it helps explain why the disease can develop in the absence of *S. mutans*, as well as why *S. mutans* can be present without lesion formation (Nyvad 1993). In spite of their ability to stick to test tube surfaces, mutans streptococci are not common early colonizers of teeth (Nyvad and Kilian 1987), not even in caries-active individuals (Nyvad and Kilian 1990). This suggests that acidogenic bacteria other than mutans streptococci contribute significantly to pH reduction in young biofilms. In fact, it has been proposed that, despite virulent production of organic acids and insoluble glucans, *S. mutans* is less competitive in clinically sound environments than are other oral streptococci (Takahashi and Nyvad 2008). The interplay between bacterial ecology and biogeochemistry is summarized in "the ecological caries hypothesis" (Fig. 4.8). *Actinomyces* spp. and non-mutans streptococci possess more diverse physiological activities, and these organisms are more adept at adjustment to minor pH perturbations, so they predominate at clinically sound sites: the dynamic stability stage (Fig. 4.8). However, any sustained moderate reduction in biofilm pH may adaptively enhance acidogenicity and acidurance in the non-mutans bacteria, and thereby drive pH even further downward. Such conditions would then favor outgrowth of the so-called "low-pH" non-mutans streptococci (van Houte et al. 1996, van Ruyven et al. 2000) and *Actinomyces* spp. via acid selection: the acidogenic stage (Fig. 4.8). Only where prolonged acidic conditions prevail do "pH-strategists" (i.e., mutans streptococci and lactobacilli) replace the low-pH non-mutans bacteria (aciduric stage, Fig. 4.8). The ecology-driven hypothesis should also take into account the "mixed-bacteria ecological approach" (Kleinberg 2002), which accentuates the ratio of acidiogenic bacteria to ammonia-producing bacteria. It has been proposed that the cascade of events within the biofilm which result in pH change is correlated with modification of hard-tissue surface texture, from smooth to rough (on enamel) and from hard to smooth (on dentin) (Takahashi and Nyvad 2008). Importantly, caries development is a dynamic process in which the reversibility of the ecological factors can be used to reverse hard-tissue changes, at least for non-cavitated lesions (Fig. 4.8).

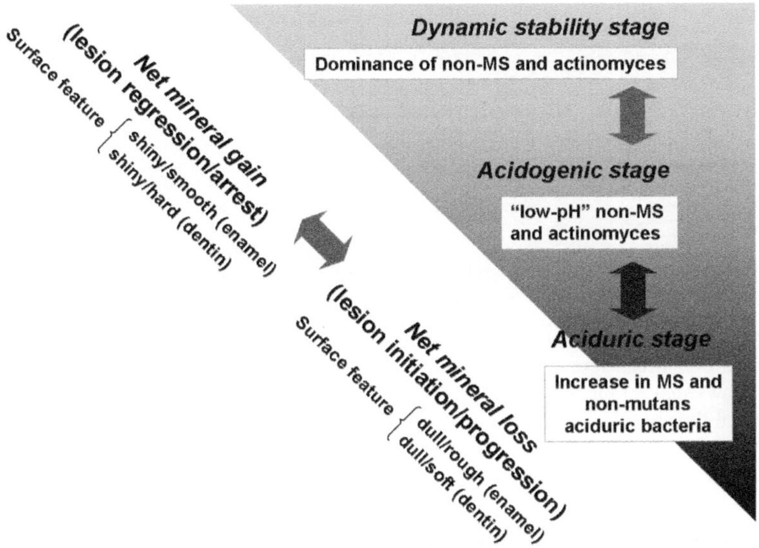

Fig. 4.8 The ecological caries hypothesis (Takahashi and Nyvad 2008). Note that dynamic changes in the composition of the biofilm are associated with changes in the surface features of the dental hard tissues. For further description, see text

In summary, bacteria respond to environmental change through phenotypic and regulatory mechanisms. Caries progression is dictated less by the presence or absence of particular bacteria than by the interplay of environmental change and biological response (Takahashi and Nyvad 2008). Acceptance of ecological hypotheses rejects a definition of caries in terms of classical infectious disease. Therefore, strategies for caries prevention and control that target specific microorganisms, such as the mutans streptococci, are likely to meet with only limited success. Vaccination, gene therapy, and antimicrobial treatment, although technologically interesting approaches, are unlikely to be of significant practical efficacy. An effective strategy should focus on control of biofilm acidification by favoring conditions that promote growth of non-mutans bacteria. Practical examples of this strategy include mechanical biofilm control (Nyvad 2008), reduction of sugary food consumption (Zero 2004), and neutralization of acid by stimulation of saliva production (Bardow et al. 2008).

4.6 Periodontal Diseases

Chronic periodontitis, the most common form of periodontitis, progresses very slowly; complete destruction of the dentition's supportive tissues can take decades (Fig. 4.9) (Flemmig 1999). The initial lesion tends to develop over the third and fourth decades of life (Heitz-Mayfield et al. 2003). Once established, the disease progresses through bursts of active destruction interspersed with periods of relative quiescence during which partial healing may occur (Lindhe et al. 1983).

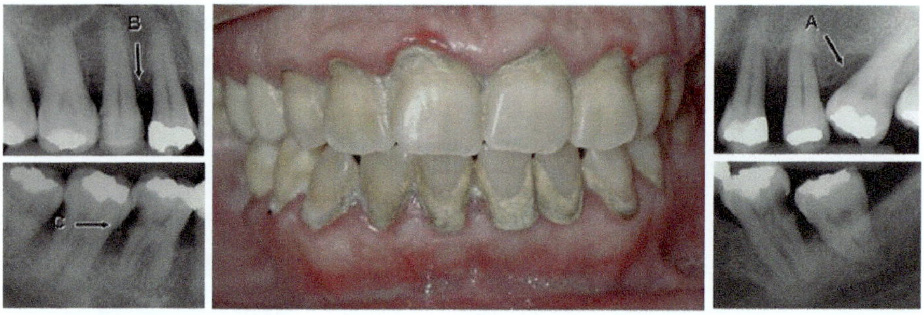

Fig. 4.9 Clinical presentation of a 45-year-old male patient diagnosed with severe generalized chronic periodontitis. Presence of supragingival biofilm and dental calculus are evident. Gingival margin shows signs of gingivitis with presence of edema and redness. Radiographs of the posterior areas reveal the presence of vertical (**a**) and horizontal (**b**) bone loss and subgingival calculus (**c**) (*arrows*)

Generally, periodontal infections have been associated with high levels of a particular group of pathogens: *P. gingivalis, T. forsythia*, *T. denticola*, and *Eubacterium nodatum* (Socransky et al. 1998, Haffajee et al. 2006b). So-called aggressive periodontitis is associated with a younger age group and tends to initiate during puberty. Localized aggressive periodontitis (LAP) has a distinct clinical manifestation; only first molars and incisors develop vertical bone defects. Interestingly, these patients present low levels of plaque and calculus, as well as little gingival inflammation (Fig. 4.10). While this form of periodontal disease has been associated primarily with leukotoxic strains of *A. actinomycetemcomitans, P. gingivalis* has also been found in high proportions (Tonetti and Mombelli 1999). Generalized forms of aggressive periodontitis affect other teeth and tend to be associated with a complex set of pathogens comprising *A. actinomycetemcomitans*, *P. gingivalis*, *T. forsythia*,

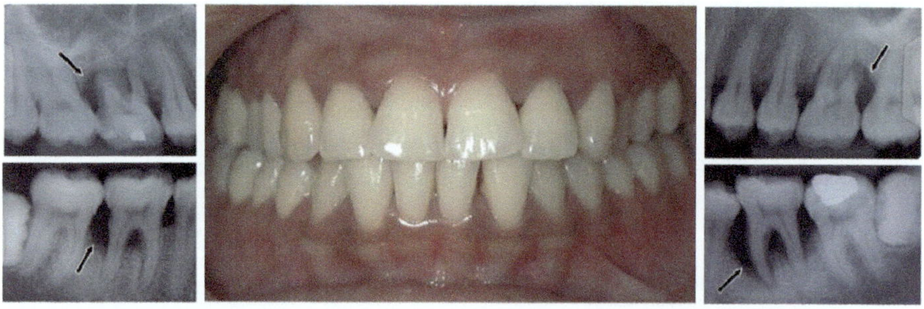

Fig. 4.10 Clinical presentation of a 21-year-old female patient diagnosed with localized aggressive periodontitis. Note the minor presence of supragingival biofilm and the healthy appearance of the gingival margin. Radiographs of the posterior area show the presence of localized vertical bone defects around the first molars (*arrows*). Furcation defects can also be detected in all first molars. Courtesy of Dr. Bruno Rescala and Dr. Wilson Rosalem Jr., Brazil

Campylobacter rectus, and *Prevotella intermedia* (Van Steenbergen et al. 1993, Tonetti and Mombelli 1999).

4.6.1 Mechanisms of Bacterial Pathogenesis

Periopathogens destroy tissue primarily through the production of proteolytic enzymes. *P. gingivalis*, for example, produces at least 10 proteases, of which the best studied are the gingipains: the Arg-X-specific enzymes RgpA and RgpB, and the Lys-X-specific enzyme Kgp. These enzymes comprise a family of isoforms that can be post-translationally modified with glycan chains and that can be presented on the cell surface or secreted into the extracellular milieu (O'Brien-Simpson et al. 2001, Gallagher et al. 2003, Potempa et al. 2003, Sheets et al. 2008). These enzymes degrade structural components of the periodontium, along with immune effectors such as cytokines, antibodies, complement components, and surface receptors. Gingipains, as well as other toxic metabolites of *P. gingivalis*, such as short-chain fatty acids, can induce apoptosis in various host cells. In LAP, *A. actinomycetemcomitans* is the consensus pathogen and, consistent with the low levels of inflammation characteristic for the disease, *A. actinomycetemcomitans* is not highly proteolytic; the main virulence factors are cytotoxins (Henderson et al. 2003). The leukotoxin of *A. actinomycetemcomitans* is a member of the RTX family of toxins and kills leukocytes through necrotic as well as apoptotic mechanisms (Fong et al. 2006). *A. actinomycetemcomitans* also produces a cytolethal distending toxin (CDT) that induces G2 arrest and apoptosis in T cells (Shenker et al. 2001). *A. actinomycetemcomitans* and *P. gingivalis*, when shed from the biofilm, can invade and survive within gingival epithelial cells (Lamont and Yilmaz 2002). Epithelial infection can impact production of innate immune effector molecules, such as cytokines, and can compromise host immunity against the biofilm as a whole. Moreover, epithelial cells with internalized *P. gingivalis* are resistant to apoptotic death and indeed accelerate through the cell cycle (Kuboniwa et al. 2008, Yilmaz et al. 2008).

4.6.2 Clinical Parameters of Periodontal Diseases

Periodontal diseases are clinically diagnosed by assessments of gingival inflammation and measurements of tissue destruction. Changes in color, shape, and texture of the gingival margin are the basis of indices used to assess gingivitis. Indices can incorporate an assessment of gingival bleeding. Damage to tooth-supporting tissue is quantified by measurements of probing pocket depth (PPD) and of clinical attachment level (CAL). PPD measurements are obtained by insertion of a periodontal probe into the gingival sulcus to determine the distance from the gingival margin to the bottom of the pocket. CAL measures the distance from a "fixed" landmark on the tooth, such as the cemento-enamel junction, to the depth of the pocket. Because the gingival margin fluctuates in response to inflammation (hyperplasia or recession), CAL is a more accurate measure of attachment loss. Bleeding after periodontal probing also constitutes an assessment of inflammation at the bottom of the pocket. Dental radiographs can be used to quantify alveolar bone loss. Because periodontal diseases are site-specific, clinical measurements are taken at several sites per tooth. The severity of the disease will be determined by the values of the PPD and CAL, while the extent of the disease (localized vs. generalized) will be established by the number of teeth or sites affected (Flemmig 1999). These clinical measurements are likewise employed to assess therapy outcome.

4.6.3 Epidemiology of Periodontal Diseases

Gingivitis is ubiquitous; most, if not all, individuals will manifest at least mild forms of gingival inflammation during certain periods of their lives. Several epidemiological surveys using full- or partial-mouth examinations and with different criteria to define periodontitis have established that chronic periodontitis is also rather common in adults. The overall prevalence of moderate forms of the disease in USA has been estimated at 44% (Papapanou 1996). More severe forms of the disease affect 10–15% of the population worldwide (Papapanou 1999). Recent data suggest that the prevalence of mild to moderate forms of the disease might be declining in developed countries due to routine implementation of brushing and flossing. However, the percentage of the population affected by severe forms of periodontitis leading to tooth loss has not changed, even in populations with a high standard of plaque control (Hugoson et al. 2008). Longer retention of teeth coupled with an aging population might account for future increases in the number of subjects affected by periodontal destruction (Papapanou 1999). Aggressive forms of periodontitis are far less common and their prevalence varies greatly because they are heavily influenced by the genetic background of the population. For instance, studies in USA have reported the prevalence of aggressive periodontitis in young (5–17 years old) Caucasian populations at less than 1%, but the prevalence reaches 2.6% in African-American children (Loe and Brown 1991).

4.6.4 Periodontal Therapy

Several features of periodontal diseases impact the outcome of anti-infective therapy. The polymicrobial biofilm lives as a mass located "outside the body," i.e., within the periodontal pocket. The exact composition of the microbiota responsible for initiation and progression of the different disease types is unknown. Periodontal infections are endogenous, and periodontal pathogens can colonize other oral surfaces besides the teeth. In addition, subjects with similar clinical manifestations can present distinct subgingival microbial profiles (Socransky and Haffajee 1997, Teles et al. 2006). Consequently specific microbial targets for therapy have been difficult to identify. Subgingival biofilms display resistance to antimicrobials and to host immune mechanisms (Socransky and Haffajee 2002). Pathogens tend to recolonize the periodontal pockets after subgingival cleaning; thus, disease recurrence is frequent. Traditional periodontal therapy relies on mechanical removal of the subgingival biofilm mass. Because sites and subjects differ in subgingival microbial composition, this "one therapy fits all" tactic sometimes fails, and a subset of patients is refractory to traditional therapy (Teles et al. 2006).

The clinical goals of therapy are to reduce inflammation, to decrease PPD and improve CAL, and to maintain these improvements over time. The microbiological goal of periodontal therapy is to reduce the level of periodontal pathogens and to create an environment conducive to establishment of a climax community compatible with periodontal health. In addition, treatment should not permit the establishment of uncommon (opportunistic) pathogens or favor antibiotic-resistant strains. Control of supragingival biofilm accumulation after therapy is essential to avoid disease recurrence (Teles et al. 2006). Supra- and subgingival biofilms are typically removed mechanically. Mineralization within supragingival biofilms (dental calculus) enhances plaque accumulation (White 1997). Dental prophylaxis and supragingival scaling remove calculus and non-mineralized deposits above the gingival margin, thereby reducing inflammation and creating conditions which assist the patient in maintaining good oral hygiene. Mineral ions present in GCF can lead to calcification within subgingival biofilms as well (White 1997). Scaling and root planing (SRP) is a procedure in which curettes are used to remove subgingival calculus and plaque from the root surface of deep pockets. Surgical procedures can be used to reach areas not otherwise accessible to SRP, such as furcation defects. After removal of supra- and subgingival biofilms, inflammation subsides and the periodontal tissues undergo partial healing. A decrease in PPD, a gain in CAL, and a reduction in the percentage of sites bleeding on probing (Badersten et al. 1984) are achieved. For these improvements to be maintained, frequent self-performed supragingival plaque removal is required together with regular visits to a dental practitioner who can re-instrument residual pockets, reinforce oral hygiene instructions, and monitor disease recurrence (Axelsson et al. 2004). Periodontal surgery results in fewer residual pockets, thereby decreasing burden on the clinician during the maintenance phase (Serino et al. 2001).

Although infection defines periodontal diseases, antibiotics are not routinely used to control these diseases. Periodontitis patients generally respond well to mechanical biofilm removal. Therefore the use of antibiotics in therapy can be justified only in cases where it results in greater clinical improvements compared to mechanical therapy alone. Periodontists have also refrained from using systemic antibiotics in therapy due to concern over emergence of antibiotic-resistant bacterial strains (Van Winkelhoff et al. 1996). Further, biofilms show enhanced resistance to antimicrobials compared to planktonic cells (Socransky and Haffajee 2002). Despite these considerations, the use of systemic antibiotics for the treatment of periodontal infections is potentially advantageous. Systemic antibiotics may reach microorganisms at the base of deep periodontal pockets, otherwise inaccessible subgingival areas such as furcations, microorganisms that reside within gingival epithelial and connective tissue, and pathogens that colonize oral soft tissues and saliva.

The use of antibiotics as adjuncts to mechanical therapy has been examined in several clinical trials, and evidence indicates that use can result in clinical and microbiological improvements beyond those accomplished with mechanical therapy alone (Herrera et al. 2002, Haffajee et al. 2003). Benefits seem particularly noticeable in deep pockets, in smokers, and in subjects with aggressive periodontitis (Haffajee et al. 2003). Furthermore, decreases occur in the number of sites that show disease progression during maintenance in subjects who have received systemic antibiotics as part of their treatment (Haffajee et al. 2006a). Interestingly, studies comparing the efficacy of systemic antibiotics alone to mechanical therapy alone have documented similar clinical and microbiological outcomes (Lopez et al. 2006). This evidence suggests that the level of antibiotic protection experienced in subgingival biofilm is not sufficient to render antibiotics ineffective, perhaps because GCF enables entrance of systemic antibiotics to the periodontal pockets. Because several periodontal pathogens colonize the subgingival habitat closer to the pocket's epithelial lining, they are readily accessible to systemic antibiotics (Socransky and Haffajee 2005).

Studies on the microbiological outcome of different periodontal therapies have demonstrated that positive clinical outcome post-therapy is accompanied by reduction in several subgingival species; reduction in *Campylobacter* sp., *Fusobacterium* sp. and *Prevotella* sp. seems essential to clinical success of mechanical therapy (Haffajee et al. 2006a). Use of systemic antibiotics, in particular amoxicillin 500 mg plus metronidazole 250 mg t.i.d. for 14 days, has resulted in further reduction in these bacterial species compared to that in subjects who did not receive adjunctive therapy (Haffajee et al. 2006a). This beneficial effect was more prominent in subjects with a high percentage of deep pockets.

Periodontal maintenance will retard progression of disease and also maintain most gains obtained during the active phase of the treatment. However, despite the best efforts of patients and clinicians, periodontal disease will tend to progress even in the presence of good home care. In fact, disease progression in subjects undergoing periodontal maintenance is higher than that in healthy subjects on a preventive program (Teles et al. 2008). This observation suggests that treatment does not bring the patients' periodontal condition to a completely healthy state. Microbiological results from that study demonstrated that elevated levels of *Porphyromonas gingivalis, Tannerella forsythia*, and *Treponema denticola* remain after 3 years of maintenance compared to the levels found in healthy subjects; the higher bacterial burden is possibly associated with increased tendency for disease progression in maintenance subjects.

4.6.5 Refractory Periodontitis

A small percentage of periodontitis patients do not respond to therapy. These refractory cases present considerable disease progression even after comprehensive periodontal treatment involving SRP and surgery, systemic antibiotics, and the practice of immaculate plaque control (Fig. 4.11). It has been hypothesized that these individuals harbor more virulent pathogen strains, or that they carry a set

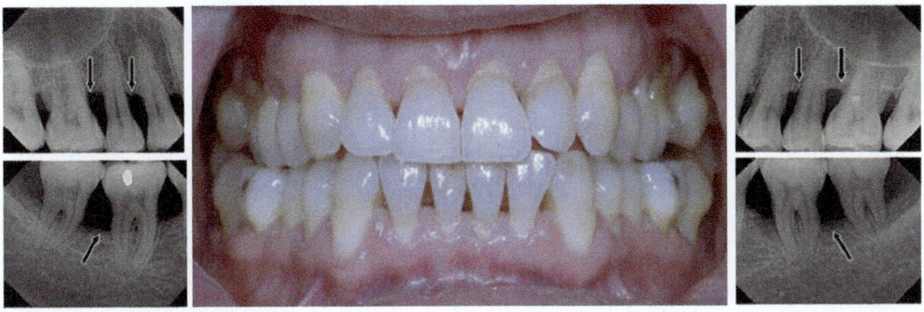

Fig. 4.11 Clinical presentation of a 32-year-old male patient diagnosed with "refractory" periodontitis. Note the immaculate oral hygiene and the healthy appearance of the gingival margin. The case is also characterized by the presence of generalized recession of the buccal and interproximal aspects of the teeth. Radiographs of the posterior area show the presence of horizontal bone loss around pre-molars and molars (*arrows*)

of pathogens not traditionally associated with periodontal infections, or that they have a greater susceptibility to low levels of periodontal pathogens. A recent study has reported positive results in cases treated with a combination of different anti-infective therapies including SRP, tetracycline delivered locally at pockets $\geq$ 4 mm, concurrent administration of amoxicillin 500 mg plus metronidazole 250 mg t.i.d. for 14 days, weekly professional removal of supragingival plaque for 3 months, and maintenance SRP every 3 months (Haffajee et al. 2004). Disease progression was arrested in 10 of 14 subjects during this 2-year study. Interestingly, the baseline levels of 40 bacterial species were even lower than those in chronic periodontitis cases after therapy, and these low levels were reduced even further by the combined therapy. The data suggest that even low levels of pathogens can sustain disease progression in refractory subjects.

4.7 A Role for Oral Biofilms in Systemic Infection?

While biofilm-induced oral diseases are usually site-specific, organisms from the biofilm can enter the circulation and cause bacterial endocarditis (see also Chapter 5). Furthermore, evidence is accumulating that implicates the oral biofilm in other serious systemic conditions, such as atherosclerosis and preterm delivery of low-birth-weight infants (Kim and Amar 2006). Transient bacteremias can occur after procedures such as professional teeth cleaning or SRP. Many oral bacteria are resistant to humoral and cellular antibacterial systems in the blood, particularly when aggregates of organisms are sloughed from the biofilm. Initiation of bacterial endocarditis begins at a defect in the heart valve that resulted from earlier disease (especially rheumatic fever), from injury, or from a developmental abnormality. Platelets become activated to repair the defect, and the adherent platelets and exposed connective tissue bind bacteria: frequently the oral streptococci (Douglas et al. 1993). The streptococci induce platelets to bind and process fibrinogen; the resultant vegetative mass of fibrin, platelets and bacteria causes valvular insufficiency with a potential for embolization and infarction of other organs (Herzberg et al. 1997). The pathogenic basis of oral bacterial biofilm causality in other systemic infections, such as atherosclerosis or preterm delivery, is less well defined. Clearly, any bacteria in the bloodstream could be carried to these remote sites and establish an infection. Other potential mechanisms (none mutually exclusive) include initiation of an inflammatory response in the periodontal tissues, or in the circulation, with systemic consequences; or induction of an auto-immune response by bacterial antigens such as heat shock proteins (Garcia et al. 2001, Kim and Amar 2006).

4.8 Immunobiology of Biofilm/Tissue Interaction

The periodontium, the host tissue that houses the teeth, is constantly exposed not only to naturally occurring dental plaque, but also to external microbial threats from air, water, and food. In fact, the tooth surface provides an excellent platform for oral bacteria and transient species to release their secreted and shed components into the adjacent periodontal tissue (Darveau et al. 1997). Our current understanding of how periodontal tissue provides a strong defensive mechanism against multiple and continued microbial assaults is that clinically healthy periodontal tissue exists in homeostasis with the dental plaque biofilm. Homeostasis represents a state of the "controlled" inflammation found in periodontal tissue. The innate host defense system is continually active in this tissue as evidenced by the orchestrated expression of select innate defense mediators. Early histological studies (Page and Schroeder 1976) of clinically healthy tissue demonstrated that a cellular infiltrate is located in juxtaposition to the colonized tooth surface. A portion of this cellular infiltrate has been described as forming a "wall" of neutrophils precisely located between bacteria and residing just outside the junctional epithelium, the epithelial cell surface closest to the dental plaque biofilm (Kornman et al. 1997). Consistent with these observations, molecular characterization of healthy periodontal tissue has demonstrated that IL-8, ICAM, and E-selectin are expressed in clinically healthy tissue (Moughal et al. 1992, Nylander et al. 1993, Gemmell et al. 1994, Tonetti et al. 1994, Tonetti 1997). These inflammatory mediators are necessary for leukocyte diapedesis from the vasculature and directed movement through tissue. E-selectin expression on endothelial cells facilitates a tethering interaction between the leukocyte and the endothelial cell wall initiating the "rolling" stage required for leukocyte exit (Springer 1994). IL-8 is a key neutrophil chemoattractant and ICAM facilitates cellular adhesion. It has been demonstrated that a gradient of IL-8 and ICAM-1 expression exists in clinically healthy tissue (Tonetti et al. 1998). IL-8 expression is greatest at the most superficial junctional epithelial cell layers and the levels of ICAM-1 increase toward areas exposed to bacterial challenges. More recent immunohistochemical and in situ studies have revealed that clinically healthy periodontal tissue also expresses human β-defensin molecules 1, 2, and 3 (Lu et al. 2004, 2005) as well as soluble (Jin and Darveau 2001) and membrane-bound CD14 (Jin et al. 2004),

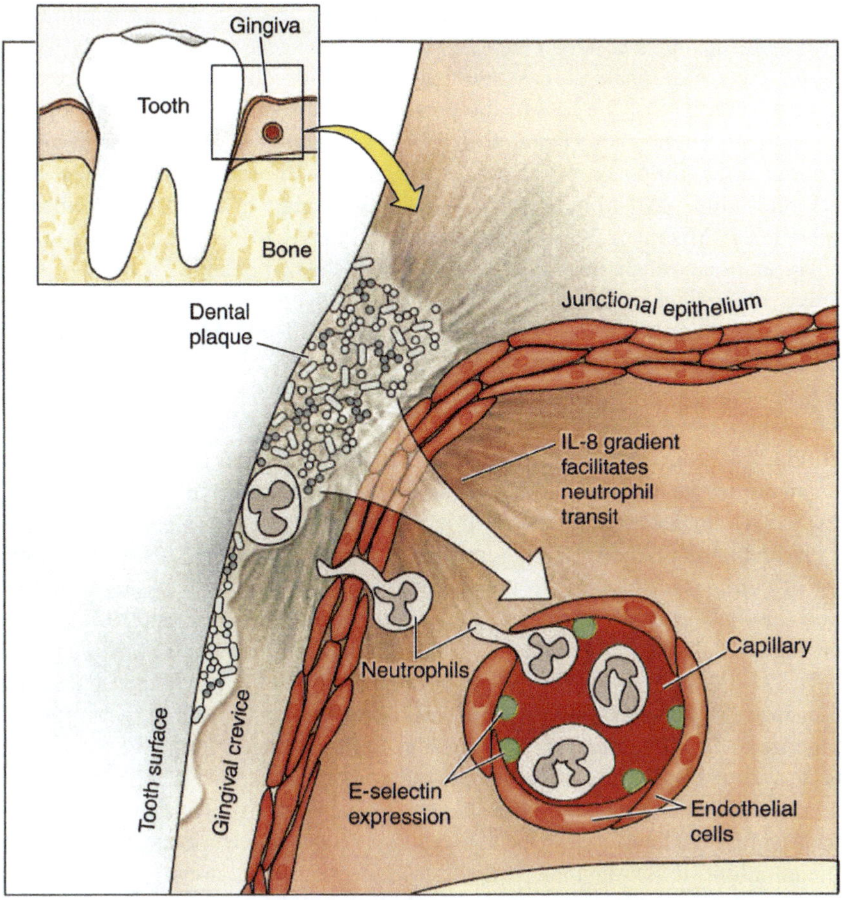

Fig. 4.12 Innate host defense status in clinically normal periodontal tissue. Clinically healthy tissue expresses low levels of E-selectin and a gradient of IL-8 (represented in shades of *red*) that facilitates the transit of neutrophils through the tissue and into the gingival crevice where they protect the host from infection. These mediators are made in response to a highly organized bacterial biofilm, termed dental plaque. Figure adapted from Whittaker et al. (1996) and Tonetti et al. (1994). Original appearance in ASM News, Vol. 69, April 2003 (American Society for Microbiology Press)

and lipopolysaccharide binding protein (LBP) (Ren et al. 2004). LBP expression is greatest in the gingival epithelium (Ren et al. 2004). These innate defense proteins function in either bacterial killing or removal, consistent with the notion that healthy periodontal tissue is "armed" by the innate host defense system to protect against bacterial infection. Clinically healthy human gingival tissue also expresses low levels of TLR2 (Ren et al. 2005, Sugawara et al. 2006). Expression of TLR4 in healthy tissue was reported in one study (Sugawara et al. 2006), but expression of this innate host defense receptor was not observed in another study (Ren et al. 2005). A diagrammatic representation of innate immune status of healthy tissue is shown in Fig. 4.12.

The innate defense status of clinically healthy tissue is highly significant for periodontal health (see also Chapter 11). The strongest evidence for the key role that the pattern of innate defense mediator expression plays in maintenance of periodontal health is that the loss of the protective neutrophilic barrier function, either by congenital deficiency (Page et al. 1987, Waldrop et al. 1987, Hart et al. 1994, Cosseau et al. 2008) or by chemical induction with anti-mitotic agents such as cyclophosamide (Attström and Schroeder 1979, Sallay et al. 1984, Hemmerle and Frank 1991, Yoshinari et al. 1994), invariably leads to disease. Furthermore, the lack of an intact innate host defense system may be responsible for the significantly increased incidence of severe periodontitis observed in diabetic patients (Type I and Type II) and tobacco users (Bergstrom et al. 1988, Macfarlane et al. 1992, Offenbacher et al. 1996, Zambon 1996, Salvi et al. 1997, Cosseau et al. 2008).

A key question is whether the commensal oral biofilm community contributes to the unique orchestrated expression of select innate defense mediators found in periodontal tissue. Several in vitro studies have shown that oral commensal bacteria can induce many innate host defense mediators (Chung and Dale 2008, Cosseau et al. 2008); however, their contribution to the selective expression pattern observed in vivo remains unknown. In contrast, much has been learned about the contribution of commensal bacteria to the immune status of intestinal tissue in germ-free mice (animals completely devoid of bacteria) (Gordon and Pesti 1971, Umesaki and Setoyama 2000, Hooper et al. 2001, Xu and Gordon 2003, Macpherson and Harris 2004). These animals are born through Caesarean section, housed aseptically in an isolator with sterile filtered air and raised using sterile food, water, and bedding. Germ-free mice are distinct from specific-pathogen-free (SPF) mice which are devoid of known mouse pathogens but contain intestinal bacteria (Macpherson and Harris 2004). Studies in germ-free mice have demonstrated that the quality and quantity of intestinal commensal microorganisms induce a state of "controlled" inflammation similar to that in the intestine of normal animals (Chadwick and Anderson 1992, Cebra 1999). Surprisingly little work using germ-free mice exists on the contribution of oral commensal bacteria to innate mediator expression in periodontal tissue. A preliminary investigation comparing innate host mediator expression in germ-free and conventionally reared mice found that IL-1β levels were significantly higher in conventionally reared mice, consistent with the notion that in the periodontium, as in the intestine, commensal bacteria actively participate in the establishment of innate mediator expression in clinically healthy or normal tissue (Dixon et al. 2004). The limited number of samples examined in the pilot study may be one reason why only IL-1β was identified as being differentially expressed in the germ-free animals. That this inflammatory mediator was elevated in healthy conventionally reared animals compared to germ-free controls may appear paradoxical because IL-1β has been associated with the development of periodontitis (Masada et al. 1990). However, the observed IL-1β expression could represent a form of indirect activation of periodontal tissue by oral commensal bacteria. Indirect activation is defined as those effects which occur when a bacterium activates one cell type and then the product from that cell acts upon another cell or cell type. In contrast, direct activation occurs when the bacteria or bacterial components directly stimulate a cell to respond (e.g., monocyte production of cytokines, chemokines, or cellular adhesion molecules). Indirect and direct activation can both occur in both myeloid and in non-myeloid cells (Heath et al. 1987, Yamazaki et al. 1992, Gemmell and Seymour 1993, Lindemann et al. 1995). Therefore, the presence of IL-1β in clinically healthy periodontal tissue may serve as a "priming" mechanism for several different cell types found in the periodontium. The contribution of the oral commensal biofilm community to innate host mediator expression is largely unexplored, and greater understanding is this area could result in rapid practical benefit.

4.9 Conclusion

Bacterial biofilms of the oral cavity present a unique opportunity for investigation of inter-bacterial interactions and the impact of those interactions on host pathology. Their accessibility has made them the first human bacterial microbiomes to be studied, with research dating back at least to the time of Antonie van Leeuwenhoek. Research on these biofilms spurred development of anaerobic culture techniques. Consistent with such a long and diverse research history, an enormous amount of physiological and taxonomic information on oral bacteria exists. Genomic information has been harder to come by, but the complete genome for several representatives of different oral taxons is either currently available in useful form or is under assembly. All these factors guarantee that oral biofilms will remain the best-characterized and most easily manipulated human microbial populations. The next frontier of biofilm biology, at least from the standpoint of human diseases that have no single causative organism, is that of host interaction. In fact, polymicrobial communities, such as those of the oral cavity and of the gut, are always effected to some extent by any disease and its subsequent treatment. Thus, the dynamics of inter-bacterial interactions and, more importantly, the outcome of those interactions on the host are critical to maintaining a state of health in human beings.

Acknowledgements Robert J. Palmer Jr. is supported by the Intramural Research Program, National Institute of Dental and Craniofacial Research, National Institutes of Health.

References

Aas JA et al (2005) Defining the normal bacterial flora of the oral cavity. J Clin Microbiol 43: 5721–5732

Aas JA et al (2008) Bacteria of dental caries in primary and permanent teeth in children and young adults. J Clin Microbiol 46:1407–1417

Årtun J, Thylstrup A (1989) A 3-year clinical and SEM study of surface changes of carious enamel lesions after inactivation. Am J Orthod Dentofacial Orthop 95:327–333

Attström R, Schroeder HE (1979) Effect of experimental neutropenia on initial gingivitis in dogs. Scand J Dent Res 87:7–23

Axelsson P et al (2004) The long-term effect of a plaque control program on tooth mortality, caries and periodontal disease in adults. Results after 30 years of maintenance. J Clin Periodontol 31:749–757

Badersten A et al (1984) Effect of nonsurgical periodontal therapy. II. Severely advanced periodontitis. J Clin Periodontol 11:63–76
Baelum V et al (2003) Application of survival analysis to carious lesion transitions in intervention trials. Community Dent Oral Epidemiol 31:252–260
Baelum V et al (2006) Dental caries paradigms in diagnosis and diagnostic research. Eur J Oral Sciv 114:263–277
Baelum V et al (2008) "for richer, for poorer, in sickness and in health..." the role of dentistry in controlling caries and periodontitis globally. In: Fejerskov O, Kidd E (ed) Dental caries: the disease and its clinical management, 2nd edn. Blackwell Munksgaard, Oxford
Banas JA (2004) Virulence properties of *Streptococcus mutans*. Front Biosci 9:1267–1277
Bardow A et al (2008) The role of saliva. In: Fejerskov O, Kidd E (ed) Dental caries: THE disease and its clinical management, 2nd edn. Blackwell Munksgaard, Oxford
Becker MR et al (2002) Molecular analysis of bacterial species associated with childhood caries. J Clin Microbiol 40:1001–1009
Bergstrom J et al (1988) Influence of cigarette smoking on vascular reaction during experimental gingivitis. Scand J Dent Res 96:34–39
Blehert DS et al (2003) Autoinducer 2 production by *Streptococcus gordonii* DL1 and the biofilm phenotype of a *luxS* mutant are influenced by nutritional conditions. J Bacteriol 185:4851–4860
Brinig MM et al (2003) Prevalence of bacteria of division TM7 in human subgingival plaque and their association with disease. Appl Environ Microbiol 69:1687–1694
Cebra JJ (1999) Influences of microbiota on intestinal immune system development. Am J Clin Nutr 69:1046S–1051S
Chadwick VS, Anderson R P (1992) Microorganisms and their products in inflammatory bowel disease. In: Macdermott RP, Stenson WF (ed) Inflammatory bowel disease edn. Elseviere, Amsterdam
Chalmers NI et al (2008) Characterization of a *Streptococcus* sp.–*Veillonella* sp. community micromanipulated from dental plaque. J Bacteriol 190:8145–8154
Chong P et al (2008) *LiaS* regulates virulence factor expression in *Streptococcus mutans*. Infect Immun 76:3093–3099
Christersson LA et al (1987) Tissue localization of *Actinobacillus actinomycetemcomitans* in human periodontitis. I. Light, immunofluorescence and electron microscopic studies. J Periodontol 58:529–539
Chung WO, Dale BA (2008) Differential utilization of nuclear factor-kappaB signaling pathways for gingival epithelial cell responses to oral commensal and pathogenic bacteria. Oral Microbiol Immunol 23:119–126
Cisar JO et al (1995) Lectin recognition of host-like saccharide motifs in streptococcal cell-wall polysaccharides. Glycobiology 5:655–662
Cosseau C et al (2008) The commensal *Streptococcus salivarius* K12 downregulates the innate immune responses of human epithelial cells and promotes host-microbe homeostasis. Infect Immun 76:4163–4175
Cotton SL et al (2009) Subgingival taxa in periodontal health and disease using HOMIM. Paper presented at the 87th General Session of the International Association for Dental Research. Miami FL, 1–4 April 2009
Darveau RP et al (1997) The microbial challenge in periodontitis. Periodontol 2000 14:12–32
Dawes C (2003) What is the critical pH and why does a tooth dissolve in acid? J Can Dent Assoc 69:722–724
Dewhirst FE, et al (2010) The human oral microbiome. J Bacteriol, doi:10.1128/JB.00542-10, PMID 20656903
Diaz PI et al (2006) Molecular characterization of subject-specific oral microflora during initial colonization of enamel. Appl Environ Microbiol 72:2837–2848
Dirks OB (1966) Posteruptive changes in dental enamel. J Dent Res 45:503–511
Dixon DR et al (2004) Modulation of the innate immune response within the periodontium. Periodontol 2000 35:53–74

Douglas CW et al (1993) Identity of viridans streptococci isolated from cases of infective endocarditis. J Med Microbiol 39:179–182

Fejerskov O et al (1981) Rational use of fluorides in caries prevention. Acta Odontologica Scandinavica 39:241–249

Fejerskov O, Manji F (1990) Risk assessment in dental caries. In: Bader JD (ed) Risk assessment in dentistry, University of North Carolina, Chapel Hill, NC

Fejerskov O (1997) Concepts of dental caries and their consequences for understanding the disease. Community Dent Oral Epidemiol 25:5–12

Fitzgerald RJ, Keyes PH (1960) Demonstration of the etiologic role of streptococci in experimental caries in the hamster. J Am Dent Assoc 61:9–19

Flemmig TF (1999) Periodontitis. Ann Periodontol 4:32–37

Fong KP et al (2006) *Actinobacillus actinomycetemcomitans* leukotoxin requires lipid microdomains for target cell cytotoxicity. Cell Microbiol 8:1753–1767

Friedman MT et al (1992) The "Plaque-free zone" in health and disease: a scanning electron microscope study. J Periodontol 63:890–896

Gallagher A et al (2003) Glycosylation of the Arg-gingipains of *Porphyromonas gingivalis* and comparison with glycoconjugate structure and synthesis in other bacteria. Curr Protein Pept Sci 4:427–441

Garcia RI et al (2001) Relationship between periodontal disease and systemic health. Periodontol 2000 25:21–36

Gemmell E, Seymour GJ (1993) Interleukin 1, interleukin 6 and transforming growth factor-beta production by human gingival mononuclear cells following stimulation with *Porphyromonas gingivalis* and *Fusobacterium nucleatum*. J Periodontal Res 28:122–129

Gemmell E et al (1994) Adhesion molecule expression in chronic inflammatory periodontal disease tissue. J Periodont Res 29:46–53

Gibbons RJ, Nygaard M (1970) Interbacterial aggregation of plaque bacteria. Arch Oral Biol 15:1397–1400

Gordon HA, Pesti L (1971) The gnotobiotic animal as a tool in the study of host microbial relationships. Bacteriol Rev 35:390–429

Groeneveld A (1985) Longitudinal study of prevalence of enamel lesions in a fluoridated and non-fluoridated area. Community Dent Oral Epidemiol 13:159–163

Haffajee a D et al (2003) Systemic anti-infective periodontal therapy. A systematic review. Ann Periodontol 8:115–181

Haffajee a D et al (2004) Clinical and microbiological changes associated with the use of combined antimicrobial therapies to treat "refractory" periodontitis. J Clin Periodontol 31: 869–877

Haffajee AD et al (2006a) The effect of periodontal therapy on the composition of the subgingival microbiota. Periodontol 2000 42:219–258

Haffajee AD et al (2006b) Association of *Eubacterium nodatum* and *Treponema denticola* with human periodontitis lesions. Oral Microbiol Immunol 21:269–282

Hart TC et al (1994) Neutrophil defects as risk factors for periodontal diseases. J Periodontol 65:521–529

Heath JK et al (1987) Bacterial antigens induce collagenase and prostaglandin E2 synthesis in human gingival fibroblasts through a primary effect on circulating mononuclear cells. Infect Immun 55:2148–2154

Heitz-Mayfield LJ et al (2003) Clinical course of chronic periodontitis. II. Incidence, characteristics and time of occurrence of the initial periodontal lesion. J Clin Periodontol 30:902–908

Hemmerle J, Frank RM (1991) Bacterial invasion of periodontal tissues after experimental immunosuppression in rats. J Biol Buccale 19:271–282

Henderson B et al (2003) Molecular pathogenicity of the oral opportunistic pathogen Actinobacillus actinomycetemcomitans. Annu Rev Microbiol 57:29–55

Herrera D et al (2002) A systematic review on the effect of systemic antimicrobials as an adjunct to scaling and root planing in periodontitis patients. J Clin Periodontol 29 Suppl 3:136–159

Herzberg MC et al (1997) Host-pathogen interactions in bacterial endocarditis: Streptococcal virulence in the host. Adv Dent Res 11:69–74

Holmen L et al (1987) Clinical and histological features observed during arrestment of active enamel carious lesions in vivo. Caries Res 21:546–554

Hooper LV et al (2001) Molecular analysis of commensal host-microbial relationships in the intestine. Science 291:881–884

Hugoson A et al (2008) Trends over 30 years, 1973–2003, in the prevalence and severity of periodontal disease. J Clin Periodontol 35:405–414

Imfeld T, Lutz F (1980) Intraplaque acid formation assessed in vivo in children and young adults. Pediatr Dent 2:87–93

Jin L, Darveau RP (2001) Soluble CD14 levels in gingival crevicular fluid of subjects with untreated adult periodontitis. J Periodontol 72:634–640

Jin L et al (2004) The in vivo expression of membrane-bound cd14 in periodontal health and disease. J Periodontol 75:578–585

Keyes PH (1960) The infectious and transmissible nature of experimental dental caries – findings and implications. Arch Oral Biol 1:304–319

Kigure T et al (1995) Distribution of *Porphyromonas gingivalis* and *Treponema denticola* in human subgingival plaque at different periodontal pocket depths examined by immunohistochemical methods. J Periodontal Res 30:332–341

Kim J, Amar S (2006) Periodontal disease and systemic conditions: a bidirectional relationship. Odontology 94:10–21

Kleinberg I (2002) A mixed-bacteria ecological approach to understanding the role of the oral bacteria in dental caries causation: an alternative to *Streptococcus mutans* and the specific-plaque hypothesis. Crit Rev Oral Biol Med 13:108–125

Kolenbrander PE (1988) Intergeneric coaggregation among human oral bacteria and ecology of dental plaque. Annu Rev Microbiol 42:627–656

Kolenbrander PE, Palmer RJ Jr (2004) Human oral bacterial biofilms. In: Ghannoum M, O'Toole GA (ed) Microbial biofilms, edn. ASM Press, Washington, DC

Kornman KS et al (1997) The host response to the microbial challenge in periodontitis: assembling the players. Periodontol 2000 14:33–53

Kreth J et al (2007) The response regulator comE in *Streptococcus mutans* functions both as a transcription activator of mutacin production and repressor of CSP biosynthesis. Microbiology (Reading, Engl) 153:1799–1807

Kuboniwa M et al (2008) *P. gingivalis* accelerates gingival epithelial cell progression through the cell cycle. Microbes Infect 10:122–128

Lamont RJ, Yilmaz O (2002) In or out: the invasiveness of oral bacteria. Periodontol 2000 30: 61–69

Larsen MJ (1990) Chemical events during tooth dissolution. J Dent Res 69:575–580

Lemos JA, Burne RA (2008) A model of efficiency: stress tolerance by *Streptococcus mutans*. Microbiology (Reading, Engl) 154:3247–3255

Lepp PW et al (2004) Methanogenic archaea and human periodontal disease. Proceedings of the National Academy of Sciences of the United States of America 101:6176–6181

Lindemann RA et al (1995) Effect of whole oral bacteria and extracted lipopolysaccharides on peripheral blood leukocyte interleukin-2 receptor expression. J Periodontal Res 30:264–271

Lindhe J et al (1983) Progression of periodontal disease in adult subjects in the absence of periodontal therapy. J Clin Periodontol 10:433–442

Listgarten MA (1994) The structure of dental plaque. Periodontol 2000 5:52–65

Lo ECM et al (1998) Arresting dentine caries in Chinese preschool children. Int J Paediatr Dent 8:253–260

Loe H, Brown LJ (1991) Early onset periodontitis in the United States of America. J Periodontol 62:608–616

Lopez NJ et al (2006) Effects of metronidazole plus amoxicillin as the only therapy on the microbiological and clinical parameters of untreated chronic periodontitis. J Clin Periodontol 33:648–660

Lu Q et al (2004) Expression of human beta-defensins-1 and -2 peptides in unresolved chronic periodontitis. J Periodontal Res 39:221–227

Lu Q et al (2005) Expression of human beta-defensin-3 in gingival epithelia. J Periodontal Res 40:474–481

Macfarlane GD et al (1992) Refractory periodontitis associated with abnormal polymorphonuclear leukocyte phagocytosis and cigarette smoking. J Periodontol 63:908–913

Macpherson AJ, Harris NL (2004) Interactions between commensal intestinal bacteria and the immune system. Nat Rev Immunol 4:478–485

Maeda K et al (2008) A *Porphyromonas gingivalis* tyrosine phosphatase is a multifunctional regulator of virulence attributes. Mol Microbiol 69:1153–1164

Mager DL et al (2003) Distribution of selected bacterial species on intraoral surfaces. J Clin Periodontol 30:644–654

Manji F et al (1991) A random effects model for some epidemiological features of dental caries. Community Dent Oral Epidemiol 19:324–328

Marsh PD (2003) Are dental diseases examples of ecological catastrophes? Microbiology (Reading, Engl) 149:279–294

Masada MP et al (1990) Measurement of interleukin-1 alpha and -1 beta in gingival crevicular fluid: implications for the pathogenesis of periodontal disease. J Periodontal Res 25: 156–163

McNab R et al (2003) LuxS-based signaling in *Streptococcus gordonii*: autoinducer 2 controls carbohydrate metabolism and biofilm formation with *Porphyromonas gingivalis*. J Bacteriol 185:274–284

Merritt J et al (2003) Mutation of *luxS* affects biofilm formation in *Streptococcus mutans*. Infect Immun 71:1972–1979

Moughal NA et al (1992) Endothelial cell leukocyte adhesion molecule-1 (ELAM-1) and intercellular adhesion molecule-1 (ICAM-1) expression in gingival tissue during health and experimentally-induced gingivitis. J Periodont Res 27:623–630

Newman P et al (1979) An in-dwelling electrode for in-vivo measurement of the pH of dental plaque in man. Arch Oral Biol 24:501–507

Noiri Y et al (1997) An immunohistochemical study on the localization of *Porphyromonas gingivalis*, *Campylobacter rectus* and *Actinomyces viscosus* in human periodontal pockets. J Periodontal Res 32:598–607

Noiri Y, Ebisu S (2000) Identification of periodontal disease-associated bacteria in the "plaque-free zone". J Periodontol 71:1319–1326

Noiri Y et al (2001) The localization of periodontal-disease-associated bacteria in human periodontal pockets. J Dent Res 80:1930–1934

Nylander K et al (1993) Expression of the endothelial leukocyte adhesion molecule-1 (ELAM-1) on endothelial cells in experimental gingivitis in humans. J Periodontol 64: 355–357

Nyvad B, Fejerskov O (1986) Active root surface caries converted into inactive caries as a response to oral hygiene. Scan J Dent Res 94:281–284

Nyvad B, Fejerskov O (1987a) Transmission electron microscopy of early microbial colonization of human enamel and root surfaces *in vivo*. Scan J Dent Res 95:297–307

Nyvad B, Fejerskov O (1987b) Scanning electron microscopy of early microbial colonization of human enamel and root surfaces *in vivo*. Scan J Dent Res 95:287–296

Nyvad B, Kilian M (1987) Microbiology of the early colonization of human enamel and root surfaces *in vivo*. Scand J Dent Res 95:369–380

Nyvad B, Kilian M (1990) Comparison of the initial streptococcal microflora on dental enamel in caries-active and in caries-inactive individuals. Caries Res 24:267–272

Nyvad B (1993) Microbial colonization of human tooth surfaces. APMIS 101:7–45

Nyvad B, Fejerskov O (1997) Assessing the stage of caries lesion activity on the basis of clinical and microbiological examination. Community Dent Oral Epidemiol 25:69–75

Nyvad B et al (1999) Reliability of a new caries diagnostic system differentiating between active and inactive caries lesions. Caries Res 33:252–260

Nyvad B et al (2003) Construct and predictive validity of clinical caries diagnostic criteria assessing lesion activity. J Dent Res 82:117–122
Nyvad B (2008) The role of oral hygiene. In: Fejerskov O, Kidd E (ed) Dental caries the disease and its clinical management, Blackwell Munksgaard, Oxford
O'Brien-Simpson NM et al (2001) Role of RgpA, RgpB, and Kgp proteinases in virulence of *Porphyromonas gingivalis* W50 in a murine lesion model. Infect Immun 69: 7527–7534
Offenbacher S et al (1996) Periodontal infection as a possible risk factor for preterm low birth weight. J Periodontol 67:1103–1113
Page RC, Schroeder HE (1976) Pathogenesis of inflammatory periodontal disease. A summary of current work. Lab Invest 33:235–249
Page RC et al (1987) Molecular basis for the functional abnormality in neutrophils from patients with generalized prepubertal periodontitis. J Periodontal Res 22:182–183
Page RC, Kornman KS (1997) The pathogenesis of human periodontitis: an introduction. Periodontol 2000 14:9–11
Palmer RJ Jr et al (2003) Coaggregation-mediated interactions of streptococci and actinomyces detected in initial human dental plaque. J Bacteriol 185:3400–3409
Papapanou PN (1996) Periodontal diseases: epidemiology. Ann Periodontol 1:1–36
Papapanou PN (1999) Epidemiology of periodontal diseases: an update. J Int Acad Periodontol 1:110–116
Paster BJ et al (1998) Identification of oral streptococci using PCR-based, reverse-capture, checkerboard hybridization. Methods Cell Sci 20:223–231
Paster BJ et al (2006) The breadth of bacterial diversity in the human periodontal pocket and other oral sites. Periodontol 2000 42:80–87
Petersen PE (2003) The World Oral Health Report 2003: continuous improvement of oral health in the 21st century – The approach of the WHO Global Oral Health Programme. Community Dent Oral Epidemiol 31:3–24
Potempa J et al (2003) Gingipains, the major cysteine proteinases and virulence factors of *Porphyromonas gingivalis*: structure, function and assembly of multidomain protein complexes. Curr Protein Pept Sci 4:397–407
Quirynen M et al (2005) Initial subgingival colonization of "pristine" pockets. J Dent Res 84: 340–344
Quivey RG Jr et al (2001) Genetics of acid adaptation in oral streptococci. Crit Rev Oral Biol Med 12:301–314
Quivey RG Jr et al (2000) Adaptation of oral streptococci to low pH. Adv Microb Physiol 42: 239–274
Ren L et al (2004) Local expression of lipopolysaccharide-binding protein in human gingival tissues. J Periodontal Res 39:242–248
Ren L et al (2005) The expression profile of lipopolysaccharide-binding protein, membrane-bound CD14, and toll-like receptors 2 and 4 in chronic periodontitis. J Periodontol 76: 1950–1959
Sallay K et al (1984) Bacterial invasion of oral tissues of immunosuppressed rats. Infect Immun 43:1091–1093
Salvi GE et al (1997) Influence of risk factors on the pathogenesis of periodontitis. Periodontol 2000 14:173–201
Senadheera D, Cvitkovitch DG (2008) Quorum sensing and biofilm formation by *Streptococcus mutans*. Adv Exp Med Biol 631:178–188
Serino G et al (2001) Initial outcome and long-term effect of surgical and non-surgical treatment of advanced periodontal disease. J Clin Periodontol 28:910–916
Shao H et al (2007) Autoinducer 2 is required for biofilm growth of *Aggregatibacter (Actinobacillus) actinomycetemcomitans*. Infect Immun 75:4211–4218
Sheets SM et al (2008) Gingipain-dependent interactions with the host are important for survival of *Porphyromonas gingivalis*. Front Biosci 13:3215–3238

Shenker BJ et al (2001) Induction of apoptosis in human T cells by *Actinobacillus actinomycetemcomitans* cytolethal distending toxin is a consequence of G2 arrest of the cell cycle. J Immunol 167:435–441

Simionato MR et al (2006) *Porphyromonas gingivalis* genes involved in community development with *Streptococcus gordonii*. Infect Immun 74:6419–6428

Socransky SS, Haffajee a D (1997) The nature of periodontal diseases. Ann Periodontol 2:3–10

Socransky SS et al (1998) Microbial complexes in subgingival plaque. J Clin Periodontol 25: 134–144

Socransky SS, Haffajee AD (2002) Dental biofilms: difficult therapeutic targets. Periodontol 2000 28:12–55

Socransky SS, Haffajee AD (2005) Periodontal microbial ecology. Periodontol 2000 38:135–187

Springer TA (1994) Traffic signals for lymphocyte recirculation and leukocyte emigration: the multistep paradigm. Cell 76:301–314

Stephan R (1940) Changes in hydrogen-ion concentration on tooth surfaces and in carious lesions. J Am Dent Assoc 27:718–723

Sugawara Y et al (2006) Toll-like receptors, NOD1, and NOD2 in oral epithelial cells. J Dent Res 85:524–529

Takahashi N, Nyvad B (2008) Caries ecology revisited: microbial dynamics and the caries process. Caries Res 42:409–418

Teles RP et al (2006) Microbiological goals of periodontal therapy. Periodontol 2000 42:180–218

Teles RP et al (2008) Disease progression in periodontally healthy and maintenance subjects. J Periodontol 79:784–794

Thylstrup A et al (1994) In vivo caries models-mechanisms for caries initiation and arrestment. Adv Dent Res 8:144–157

Tonetti MS et al (1994) Localized expression of mRNA for phagocyte-specific chemotactic cytokines in human periodontal infections. Infect Immun 62:4005–4014

Tonetti MS (1997) Molecular factors associated with compartmentalization of gingival immune responses and transepithelial neutrophil migration. J Periodontal Res 32:104–109

Tonetti MS et al (1998) Neutrophil migration into the gingival sulcus is associated with transepithelial gradients of interleukin-8 and ICAM-1. J Periodontol 69:1139–1147

Tonetti MS, Mombelli A (1999) Early-onset periodontitis. Ann Periodontol 4:39–52

Umesaki Y, Setoyama H (2000) Structure of the intestinal flora responsible for development of the gut immune system in a rodent model. Microbes Infect 2:1343–1351

van Houte J et al (1996) The final pH of bacteria comprising the predominant flora on sound and carious human root and enamel surfaces. J Dent Res 75:1008–1014

van Ruyven FO J et al (2000) Relationship among mutans streptococci, "low-pH" bacteria, and lodophilic polysaccharide-producing bacteria in dental plaque and early enamel caries in humans. J Dent Res 79:778–784

Van Steenbergen TJ et al (1993) Microbiological and clinical monitoring of non-localized juvenile periodontitis in young adults: a report of 11 cases. J Periodontol 64:40–47

Van Winkelhoff AJ et al (1996) Systemic antibiotic therapy in periodontics. Periodontol 2000 10:45–78

Waldrop TC et al (1987) Periodontal manifestations of the heritable Mac-1, LFA-1, deficiency syndrome. Clinical, histopathologic and molecular characteristics. J Periodontol 58: 400–416

Wecke J et al (2000) A novel technique for monitoring the development of bacterial biofilms in human periodontal pockets. FEMS Microbiol Lett 191:95–101

White DJ (1997) Dental calculus: recent insights into occurrence, formation, prevention, removal and oral health effects of supragingival and subgingival deposits. Eur J Oral Sci 105: 508–522

Whittaker CJ, Klier CM, Kolenbrander PE (1996) Annual Review of Microbiology 50:513–552

Xie H et al (2000) Intergeneric communication in dental plaque biofilms. J Bacteriol 182: 7067–7069

Xie H et al (2007) Identification of a signalling molecule involved in bacterial intergeneric communication. Microbiology (Reading, Engl) 153:3228–3234
Xu J, Gordon JI (2003) Inaugural article: honor thy symbionts. Proc Natl Acad Sci USA 100:10452–10459
Yamazaki K et al (1992) Direct and indirect effects of *Porphyromonas gingivalis* lipopolysaccharide on interleukin-6 production by human gingival fibroblasts. Oral Microbiol Immunol 7:218–224
Yilmaz O et al (2008) ATP scavenging by the intracellular pathogen *Porphyromonas gingivalis* inhibits P2X7-mediated host-cell apoptosis. Cell Microbiol 10:863–875
Yoshinari N et al (1994) Effect of long-term methotrexate-induced neutropenia on experimental periodontal lesion in rats. J Periodontal Res 29:393–400
Zambon JJ (1996) Periodontal diseases: microbial factors. Ann Periodontol 1:879–925
Zero DT (2004) Sugars – the arch criminal? Caries Res 38:277–285

Chapter 5
Implant-Associated Infection

Werner Zimmerli and Andrej Trampuz

5.1 Introduction

5.1.1 Epidemiology of Implant-Associated Infection

During the last 50 years, medical devices have gained growing importance (Darouiche 2004). On the one side, there is an increasing need due to the higher median age of the population suffering from degenerative diseases. On the other side, technology evolved and many functions can be replaced by novel medical devices, such as cochlear implants, brain stimulators, and ventricular assist devices (Waldvogel and Bisno 2000). In addition, the broader availability of health insurance in some countries led to an increased use especially of orthopedic and cardiac devices. As an example, the annual number of inserted joint prostheses is 600,000, and the one of inserted pacemaker-defibrillators is 300,000 in USA (Darouiche 2004).

Unfortunately, implantation of medical devices is an important source of nosocomial infections during the perioperative period (Roumbelaki et al. 2008). In addition, implanted devices can be infected by the hematogenous route as long as they remain in the host (Zimmerli et al. 1985, Trampuz and Zimmerli 2008). Thus, the benefit of improved function or alleviated pain has to be paid with infectious complications. Even if the average infection rate is low in most totally implanted devices, infection represents a large personal and economic burden. Its treatment often requires several surgical interventions and a long treatment period for the patient.

5.1.2 Overview of Selected Permanent Implants Used in Medicine

Medical devices can be used either transiently or permanently. Most patients in intensive care units have several removable devices which will not be discussed

W. Zimmerli (✉)
Department of Infectious Diseases and Internal Medicine, Medical University Clinic, Kantonsspital, Rheinstrasse 26, CH-4410 Liestal, Switzerland
e-mail: werner.zimmerli@unibas.ch

T. Bjarnsholt et al. (eds.), *Biofilm Infections*, DOI 10.1007/978-1-4419-6084-9_5,

in this chapter (see Chapter 6). We will exclusively deal with permanent implants which are increasingly used in different areas of surgery. They can be classified according to their localization in the organism as intravascular or extravascular devices.

5.1.2.1 Intravascular Devices

Artificial heart valves are used to correct valvular stenosis or regurgitation. Not only mechanical, but also biological devices of non-vascularized dead tissue (e.g., porcine bioprostheses) are prone to infection (Geissdorfer et al. 2007). In a randomized prospective trial comparing mechanical heart valves with porcine bioprostheses, the risk for endocarditis was not statistically different (Hammermeister et al. 2000, Oxenham et al. 2003). Thus, avascular biological material seems to behave like synthetic material in terms of susceptibility to infection.

Vascular prostheses are used either as arterio-venous shunts in patients undergoing hemodialysis or for arterial replacement in case of vascular occlusive disease (Antonios et al. 2006, Berardinelli 2006). Whereas the former is repetitively endangered for exogenous infection during puncture, the latter is prone to exogenous infection exclusively during the perioperative period.

The most frequent permanent intravascular implants are *venous access devices* with a subcutaneous reservoir which are routinely used in oncology for chemotherapy, or in infectious diseases for outpatient intravenous antibiotic therapy (Maki 2006, Raad 2007). These devices have an extravascular (port and proximal catheter) and an intravascular (distal catheter) part. Since the port is repetitively manipulated, it is prone to exogenous infection. In contrast, the intravascular part is mainly endangered by circulating microorganisms and, to a lesser extent, by contaminated infusion material. Similarly, the intravascular parts of *electrophysiologic devices* (cardiac pacemakers and implantable cardioverter defibrillator) can also be colonized during bloodborne infection (Marrie and Costerton 1984, Chua et al. 2000, Uslan et al. 2007). However, the risk of exogenous infection is almost exclusively limited to the perioperative period. *Intravascular stents* are commonly used in interventional cardiology. For unknown reasons, infection of coronary stents is an extremely rare event. Such infections are more frequent in stents of larger vessels, such as the aorta (Antonios and Baddour 2004).

5.1.2.2 Intracorporeal Extravascular Devices

These devices have no direct contact to the bloodstream, but are in different compartments of the body. Orthopedic implants, such as *artificial joints* and *internal fixation devices*, are by far the most important implants in human medicine (Steckelberg and Osmon 2000, Zimmerli 2004, Trampuz and Zimmerli 2008). The infection rate is between 0.5 and 5% according to the joint, with the lowest rate after total hip arthoplasty and the highest rate after elbow joint replacement (Yamaguchi 1998, Trampuz and Zimmerli 2008). The infection rate is even higher after internal fixation of open fractures (5–10%) (Ochsner 2006). Interestingly, even these extravascular devices are endangered during episodes of bacteremia

(Zimmerli 1985). *Reconstructive and aesthetic implants*, such as breast-, nose-chin and testis-prostheses, have a low infection rate (<2%) and are mainly endangered by microorganisms of the skin flora. These infections, caused by low-virulence bacteria, such as *Propionibacterium acnes* or coagulase-negative staphylococci, generally do not cause frank infection, but capsule fibrosis, chronic inflammation, or pain (Pittet 2005, Rieger et al. 2009).

Neurosurgical devices (e.g., internal ventricular drainage devices, deep brain stimulators, and intrathecal opiate pumps) are a special problem, since part of these devices is localized in a compartment with ineffective host defense mechanisms, namely brain or cerebrospinal fluid (Conen et al. 2008, Lietard et al. 2008). Thus, the manifestation of these infections may be subtle especially in case of microorganisms. *Chochlear implants* are increasingly used especially in children with hearing deficiency. Part of the implant is integrated in bone, part in the middle ear. Thus, exogenous microorganisms may penetrate along the electrode and the auditory nerve into the central nervous system (Wei et al. 2008).

In *hernia repair*, prosthetic material is now routinely used. Interestingly, according to a recent Cochrane analysis, the infection rate was not higher in patients with hernioplasty including an artificial mesh graft (1.4%) as compared to those with herniorrhaphy without implant (3.5%) (Perez 2005, Sanchez-Manuel et al. 2007). Thus, in this situation the presence of the foreign device seems not to increase the risk of infection. However, once infected, for efficacious therapy, the mesh graft has to be removed in most cases.

5.1.2.3 Intracorporeal Devices with Direct Extra-corporeal Connection

Implants with direct extracorporeal connection are continuously exposed to exogenous microorganisms from the skin or mucous membranes. Therefore, their risk for infection is higher than the one of totally implanted devices. *Dental implants* are among the most frequently used foreign devices. Despite their anatomic position in an environment with a very high bacterial load (oral cavity), the risk for peri-implantitis is quite low, i.e., ~5% (Klinge et al. 2005, Zimmerli 2008). It is conceivable that the high number of exudate granulocytes in the periodontal region inhibits bacterial growth from the biofilm covering the bone skrew. The *cardiac assist device* (artificial heart) is implanted to bridge the time until a transplant organ is available for a patient with terminal cardiac failure. In modern devices such as HeartMate II®, the inflow conduit, the outflow graft, the blood pump, and a microprocessor-based system controller are totally implanted (Hammermeister et al. 2000). However, the electrical lead is tunneled subcutaneously and exits through the abdominal wall (Frazier et al. 2007). This limits the lifetime of these devices due to frequent exogenous infection. *Peritoneal dialysis catheters* (Tenkhoff catheters) are used as method for kidney replacement. Exogenous catheter-associated infections and especially peritonitis are frequent, i.e., one episode per 16–30 patient-months, even in patients with excellent aseptic technique of manipulation (Jassal and Lok 2008). In oncology, different types of intravenous devices, not only totally implanted but also tunneled catheters with percutaneous exit, are used. The risk for infection is greater with tunneled (1.6/1000 days) than with totally implanted catheters (0.1/1000 days) (Maki et al. 2006).

5.1.2.4 Extracorporeal Devices

Even extracorporeal devices may harm the body by invasive biofilm infection. In general, *suprapubic bladder catheters* do not result in invasive device-associated infection, despite the biofilm on the catheter (Jacobsen et al. 2008). In contrast, *intrauterine devices* may result in invasive local infection, especially if they are covered with a biofilm of *Actinomyces* spp. (Akhan et al. 2008).

5.2 Pathogenesis

The pathogenesis of implant-associated infection involves interactions between the microorganism, the implant and the host (Darouiche 2001). The interaction of the microorganism with the device results in the formation of a biofilm, whereas interactions of host defense mechanisms with the implant result in activated and exhausted phagocytes (Zimmerli et al. 1984) (see below).

5.2.1 Route of Infection

Each implant can be colonized either by the exogenous or by the endogenous route. Exogenous infections mainly originate during the perioperative period. However, it may occur at any time after implantation, if the implant is adjacent to an infectious focus (e.g., appendicitis in a patient with a ventriculo-peritoneal shunt) or if it is exposed to the exogenous surface (e.g., pressure necrosis of a pacemaker battery pocket). Endogenous infection occurs at any time after implantation either by the hematogenous or by the lymphogenous route. Experimentally, hematogenous seeding with *S. aureus* occurs at a bacterial density of 100–1000 cfu/ml blood (Zimmerli et al. 1985). Thus, episodes of bacteremia during common activities such as tooth brushing or extraction of non-infected teeth do not endanger implants, since the density of bacteria in the bloodstream during dental extraction is as low as 1 and 28 cfu/ml blood (Lucas et al. 2002). Implants are mainly endangered during skin infection (Maderazo et al. 1988). This indicates that some of these infections may occur by the lymphogenic route.

5.2.2 Mechanisms of Persistence

5.2.2.1 Microbiology of Implant-Associated Infections

Coagulase-negative staphylococci, *S. aureus*, beta-hemolytic streptococci, and enterobacteriaceae are the most important microorganisms in all type of implant-associated infections. In low-grade infections, microorganisms of the skin flora are frequent. Among them, coagulase-negative staphylococci, *P. acnes* and

Corynebacterium ssp. are the most typical ones. These microorganisms have the ability to persist as biofilm (see below) (Hall-Stoodley et al. 2004).

5.2.2.2 Intracellular Persistence

Implant-associated infections are characterized by persistence and recurrence. One of the mechanisms of persistence is the localization of bacteria in host cells that are not naturally phagocytic (epithelial and endothelial cells), and therefore not able to kill bacteria (Garzoni and Kelley 2009, Sendi and Proctor 2009). Microorganisms with normal phenotypic appearance, as well as so-called small colony variants (SCV), can persist intracellularly (Fig. 5.1). SCV have been reported for *S. aureus* and other staphylococcal species, as well as from Gram-negative bacilli such as *E. coli* and *Pseudomonas aeruginosa* (von Eiff et al. 2006). As compared to the normal phenotype, SCV of *S. aureus* show increased uptake by host cells, resistance to intracellular defenses, and reduced stimulation of host defenses. We described five patients with *S. aureus* SCV in prosthetic joint infection (Sendi et al. 2006). All of them were previously treated with a prolonged course of antibiotics. Thus, intracellular localization and/or transformation in SCV phenotype seem to play an important role in device-associated infection.

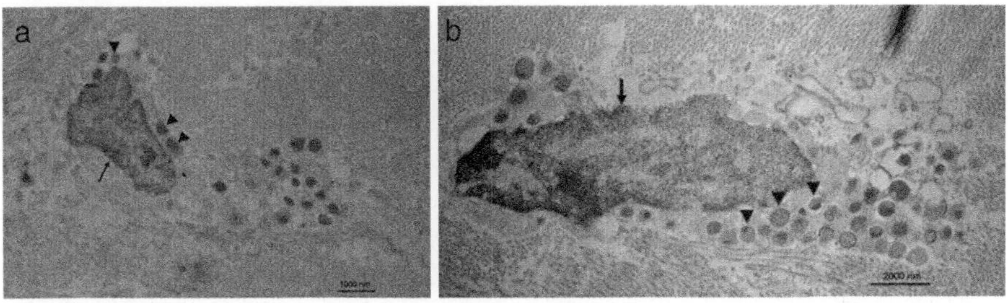

Fig. 5.1 Transmission electron micrography of sample obtained from the hip joint capsule of a patient with a pretreated periprosthetic joint infection caused by a small-colony variant of *S. aureus*. Intracellular cocci in a fibroblast from a periprosthetic biopsy. Two different magnifications according to the scale on the figures. Reproduced from Sendi et al. (2006) (© 2006 by Journals Division, University of Chicago Press, Chicago) ⟶ Nucleus of fibroblast; ▼ Intracellular cocci

5.2.2.3 Microbial Biofilms on Implants

The ability to form biofilms is a virulence factor of many microorganisms formerly considered as non-pathogenic. This is particularly true for coagulase-negative staphylococci, but also for *P. acnes* (Ramage et al. 2003). Implant-associated infections are typically caused by microorganisms growing in biofilms. Formation of biofilms involves several steps, starting with rapid surface attachment and followed by multilayered cellular proliferation and intercellular adhesion in an extracellular polysaccharide matrix (slime) excreted by the bacteria (Gotz 2002, Trampuz et al. 2003, Yao et al. 2005). Adherence of *S. epidermidis* to the surface of the device involves rapid attachment mediated by non-specific factors, such as surface tension, hydrophobicity, and electrostatic forces, or by specific adhesins (Williams et al. 2002). Adherence of *S. aureus* is more dependent on the presence of host-tissue ligands, such as fibronectin, fibrinogen, and collagen. This initial phase of adherence is followed by an accumulative phase, during which bacterial cells adhere to each other and create a biofilm. This process is mediated by the polysaccharide intercellular adhesin (PIA) encoded by the *ica* operon. Adherence and accumulation (cell–cell adhesion and proliferation) are separately controlled. It has been shown that *S. epidermidis* RP 62A and its isogenic slime-negative mutant M7 differ in their property to accumulate. Both adhere, but only the slime-producing strain is able to accumulate on implants (Konig et al. 1998). Interestingly, this parent strain is more resistant to antimicrobial agents than the mutant strain in vitro as well as in an animal model, indicating the pathogenic role of accumulation regarding persistence (Schwank et al. 1998). Release of cell-to-cell signaling molecules (quorum sensing) induces bacteria in a population to respond in concert by changing patterns of gene expression involved in biofilm differentiation (Davies et al. 1998). In a recent analysis of gene expression in *Staphylococcus epidermidis* biofilms, it could be shown that genes for translation, transcription, and global regulatory systems are down-regulated, whereas those for chaperones, stress factors, resistance factors, and osmoprotection are up-regulated (Yao et al. 2005). Thus, the pattern of gene expression in *S. epidermidis* biofilms leads to a non-aggressive and protected form of bacterial growth with low metabolic activity allowing long-term survival during chronic infection.

5.2.3 Host Defense Around Implants

Implanted devices have a high risk to be colonized by bacterial and fungal agents. It has been shown that the presence of a foreign body decreases the minimal infecting dose of *S. aureus* leading to a permanent abscess >100,000-fold (Elek and Conen 1957, Zimmerli et al. 1982). In an animal model, 100 cfu *S. aureus* were sufficient to infect 95% of the subcutaneous implants. Therefore, even a minimal contamination of the implant either during surgery or in the early perioperative period may result in a permanent device-associated infection. In contrast to wound infections in the absence of foreign material, such infections never spontaneously heal (Zimmerli et al. 1982). Implants are not only endangered by the exogenous route during the perioperative period, but also at any time by the hematogenous route (Zimmerli et al. 1985, Zimmerli et al. 2004).

The most important microorganisms causing implant associated infection are *S. aureus* and coagulase-negative staphylococci (Giulieri et al. 2004, Laffer et al. 2006). Thus, even a low number of these skin-derived bacteria can colonize an implant during surgery. Obviously, granulocytes around an implant are not able to clear even a small quantity of microorganisms. This is due to a local granulocyte defect induced by the non-phagocytosable foreign body. Granulocytes accumulating around an implant are partially degranulated and have a decreased superoxide production and an impaired killing of *S. aureus* (Zimmerli et al. 1982, Zimmerli et al. 1984). Once a biofilm is established, adherent staphylococci can persist because they resist even to functionally intact granulocytes, as shown in an in vitro model (see also Chapter 11).

5.3 Selected Implant-Associated Infections

5.3.1 Prosthetic Joint Associated

Infections associated to prosthetic joints are classified according to their time of manifestation. Early infections occur during the first 2 months after implantation. Delayed infections are diagnosed after the second month up to 2 years, and late infections occur lifelong after the second postoperative year (Zimmerli et al. 2004). Early infections are exogenous and typically caused by microorganisms with high virulence, such as *S. aureus* or *E. coli* (Steckelberg and Osmon 2000, Zimmerli et al. 2004). Delayed infections are also perioperatively acquired, but caused by microorganisms of lower virulence such as coagulase-negative staphylococci and *Propoionibacterium acnes*. Late infections are hematogenous. The main primary foci are skin infections (*S. aureus*, group-A streptococci), respiratory tract infections (*Streptococcus pneumoniae*), or urinary tract (*E. coli*) infections (Maderazo et al. 1988).

Signs and symptoms vary according to the type of infection. In early infection, signs of wound infection, such as erythema and prolonged wound secretion or synovitis, are easy to recognize. In contrast, in delayed infection, signs and symptoms are not specific for infection. In general, the patient suffers from persisting pain after surgery without systemic signs of infection. After a prolonged time, signs of implant loosening predominate (Zimmerli et al. 2004). The main challenge in late infection is the fact that signs of the primary infection (skin infection, pneumonia, etc.) predominate, and symptoms of arthritis must be actively searched (Maderazo et al. 1988).

Before starting antimicrobial therapy, the microbiological diagnosis is strictly required. Therefore, in case of suspicion, arthrotomy with appropriate diagnostic sampling (synovia and synovial biopsies) should be performed. In patients with infection associated to total knee arthroplasty, cytology of synovia confirms infection, if leukocytes are >1700/ul and/or the granulocyte fraction is >65% with excellent sensitivity (94 and 97%) (Trampuz et al. 2004). In other joints, cytology has not been tested. The sensitivity of the culture of synovia or even biopsies is limited (45–100%) especially in case of previous antimicrobial therapy or in delayed infection. Sonication of devices (screws, inlays and prosthesis) improves the sensitivity especially in patients with recent antimicrobial therapy (Trampuz et al. 2007).

According to the traditional rules, standard treatment includes two-stage exchange with meticulous removal of all foreign material combined with a 6-week course of intravenous antibiotics (Brause 2005). This approach is fastidious, time-consuming, and the functional result may be suboptimal due to repetitive surgery resulting in bone and muscle destruction. Alternatively, lifelong suppressive oral antimicrobial treatment without surgical intervention is suggested by some authors. However, this approach does not eradicate the infection, but only suppress clinical symptoms. In order to reach the ultimate goal of a successful therapy, namely a long-term pain-free functional joint by complete eradication of infection, a combination of both, an appropriate surgical procedure and an antimicrobial treatment acting on adherent bacteria is needed (Widmer et al. 1990, Widmer et al. 1991, Zimmerli et al. 1994). The cornerstone of successful treatment is early diagnosis. Since treatment is less invasive (see below) in patients with a short history of infection (<3 weeks), delay in diagnosis should be avoided. This can be done by a high degree of suspicion and by strict avoidance of empirical antibiotic therapy without unequivocal diagnosis.

In implant-associated infection, antimicrobial treatment without any surgical intervention usually fails. We recently published an algorithm (Fig. 5.2) which allows to choose the optimal surgical intervention according to well different criteria (Zimmerli and Ochsner 2003, Zimmerli et al. 2004, Trampuz and Zimmerli 2005).

Debridement with retention. The success rate with this procedure is widely underestimated, because in many reports it has been used without proper selection of eligible patients. Not only in older (Brandt et al. 1997, Crockarell et al. 1998, Tattevin et al. 1999) but also in very recent publications (Berbari et al. 2006, Marculescu et al. 2006), success rates below 35% have been reported. Therefore, this procedure should be reserved for a well-defined population of patients. If the following conditions are fulfilled, the success rate is similar to the one for exchange surgery, namely >80% (Zimmerli et al. 1998, Giulieri et al. 2004, Byren et al. 2009). The requirements are (1) a stable implant, (2) a pathogen with susceptibility to antimicrobial agents active against surface-adhering microorganisms, (3) absence of a sinus tract or an abscess, and (4) duration of symptoms of infection less than 3 weeks.

One-stage (*direct*) exchange includes the removal and implantation of a new prosthesis during the same surgical procedure. Patients with intact or only slightly compromised soft tissue qualify for this procedure. A success rate of 86–100% can be expected in appropriately selected patients (Raut et al. 1994, Ure et al. 1998). If resistant or difficult-to-treat microorganisms are causing the infection, such as methicillin-resistant *S. aureus* (MRSA), small-colony variants of staphylococci, enterococci, quinolone-resistant *P. aeruginosa* or fungi, a two-stage revision should be preferred.

Two-stage (*staged*) *exchange* includes removal of the prosthesis with implantation of a new prosthesis during a later surgical procedure. If no difficult-to-treat microorganisms are isolated, a short interval until reimplantation (2–4 weeks) and a temporary antimicrobial-impregnated bone cement spacer may be used. If difficult-to-treat microorganisms are isolated, a longer interval (8 weeks) without a spacer is preferred. The two-stage procedure can be used for every patient and has success

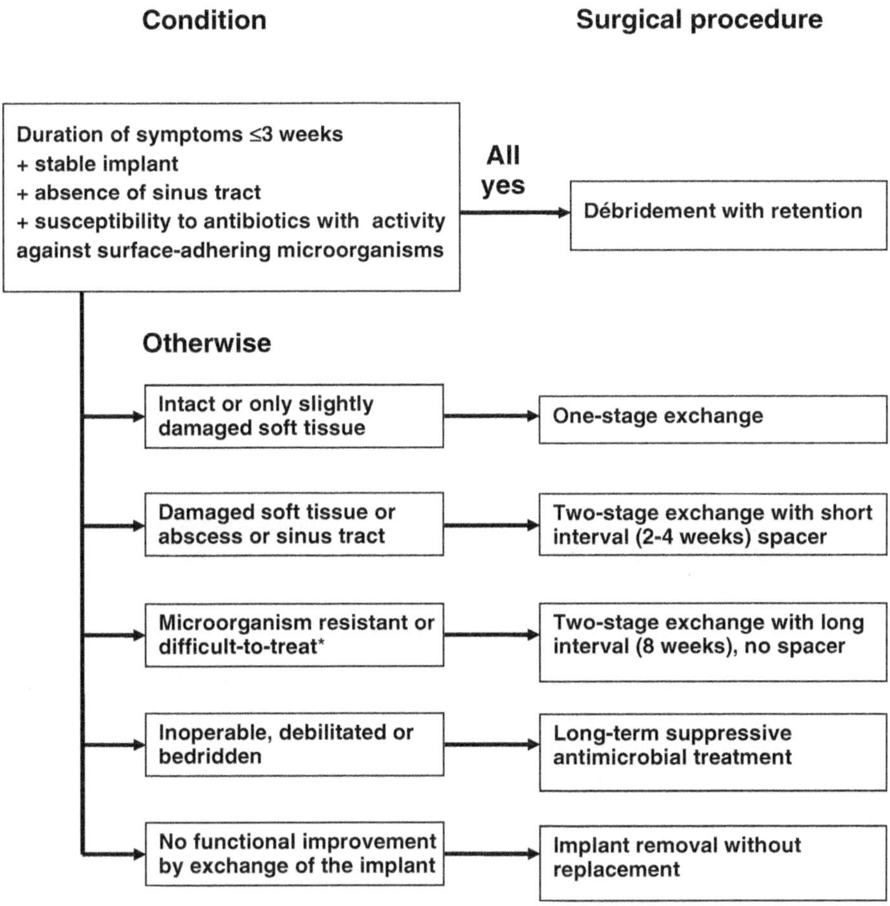

Fig. 5.2 Surgical treatment algorithm for prosthetic joint infections (modified according to Trampuz and Zimmerli (2005). *Difficult-to-treat microorganisms include methicillin-resistant *S. aureus* (MRSA), small-colony variants of staphylococci, enterococci, quinolone-resistant *P. aeruginosa* and fungi

rate generally exceeding 90% (Colyer and Capello 1994, Langlais 2003, Zimmerli and Ochsner 2003, Zimmerli et al. 2004). However, the expenditure for the patient and the surgeon is higher than for other surgical options.

Permanent removal of the device is usually reserved for patients with a high risk of reinfection (e.g., severe immunosuppression, active intravenous drug use) or when no functional improvement after reimplantation is expected. Alternatively, long-term antimicrobial suppression may be chosen, if the patient is inoperable, bedridden, or debilitated. However, suppressive therapy only controls clinical symptoms rather than curing the infection. Therefore, infection relapses occur in most patients (>80%) when antimicrobials are discontinued.

5.3.2 Mammary Implants

Infection following breast reconstruction with implants is a rare but feared complication. The infection rate is variable according to the type of surgical procedure performed. It is influenced by local factors in the wound, such as decreased skin vascularization, the presence of necrotic tissue or hematoma, as well as host defense mechanisms of the patient. In case of simple augmentation, the infection rate should not exceed 2% (Gabriel et al. 1997, Darouiche 2004), whereas in patients with mastectomy for breast cancer combined with immediate implant, it can be around 12% (Olsen et al. 2008). The microbiology of acute mammary implant associated infection is similar to the one of other surgical site infections, i.e., the main microorganism are *S. aureus*, enterobacteriaceae, and *P. aeruginosa* (Pittet et al. 1999).

Early acute infection is characterized by swelling, tenderness, pain, delayed wound healing, and fever. The clinical picture of late infection is different. Since many years, low-grade infection has been suspected to cause capsular contracture (Shah et al. 1981). In the meantime, several studies showed evidence for the bacterial etiology of capsular contracture. In two studies, sonication of removed breast implants of patients with contracture revealed a significant amount of bacteria in 33 and 41% of the cases (Del Pozo et al. 2009, Rieger et al. 2009) reported that the mean indwelling time was 10.4 years. Thus, obviously, this type of biofilm infections does not manifest with overt signs and symptoms of infection.

The classical treatment of periprosthetic infection includes breast implant removal and delayed reinsertion, combined with long-term antibiotic therapy (Spear et al. 2004). In case of unilateral infection, at time of reimplantation, the contralateral implant is also removed, and a new pair of implants are inserted in order to achieve symmetry (Darouiche 2004). However, direct exchange with implant removal, pocket curettage, capsulectomy, debridement, placement of a new implant, and postoperative antibiotics may also be successful in selected cases (Spear et al. 2004, Chun and Schulman 2007).

5.3.3 Cardiac Pacemakers and Cardioverter-Defibrillators (ICD)

Cardiac pacemakers and ICD are increasingly used due to the higher median age of the population and the ever broadening indications for ICD. Nowadays, the infection rate should not exceed 1–2% (Chua et al. 2000, Klug et al. 2007). Exogenous infection generally occurs within 1–2 months after implantation. However, it can also be a late event in case of erosion of the battery pocket due to pressure of the device. Hematogenous infection may result during prolonged bacteremia. It affects the electrode and is hard to diagnose due to the lack of local signs of infection. In a series of 123 patients, the most common pathogens were coagulase-negative staphylococci (68%), *S. aureus* (24%), and enterobacteriaceae (17%). Thirteen percent of the infections were polymicrobial (Chua et al. 2000).

The clinical presentation of pacemaker- and ICD-associated infection is similar (Chua et al. 2000). It includes signs of pocket inflammation, such as erythema, swelling, warmth, pain, erosion, and sinus tract. These signs are typical for exogenous infection and mostly lacking in case of hematogenous infection. Systemic signs of infection, including fever, chills, tachycardia, malaise, and anorexia are less frequent. These signs, as well as pacing threshold elevation, are suggestive for electrode-associated infection. Echocardiography should be performed in order to look for vegetations on valves or electrode which can be found in about 10% of the cases.

Treatment includes removal of device and all lead material combined with antimicrobial therapy. In case of late exogenous infection due to pocket necrosis, the generator is sometimes exchanged without removal of the electrodes. However, this should be the exception, because the risk of recurrence is high. In case of diagnosis of electrode-associated infection, the exchange of all hardware should be the rule (Bucher et al. 2000). This is especially important if the diagnosis was delayed and the biofilm was fully established on the electrode.

5.3.4 Vascular Prostheses

Infection associated with vascular prostheses is rare but dreaded, because it is a life- or limb-endangering event. The most frequent microorgansims are *S. aureus, S. epidermidis*, and group B-streptococci. Polymicrobial infection is mainly observed in patients with enteric ischemia after aortic replacement. The incidence is <1% in aortic up to 6% in lower extremity grafts (Antonios and Baddour 2004). The risk of hematogenous seeding indirectly correlated with the integrity of the neoendothelium (Lanzetta and Owen 1995). Graft-associated infections have a high risk for mortality (up to 75% for aortic grafts) and morbidity (up to 70% limb amputation for lower extremity grafts).

The main signs and symptoms are fever and signs of wound infection such as erythema, swelling, heat, and wound healing disturbances with a protracted wet wound. After aortic replacement, the symptoms are non-specific. Sometimes, prolonged postoperative low-grade fever is the only sign. A special manifestation is the acute gastrointestinal bleeding occurring in case of erosion with connection between graft anastomosis and intestines (Valentine et al. 2008).

The microbiological diagnosis is important. Therefore, several blood cultures should be drawn before any antibiotic therapy. In addition, fluid collections should be punctured under CT-guidance. In the early postoperative period, the diagnosis of vascular graft-associated infection is not easy to diagnose because of the normal perigraft hematoma (Qvarfordt et al. 1985). After about 1–2 weeks, perigraft air is suggestive for infection. In unclear cases, a technetium-99m-HMPAO-labeled leukocyte scan allows to reliably detect infection with a sensitivity of 100% and a specificity of 92.5% (Liberatore et al. 1998).

The aims of treatment are control of sepsis, complete elimination of device-associated biofilm bacteria, and an adequate vascular supply of the affected peripheral organs. The decision whether a treatment is conservative (debridement, drainage and antibiotics) or radically surgical (excision and extraanatomic bypass) should be discussed in an experienced team of vascular surgeons and infectious disease specialists (Swain 2004). In case of sepsis after aortic replacement, antibiotic therapy alone is efficacious, as long as the patient has no periprosthetic abscess. However, in case of periprothetic abscess or sinus tract, surgical intervention is required for cure of infection. For specific antimicrobial therapy, see Table 5.1.

Table 5.1 Treatment of prosthetic joint infections. Adapted from Zimmerli et al. (2004)

Microorganism	Antimicrobial Agent[a]	Dose[b]	Route
Staphylococcus aureus or coagulase-negative staphylococci			
Methicillin-susceptible	Rifampin plus	450 mg every 12 h	PO/IV
	(flu)cloxacillin[c]	2 g every 6 h	IV
	for 2 weeks, followed by		
	Rifampin plus	450 mg every 12 h	PO
	ciprofloxacin or	750 mg every 12 h	PO
	levofloxacin	750 mg every 24 h to	PO
		500 mg every 12 h	PO
Methicillin-resistant	Rifampin plus	450 mg every 12 h	PO/IV
	vancomycin	1 g every 12 h	IV
	for 2 weeks, followed by		
	Rifampin plus	450 mg every 12 h	PO
	ciprofloxacin[d] or	750 mg every 12 h	PO
	levofloxacin[d] or	750 mg every 24 h to	PO
		500 mg every 12 h	PO
	teicoplanin[e] or	400 mg every 24 h	IV/IM
	fusidic acid or	500 mg every 8 h	PO
	cotrimoxazole or	1 forte tablet every 8 h	PO
	minocycline	100 mg every 12 h	PO
Streptococcus spp.	Penicillin G[c] or	5 million U every 6 h	IV
	Ceftriaxone	2 g every 24 h	IV
	for 4 weeks, followed by		
	Amoxicillin	750–1000 mg every 8 h	PO
Enterococcus spp. (penicillin-susceptible)	Penicillin G or	5 million U every 6 h	IV
	ampicillin or amoxicillin	2 g every 4–6 h	IV
	plus aminoglycoside[e]		IV
	For 6 weeks after removal of all foreign material		
Enterobacteriaceae (quinolone-susceptible)	Ciprofloxacin	750 mg every 12 h	PO
Nonfermenters (e.g., *P. aeruginosa*)	Cefepime or ceftazidime	2 g every 8 h	IV
	plus aminoglycoside[f]		IV
	for 2–4 weeks, followed by		
	Ciprofloxacin	750 mg every 12 h	PO
Anaerobes[g]	Clindamycin	600 mg every 6–8 h	IV
	for 2–4 weeks, followed by		
	Clindamycin	300 mg every 6 h	PO

Table 5.1 (continued)

Microorganism	Antimicrobial Agent[a]	Dose[b]	Route
Mixed infections (without methicillin-resistant staphylococci)	Amoxicillin/clavulanic acid	2.2 g every 8 h	IV
	or piperacillin/tazobactam	4.5 g every 8 h	IV
	or imipenem	500 mg every 6 h	IV
	or meropenem	1 g every 8 h	IV
	for 2–4 weeks, followed by individual regimens according to antimicrobial susceptibility		

PO, orally; IV, intravenously; IM, intramuscularly, forte tablet: trimethoprim 160 mg plus sulfamethoxazole 800 mg

[a]If implant retention or one-stage exchange is performed, the total duration of antimicrobial treatment is 3 months for hip prosthesis and 6 months for knee prosthesis. For two-stage exchange see text.

[b]All dosages are for adults assuming normal renal function.

[c]In patients with delayed hypersensitivity, cefazolin (2 g every 8 h IV) can be administered. In patients with immediate hypersensitivity, penicillin should be replaced by vancomycin (1 g every 12 h IV).

[d]Methicillin-resistant *S. aureus* should not be treated with quinolones since antimicrobial resistance may emerge during treatment.

[e]The first day, a loading dose of 800 mg IV in one or two doses should be given.

[f]Aminoglycosides can be administered in a single daily dose.

[g]Alternatively, penicillin G (5 million U every 6 h IV) or ceftriaxone (2 g every 24 h IV) can be used for Gram-positive anaerobes (e.g., *Propionibacterium acnes*) and metronidazole (500 mg every 8 h IV or PO) for Gram-negative anaerobes (e.g., *Bacteroides* spp.).

5.3.5 Dental Implants

Dental implants (see also Chapter 4) are exposed to an extremely high concentration of bacteria in saliva. Therefore, colonization by microorganisms from normal saliva flora or by putative periodontal pathogens is the rule. Interestingly, this biofilm on the screw does usually not harm. However, in about 5–8% peri-implantitis occurs. This is characterized by an inflammation leading to bone loss around the implant and consecutive loosening of the screw. Epidemiologically, implants colonized by *Streptococcus oralis, Eubacterium* sp., and *Veillonella parvula* remain asymptomatic, whereas colonization with *Streptococcus intermedius, Bacteroides forsythus, Fusobacterium nucleatum*, and *Porphyromonas gingivalis* generally result in implant loosening (Leonhardt et al. 2003). In case of severe peri-implantitis, antibiotics such as amoxicilin/clavulanic acid or clindamycin should be given. The duration of antimicrobial therapy is not well defined. In addition, the use of laser- or ultrasonic devices for detatching biofilms from the implant surfaces in peri-implantitis has been proposed (Schwarz et al. 2004, Sculean et al. 2005). However, this treatment modality is still experimental.

5.3.6 Peritoneal Dialysis Catheters

About 5–10% of the patients with end stage renal disease are maintained on chronic ambulatory peritoneal dialysis (CAPD) therapy. The problem of this technique is the high rate of CAPD-associated peritonitis. With improved technology, the infection rate decreased from one infection per 12 patient-months to one episode in 60 patient-months (Troidle and Finkelstein 2006). The main microorganisms causing peritonitis are *S. epidermidis* and *S. aureus*, less frequently *P. aeruginosa* and occasionally fungi (Troidle and Finkelstein 2006, Jassal and Lok 2008).

Risk factors for infection are older systems requiring more manipulation, nasal colonization with *S. aureus*, and depression (Troidle and Finkelstein 2006). Episodes are classified as recurrent (episode within 4 weeks after therapy with another microorganism), relapsing episode (same microorganism and susceptibility pattern within the 4-week period after antimicrobial therapy), repeat episode (same microorganism and susceptibility pattern greater than 4 weeks after therapy), refractory (persistence of cloudy efflux after 5 days of therapy), and catheter-related peritonitis. CAPD-associated peritonitis is not obligatory accompanied by abdominal pain and fever. The diagnosis of peritonitis requires the presence of 100 leukocytes per μl cloudy peritoneal effluent with >50% granulocytes. Even with adequate culturing methods, up to 20% of the episodes remain without microbial growth. Exit-site and tunnel infection are characterized by erythema, edema, and tenderness over the subcutaneous pathway (Piraino et al. 2005). Diagnostic work-up includes cytology of the peritoneal effluent and culture of effluent directly and in a blood-culture bottle.

Despite the fact that a biofilm covers the peritoneal catheter in case of peritonitis, the catheter must rarely be removed. In general, antibiotics are given by the intraperitoneal route and continued for at least 2 weeks. Details of the antimicrobial therapy are given in the guidelines and can also be looked at by contacting http://www.ispd.org (Piraino et al. 2005). Catheter removal should be considered in refractory, relapsing, or fungal peritonitis. In addition, it is also indicated in refractory exit-site and tunnel infection. There are several protocols for preventing exit-site infections, including exit-site mupirocin, intranasal mupirocin, or exit-site gentamicin cream (Piraino et al. 2005).

5.4 Prevention

5.4.1 Perioperative Antimicrobial Prophylaxis

In general, perioperative antimicrobial prophylaxis is indicated for clean-contaminated or contaminated, but not for clean wounds, because the risk for surgical wound infection is high in the former two, but low in the latter situation (Waddell and Rotstein 1994). However, since the presence of a foreign body increases the susceptibility for infection >100,000 fold, prophylaxis is indicated in each type of surgery involving permanent devices (e.g., orthopedic, cardiovascular, pacemaker, and neurosurgical surgery) (Zimmerli et al. 1982, Haas and Kaiser 2000, Trampuz and Zimmerli 2006).

Drugs for prophylaxis should be active against the most common pathogens involved in implant-associated bone infection, namely staphylococci, streptococci, and Gram-negative bacilli (Laffer et al. 2006, Trampuz and Zimmerli 2006). Since the susceptibility of these microorganisms differs, each hospital needs to continuously update the resistance pattern of own surgical site isolates. A first-generation or second-generation cephalosporin, such as cefazolin, cefamandole, or cefuroxime, is a rational choice. If the patient is allergic to cephalosporins, or in settings with high prevalence of methicillin-resistant *S. aureus* (MRSA), vancomycin or teicoplanin are alternative options. However, even in countries with a high prevalence of MRSA, no evidence of superiority of a glycopeptide prophylaxis exists. For example, in an Italian study comparing a single dose of teicoplanin with two doses of cefamandole, both prophylactic regimens were equal efficient in 496 patients with total hip replacement (Suter et al. 1994).

5.4.2 Prophylaxis of Late Hematogenous Infections in Patients with Orthopedic Devices

Hematogenous prosthetic joint infection occurs at any time after surgery. It is experimentally and clinically well established (Ahlberg et al. 1978). In a recent cohort study of 40 consecutive episodes of prosthetic knee associated infection, the fraction of hematogenous infections has been estimated to 38% (Laffer et al. 2006). Hematogenous seeding occurs during the whole life; however, the risk is highest early after implantation. However, this does not implicate that routine antibiotic prophylaxis is needed during procedures potentially causing bacteremia. To our best knowledge, the molecular proof of hematogenously caused periprosthetic joint infection is still lacking (Pallasch and Wahl 2003). The infection rate by oral microorganisms of prosthetic joints has been estimated to 0.05% (Steckelberg and Osmon 2000). Since the risk of a bacteremia from periodontitis, oral hygiene, and mastication is considerably higher than from dental treatment, the routine use of antibiotic prophylaxis during procedures with potential bacteremia is not recommended (Pallasch and Wahl 2003). Accordingly, the American Dental Association/Academy of Orthopaedic surgeons (AAOS) has published advisory statements which states that antibiotic prophylaxis is not mandatory during dental procedure, but it should be considered in patients with increased risk of hematogenous prosthetic joint infection, such as early after-joint replacement (within 1 year after implantation) in immunocompromised patients and in those with severe comorbidities (especially in persons with underlying inflammatory arthropathies) (American Academy of Orthopaedic and Surgeons, document 1033.). Very recently, Berbari et al. (Berbari et al. 2010) published a case-control study to examine the association between dental procedures with or without antibiotic prophylaxis and periprosthetic joint infection. The authors found no increased risk of periprosthetic joint infection following dental procedures. In addition, antibiotic prophylaxis was not associated with risk reduction.

5.5 General Principles of Antimicrobial Therapy

Antimicrobial therapy against an established biofilm often fails without removal of the involved implant (Gristina et al. 1989). Obviously, standard susceptibility tests are not predictive of the therapeutic outcome of device-associating infections (Widmer et al. 1990). Therefore, the impact of bacterial adherence and the formation of biofilms have to be considered in the treatment of such infections. In vitro and animal experiments with two isogenic strains of *S. epidermidis*, with and without accumulative growth on surfaces, showed phenotypic resistance to antibiotic treatment of the biofilm-forming strain (Schwank et al. 1998). We previously analyzed which susceptibility test could accurately predict the outcome of biofilm infections (Zimmerli et al. 1994). In brief, we found that the in vitro bactericidal effect of antimicrobial agents on non-growing and adherent bacteria best predicts treatment outcome in experimental implant-associated infections (Widmer et al. 1990, 1991, Zimmerli et al. 1994, Drancourt et al. 1997, Schwank et al. 1998). Rifampin best fulfills these requirements against staphylococci and quinolones against Gram-negative bacilli (Widmer et al. 1991, Zimmerli et al. 1994). Optimal antimicrobial therapy is best defined in staphylococcal implant infections. Rifampin has proven its activity in vitro in animal models and in several clinical studies (Zimmerli et al. 1994, Schwank et al. 1998, Giulieri et al. 2004, Laffer et al. 2006). It must always be combined with another drug to prevent emergence of resistance in staphylococci. Quinolones are excellent combination drugs because of their good bioavailability, activity, and safety. Newer quinolones were studied in experimental bone infections (Shirtliff et al. 2001), but only anecdotal clinical data exist with these new drugs (Frippiat et al. 2004).

Table 5.1 summarized the choice of antimicrobial agents according to the pathogen (Zimmerli et al. 2004). The correct duration of treatment has never be tested in comparative studies, and therefore remains arbitrary. In the presence of a device, recurrence is frequent. Therefore, like in tuberculosis, killing of all microorganisms is needed. In patients with orthopedic devices, treatment has been better studied than in other implant-associated infections. In a controlled trial in patients with orthopedic implant-associated infection, a duration of 3 months has been chosen for each type of implant except knee prostheses which are treated 6 months (Zimmerli et al. 1998). Intravenous treatment should be administered for the first 2 weeks, followed by oral therapy if a drug with good oral bioavailability is available.

Resistance of staphylococci to quinolones is an increasing problem. Therefore, other anti-staphylococcal drugs have been combined with rifampin, such as cotrimoxazole, minocycline, fusidic acid (Drancourt et al. 1997), or linezolid (Baldoni et al. 2009). Quinopristin-dalfopristin is active against *Enterococcus faecium* (including vancomycin-resistant strains) and *S. aureus* (including MRSA), but not against *E. faecalis*. In a study of 40 patients with orthopedic infections with MRSA, clinical success was reported in 78% and microbial eradication in 69% (Drew et al. 2000). Linezolid is active against virtually all Gram-positive cocci, including methicillin-resistant staphylococci and vancomycin-resistant enterococci

(VRE). Twenty consecutive patients treated with linezolid for orthopedic infections (15 of whom had an orthopedic device) were retrospectively evaluated (Razonable 2004). At a mean follow-up of 276 days, 55% achieved clinical cure and 35% had clinical improvement but received long-term antimicrobial suppressive therapy. Unfortunately, adverse events such as reversible myelosuppression (40%), and irreversible peripheral neuropathy (5%) were quite common. In another review, long-term use of linezolid (>28 days) was associated with severe peripheral and optic neuropathy. In most cases, optic neuropathies resolved after stopping linezolid but peripheral neuropathies did not (Bressler et al. 2004). Daptomycin is active against several Gram-positive bacteria, including MRSA, vancomycin-resistant *S. aureus*, and VRE (Carpenter and Chambers 2004). Its efficacy has been tested in an animal model of implant-associated infections, where it showed advantage in combination with rifampin as compared to vancomycin or linezolid (John et al. 2009).

5.6 Outlook

The apparent incidence of device-associated infections will increase due to better detection methods for microbial biofilms (sonication method), the growing use of permanent implants (demographic structure of population), and the longer residency time of implanted devices (better life expectation) which are at lifelong risk for hematogenous infection. There is a need for new antimicrobial agents with activity on surface-adhering microorganism (see also Chapter 14). A novel treatment mechanism could be the modulation of quorum-sensing resulting in disintegration of the biofilm. Another strategy could be the use of bacteriophages with activity on staphylococci. However, since the specificity may be to narrow and the penetration of bacteriophages in biofilms limited, bacteriophage-encoded endolysins might be an alternative treatment option (van der Ploeg 2008).

References

Ahlberg A, Carlsson AS, Lindberg L (1978) Hematogenous infection in total joint replacement. Clin Orthop Relat Res (137):69–75

Akhan SE, Dogan Y, Akhan S et al (2008) Pelvic actinomycosis mimicking ovarian malignancy: three cases. Eur J Gynaecol Oncol 29(3):294–297

American Academy of Orthopaedic Surgeons. Antibiotic prophylaxis for bacteremia in patients with total joint replacements. Document no 1033. http://www.aaos.org/about/papers/advistmt/1033.asp. Accessed 3 August 2010

Antonios VS, Baddour LM (2004) Intra-arterial device infections. Curr Infect Dis Rep 6(4): 263–269

Antonios VS, Noel AA, Steckelberg JM et al (2006) Prosthetic vascular graft infection: a risk factor analysis using a case-control study. J Infect 53(1):49–55

Baldoni D, Haschke M, Rajacic Z et al (2009) Linezolid alone or combined with rifampin against methicillin-resistant Staphylococcus aureus in experimental foreign-body infection. Antimicrob Agents Chemother 53(3):1142–1148

Berardinelli L (2006) Grafts and graft materials as vascular substitutes for haemodialysis access construction. Eur J Vasc Endovasc Surg 32(2):203–211
Berbari EF, Osmon DR, Duffy MC et al (2006) Outcome of prosthetic joint infection in patients with rheumatoid arthritis: the impact of medical and surgical therapy in 200 episodes. Clin Infect Dis 42(2):216–223
Berbari EF, Osmon DR, Carr A et al (2010) Dental procedures as risk factors for prosthetic hip or knee infection: a hospitals based prospective case-control study. Clin Infect Dis 50(1):8–16
Brandt CM, Sistrunk WW, Duffy MC et al (1997) Staphylococcus aureus prosthetic joint infection treated with debridement and prosthesis retention. Clin Infect Dis 24(5):914–919
Brause B (2005) Infections with prostheses in bones and joints. In: Mandell GL, Bennet JE, Dolin R (eds) Principles and practice of infectious diseases, 6th edn. WB Saunders, Washington, DC
Bressler AM, Zimmer SM, Gilmore JL et al (2004) Peripheral neuropathy associated with prolonged use of linezolid. Lancet Infect Dis 4(8):528–531
Bucher E, Trampuz A, Donati L et al (2000) Spondylodiscitis associated with bacteraemia due to coagulase-negative staphylococci. Eur J Clin Microbiol Infect Dis 19(2):118–120
Byren I, Bejon P, Atkins BL et al (2009) One hundred and twelve infected arthroplasties treated with "DAIR" (debridement, antibiotics and implant retention): antibiotic duration and outcome. J Antimicrob Chemother 63(6):1264–1271
Carpenter CF, Chambers HF (2004) Daptomycin: another novel agent for treating infections due to drug-resistant gram-positive pathogens. Clin Infect Dis 38(7):994–1000
Chua JD, Wilkoff BL, Lee I et al (2000) Diagnosis and management of infections involving implantable electrophysiologic cardiac devices. Ann Intern Med 133(8):604–608
Chun JK, Schulman MR (2007) The infected breast prosthesis after mastectomy reconstruction: successful salvage of nine implants in eight consecutive patients. Plast Reconstr Surg 120(3):581–589
Colyer RA, Capello WN (1994) Surgical treatment of the infected hip implant. Two-stage reimplantation with a one-month interval. Clin Orthop Relat Res (298):75–79
Conen A, Walti LN, Merlo A et al (2008) Characteristics and treatment outcome of cerebrospinal fluid shunt-associated infections in adults: a retrospective analysis over an 11-year period. Clin Infect Dis 47(1):73–82
Crockarell JR, Hanssen AD, Osmon DR et al (1998) Treatment of infection with debridement and retention of the components following hip arthroplasty. J Bone Joint Surg Am 80(9):1306–1313
Darouiche RO (2001) Device-associated infections: a macroproblem that starts with microadherence. Clin Infect Dis 33(9):1567–1572
Darouiche RO (2004) Treatment of infections associated with surgical implants. N Engl J Med 350(14):1422–1429
Davies DG, Parsek MR, Pearson JP et al (1998) The involvement of cell-to-cell signals in the development of a bacterial biofilm. Science 280(5361):295–298
Del Pozo JL, Tran NV, Petty PM et al (2009) Pilot study of association of bacteria on breast implants with capsular contracture. J Clin Microbiol 47(5):1333–1337
Drancourt M, Stein A, Argenson JN et al (1997) Oral treatment of Staphylococcus spp. infected orthopaedic implants with fusidic acid or ofloxacin in combination with rifampicin. J Antimicrob Chemother 39(2):235–240
Drew RH, Perfect JR, Srinath L et al (2000) Treatment of methicillin-resistant staphylococcus aureus infections with quinupristin-dalfopristin in patients intolerant of or failing prior therapy. For the Synercid Emergency-Use Study Group. J Antimicrob Chemother 46(5):775–784
Elek SD, Conen PE (1957) The virulence of Staphylococcus pyogenes for man; a study of the problems of wound infection. Br J Exp Pathol 38(6):573–586
Frazier OH, Gemmato C, Myers TJ et al (2007) Initial clinical experience with the HeartMate II axial-flow left ventricular assist device. Tex Heart Inst J 34(3):275–281
Frippiat F, Meunier F, Derue G (2004) Place of newer quinolones and rifampicin in the treatment of Gram-positive bone and joint infections. J Antimicrob Chemother 54(6):1158; author reply 1159

Gabriel SE, Woods JE, O'Fallon WM et al (1997) Complications leading to surgery after breast implantation. N Engl J Med 336(10):677–682

Garzoni C, Kelley WL (2009) Staphylococcus aureus: new evidence for intracellular persistence. Trends Microbiol 17(2):59–65

Geissdorfer W, Tandler R, Schlundt C et al (2007) Fatal bioprosthetic aortic valve endocarditis due to Cardiobacterium valvarum. J Clin Microbiol 45(7):2324–2326

Giulieri SG, Graber P, Ochsner PE et al (2004) Management of infection associated with total hip arthroplasty according to a treatment algorithm. Infection 32(4):222–228

Gotz F (2002) Staphylococcus and biofilms. Mol Microbiol 43(6):1367–1378

Gristina AG, Jennings RA, Naylor PT et al (1989) Comparative in vitro antibiotic resistance of surface-colonizing coagulase-negative staphylococci. Antimicrob Agents Chemother 33(6):813–816

Haas D, Kaiser A (2000) Antimicrobial prophyloxis of infections associated with foreign bodies. In: Bisno AL, Waldvogel FA, Infections associated with indwelling medical devices, 3rd ed, Washington, DC

Hall-Stoodley L, Costerton JW, Stoodley P (2004) Bacterial biofilms: from the natural environment to infectious diseases. Nat Rev Microbiol 2(2):95–108

Hammermeister K, Sethi GK, Henderson WG et al (2000) Outcomes 15 years after valve replacement with a mechanical versus a bioprosthetic valve: final report of the Veterans Affairs randomized trial. J Am Coll Cardiol 36(4):1152–1158

Jacobsen SM, Stickler DJ, Mobley HL et al (2008) Complicated catheter-associated urinary tract infections due to Escherichia coli and Proteus mirabilis. Clin Microbiol Rev 21(1):26–59

Jassal SV, Lok CE (2008) A randomized controlled trial comparing mupirocin versus Polysporin Triple for the prevention of catheter-related infections in peritoneal dialysis patients (the MP3 study). Perit Dial Int 28(1):67–72

John AK, Baldoni D, Haschke M et al (2009) Efficacy of daptomycin in implant-associated infection due to methicillin-resistant staphylococcus aureus (MRSA): the importance of combination with rifampin. Antimicrob Agents Chemother 53:2719–2724

Klinge B, Hultin M, Berglundh T (2005) Peri-implantitis. Dent Clin North Am 49(3):661–676, vii–viii

Klug D, Balde M, Pavin D et al (2007) Risk factors related to infections of implanted pacemakers and cardioverter-defibrillators: results of a large prospective study. Circulation 116(12): 1349–1355

Konig DP, Perdreau-Remington F, Rutt J et al (1998) Slime production of Staphylococcus epidermidis: increased bacterial adherence and accumulation onto pure titanium. Acta Orthop Scand 69(5):523–526

Laffer RR, Graber P, Ochsner PE et al (2006) Outcome of prosthetic knee-associated infection: evaluation of 40 consecutive episodes at a single centre. Clin Microbiol Infect 12(5): 433–439

Langlais F (2003) Can we improve the results of revision arthroplasty for infected total hip replacement? J Bone Joint Surg Br 85(5):637–640

Lanzetta M, Owen ER (1995) Neo-endothelialisation of PTFE microvascular grafts: a five-year experience. Microsurgery 16(6):404–411

Leonhardt A, Dahlen G, Renvert S (2003) Five-year clinical, microbiological, and radiological outcome following treatment of peri-implantitis in man. J Periodontol 74(10):1415–1422

Liberatore M, Iurilli AP, Ponzo F et al (1998) Clinical usefulness of technetium-99m-HMPAO-labeled leukocyte scan in prosthetic vascular graft infection. J Nucl Med 39(5): 875–879

Lietard C, Thebaud V, Besson G et al (2008) Risk factors for neurosurgical site infections: an 18-month prospective survey. J Neurosurg 109(4):729–734

Lucas VS, Lytra V, Hassan T et al (2002) Comparison of lysis filtration and an automated blood culture system (BACTEC) for detection, quantification, and identification of odontogenic bacteremia in children. J Clin Microbiol 40(9):3416–3420

Maderazo EG, Judson S, Pasternak H (1988) Late infections of total joint prostheses. A review and recommendations for prevention. Clin Orthop Relat Res (229):131–142

Maki DG, Kluger DM, Crnich CJ (2006) The risk of bloodstream infection in adults with different intravascular devices: a systematic review of 200 published prospective studies. Mayo Clin Proc 81(9):1159–1171

Marculescu CE, Berbari EF, Hanssen AD et al (2006) Outcome of prosthetic joint infections treated with debridement and retention of components. Clin Infect Dis 42(4):471–478

Marrie TJ, Costerton JW (1984) Morphology of bacterial attachment to cardiac pacemaker leads and power packs. J Clin Microbiol 19(6):911–914

Ochsner P, Sirkin M, Trampuz A (2006) Acute infection. In: Ruedi RP, Murphy WM (eds) AO principles of fracture. Management, Stuttgart, Thieme

Olsen MA, Chu-Ongsakul S, Brandt KE et al (2008) Hospital-associated costs due to surgical site infection after breast surgery. Arch Surg 143(1):53–60; discussion 61

Oxenham H, Bloomfield P, Wheatley DJ et al (2003) Twenty year comparison of a Bjork-Shiley mechanical heart valve with porcine bioprostheses. Heart 89(7):715–721

Pallasch T, Wahl M (2003) Focal infection: new age or ancient history. Endodontic Topics 4:32–45

Perez AR, Roxas MF, Hilvano SS (2005) A randomized, double-blind, placebo-controlled trial to determine effectiveness of antibiotic prophylaxis for tension-free mesh herniorrhaphy. J Am Coll Surg 200(3):393–397; discussion 397–398

Piraino B, Bailie GR, Bernardini J et al (2005) Peritoneal dialysis-related infections recommendations: 2005 update. Perit Dial Int 25(2):107–131

Pittet B, Montandon D, Pittet D (2005) Infection in breast implants. Lancet Infect Dis 5(2): 94–106

Pittet D, Harbarth S, Ruef C et al (1999) Prevalence and risk factors for nosocomial infections in four university hospitals in Switzerland. Infect Control Hosp Epidemiol 20(1):37–42

Qvarfordt PG, Reilly LM, Mark AS et al (1985) Computerized tomographic assessment of graft incorporation after aortic reconstruction. Am J Surg 150(2):227–231

Raad I, Hanna H, Maki D (2007) Intravascular catheter-related infections: advances in diagnosis, prevention, and management. Lancet Infect Dis 7(10):645–657

Ramage G, Tunney MM, Patrick S et al (2003) Formation of Propionibacterium acnes biofilms on orthopaedic biomaterials and their susceptibility to antimicrobials. Biomaterials 24(19): 3221–3227

Raut VV, Siney PD, Wroblewski BM (1994) One-stage revision of infected total hip replacements with discharging sinuses. J Bone Joint Surg Br 76(5):721–724

Razonable RR, Osmon DR, Steckelberg JM (2004) Linezolid therapy for orthopedic infections. Mayo Clin Proc 79(9):1137–1144

Rieger UM, Pierer G, Luscher NJ et al (2009) Sonication of removed breast implants for improved detection of subclinical infection. Aesthetic Plast Surg 33(3):404–408

Roumbelaki M, Kritsotakis EI, Tsioutis C et al (2008) Surveillance of surgical site infections at a tertiary care hospital in Greece: incidence, risk factors, microbiology, and impact. Am J Infect Control 36(10):732–738

Sanchez-Manuel FJ, Lozano-Garcia J, Seco-Gil JL (2007) Antibiotic prophylaxis for hernia repair. Cochrane Database Syst Rev (3):CD003769

Schwank S, Rajacic Z, Zimmerli W et al (1998) Impact of bacterial biofilm formation on in vitro and in vivo activities of antibiotics. Antimicrob Agents Chemother 42(4):895–898

Schwarz F, Bieling K, Sculean A et al (2004) [Treatment of periimplantitis with laser or ultrasound. A review of the literature]. Schweiz Monatsschr Zahnmed 114(12):1228–1235

Sculean A, Schwarz F, Becker J (2005) Anti-infective therapy with an Er:YAG laser: influence on peri-implant healing. Expert Rev Med Devices 2(3):267–276

Sendi P, Proctor RA (2009) Staphylococcus aureus as an intracellular pathogen: the role of small colony variants. Trends Microbiol 17(2):54–58

Sendi P, Rohrbach M, Graber P et al (2006) Staphylococcus aureus small colony variants in prosthetic joint infection. Clin Infect Dis 43(8):961–967

Shah Z, Lehman JA Jr, Tan J (1981) Does infection play a role in breast capsular contracture? Plast Reconstr Surg 68(1):34–42
Shirtliff ME, Calhoun JH, Mader JT (2001) Comparative evaluation of oral levofloxacin and parenteral nafcillin in the treatment of experimental methicillin-susceptible Staphylococcus aureus osteomyelitis in rabbits. J Antimicrob Chemother 48(2):253–258
Spear SL, Howard MA, Boehmler JH et al (2004) The infected or exposed breast implant: management and treatment strategies. Plast Reconstr Surg 113(6):1634–1644
Steckelberg JM, Osmon DR (2000) Prosthetic joint infection. In: Waldvogel FA (ed) 3rd edn ASM Press Washington, DC
Suter F, Avai A, Fusco U et al (1994) Teicoplanin versus cefamandole in the prevention of infection in total hip replacement. Eur J Clin Microbiol Infect Dis 13(10):793–796
Swain TW, 3rd, Calligaro KD, Dougherty MD (2004) Management of infected aortic prosthetic grafts. Vasc Endovascular Surg 38(1):75–82
Tattevin P, Cremieux AC, Pottier P et al (1999) Prosthetic joint infection: when can prosthesis salvage be considered? Clin Infect Dis 29(2):292–295
Trampuz A, Hanssen AD, Osmon DR et al (2004) Synovial fluid leukocyte count and differential for the diagnosis of prosthetic knee infection. Am J Med 117(8):556–562
Trampuz A, Osmon DR, Hanssen AD et al (2003) Molecular and antibiofilm approaches to prosthetic joint infection. Clin Orthop Relat Res (414):69–88
Trampuz A, Piper KE, Jacobson MJ et al (2007) Sonication of removed hip and knee prostheses for diagnosis of infection. N Engl J Med 357(7):654–663
Trampuz A, Zimmerli W (2005) Prosthetic joint infections: update in diagnosis and treatment. Swiss Med Wkly 135(17–18):243–251
Trampuz A, Zimmerli W (2006) Antimicrobial agents in orthopaedic surgery: Prophylaxis and treatment. Drugs 66(8):1089–1105
Trampuz A, Zimmerli W (2008) Diagnosis and treatment of implant-associated septic arthritis and osteomyelitis. Curr Infect Dis Rep 10(5):394–403
Troidle L, Finkelstein F (2006) Treatment and outcome of CPD-associated peritonitis. Ann Clin Microbiol Antimicrob 5:6
Ure KJ, Amstutz HC, Nasser S et al (1998) Direct-exchange arthroplasty for the treatment of infection after total hip replacement. An average ten-year follow-up. J Bone Joint Surg Am 80(7):961–968
Uslan DZ, Sohail MR, St Sauver JL et al (2007) Permanent pacemaker and implantable cardioverter defibrillator infection: a population-based study. Arch Intern Med 167(7):669–675
Valentine RJ, Timaran CH, Modrall GJ et al (2008) Secondary aortoenteric fistulas versus paraprosthetic erosions: is bleeding associated with a worse outcome? J Am Coll Surg 207(6):922–927
van der Ploeg JR (2008) Characterization of Streptococcus gordonii prophage PH15: complete genome sequence and functional analysis of phage-encoded integrase and endolysin. Microbiology 154(Pt 10):2970–2978
von Eiff C, Peters G, Becker K (2006) The small colony variant (SCV) concept—the role of staphylococcal SCVs in persistent infections. Injury 37 Suppl 2:S26–33
Waddell TK, Rotstein OD (1994) Antimicrobial prophylaxis in surgery. Committee on Antimicrobial Agents, Canadian Infectious Disease Society. CMAJ 151(7):925–931
Waldvogel F, Bisno A (2000) Infections associated with indwellling medical devices 3rd edn. ASM Press Washington
Wei BP, Robins-Browne RM, Shepherd RK et al (2008) Can we prevent cochlear implant recipients from developing pneumococcal meningitis? Clin Infect Dis 46(1):e1–7
Widmer AF, Frei R, Rajacic Z et al (1990) Correlation between in vivo and in vitro efficacy of antimicrobial agents against foreign body infections. J Infect Dis 162(1):96–102
Widmer AF, Wiestner A, Frei R et al (1991) Killing of nongrowing and adherent Escherichia coli determines drug efficacy in device-related infections. Antimicrob Agents Chemother 35(4):741–746

Williams RJ, Henderson B, Sharp LJ et al (2002) Identification of a fibronectin-binding protein from Staphylococcus epidermidis. Infect Immun 70(12):6805–6810

Yamaguchi K, Adams RA, Morrey BF (1998) Infection after total elbow arthroplasty. J Bone Joint Surg Am 80(4):481–491

Yao Y, Sturdevant DE, Otto M (2005) Genomewide analysis of gene expression in Staphylococcus epidermidis biofilms: insights into the pathophysiology of S. epidermidis biofilms and the role of phenol-soluble modulins in formation of biofilms. J Infect Dis 191(2):289–298

Zimmerli W (2008) Antibiotic Therapy. In: Baltensperger M, Eyrich G (eds) Osteomyelitis of the Jaws, 1st edn. Springer, Berlin

Zimmerli W, Frei R, Widmer AF et al (1994) Microbiological tests to predict treatment outcome in experimental device-related infections due to Staphylococcus aureus. J Antimicrob Chemother 33(5):959–967

Zimmerli W, Lew PD, Waldvogel FA (1984) Pathogenesis of foreign body infection. Evidence for a local granulocyte defect. J Clin Invest 73(4):1191–1200

Zimmerli W, Ochsner PE (2003) Management of infection associated with prosthetic joints. Infection 31(2):99–108

Zimmerli W, Trampuz A, Ochsner PE (2004) Prosthetic-joint infections. N Engl J Med 351(16):1645–1654

Zimmerli W, Waldvogel FA, Vaudaux P et al (1982) Pathogenesis of foreign body infection: description and characteristics of an animal model. J Infect Dis 146(4):487–497

Zimmerli W, Widmer AF, Blatter M et al (1998) Role of rifampin for treatment of orthopedic implant-related staphylococcal infections: a randomized controlled trial. Foreign-Body Infection (FBI) Study Group. JAMA 279(19):1537–1541

Zimmerli W, Zak O, Vosbeck K (1985) Experimental hematogenous infection of subcutaneously implanted foreign bodies. Scand J Infect Dis 17(3):303–310

Chapter 6
The Role of Bacterial Biofilms in Infections of Catheters and Shunts

Trine Rolighed Thomsen, Luanne Hall-Stoodley, Claus Moser, and Paul Stoodley

6.1 Catheters and Biofilm

6.1.1 What Are Catheters and Why Are They Used?

Catheters and shunts are tubes which are used to manage the flow of fluids into, within, and out of the body. Intravascular catheters deliver fluids and medications directly into the bloodstream, while urinary catheters drain waste fluids. In some cases devices such as cerebral ventricular shunts drain fluid from the brain, to another part of the body, such as the heart or stomach where the fluids are processed internally. There are numerous indications for catheter use in a wide variety of anatomic sites and for treating many chronic diseases. Urinary or IV catheters might be placed for a few hours for patient management during a surgical procedure, radiological scan during a surgical procedure, or during an imaging examination. Some catheters may have an anticipated lifetime of months or years in certain patients. Catheters can be transcutaneous, where by definition, they breach the protective skin barrier (i.e., IV catheters and external ventricular drains), can be inserted into natural orifices (i.e., urinary and nasogastric catheters) or can be completely indwelling (i.e., cerebral ventricular shunts). In the broadest sense, endotracheal tubes and ear tubes might be considered catheters since they are tubes designed to facilitate the transport of fluids (air or liquid) into and out of the body. Some of the more common catheters are listed in Table 6.1. In addition to catheters, stents and cannulae are often used for the transport of fluids into and out of the body. A stent is a "tube" inserted into a vessel or other conduit in the body to prevent or treat stenosis, or constriction, e.g., of a vessel, esophagus, or ductus choledocus.

T.R. Thomsen (✉)
Department of Biotechnology, Chemistry, and Environmental Engineering,
Aalborg University, Sohngaardsholmsvej 49, DK-9000, Aalborg, Denmark;
Life Science division, The Danish Technological Institute,
Kongsvang Allé, DK-8000 Århus C, Denmark
e-mail: trt@bio.aau.dk

T. Bjarnsholt et al. (eds.), *Biofilm Infections*, DOI 10.1007/978-1-4419-6084-9_6,

Table 6.1 Different types of medical fluid management devices

Catheter type	Use
Central venous catheter (CVC)	Transcutaneous. Administer medication, fluids and nutrients, obtain blood for tests and directly obtain cardiovascular measurements such as the central venous pressure. Usually the CVC is passed through a vein to end up in the thoracic portion of the vena cava or in the right atrium of the heart
Urinary catheters	External intubation. Drainage of urine. Inserted directly into the urethra, generally with the aid of lubricants. Can be used for incontinence, disability or short term use during medical procedures
Dialysis catheters	Transcutaneous. Exchange of blood to and from a hemodialysis machine and the patient
Cardiac shunt	Internal. Allows blood flow between heart chambers and cardiac vessels. Shunts can be "one-way" or bidirectional
Cerebral or ventricular shunt	Internal or transcutaneous drainage. In cases of hydrocephalus, a one-way valve is used to drain excess cerebrospinal fluid (CSF) from the brain and carry it to other parts of the body
External ventricular drain	Used for external drainage of CSF. Often used for monitoring CSF in cases of suspected infection
Endotracheal tube	External intubation. To assist breathing or protect the airway
Nasogastric Catheter/ Shunt	External inturbation. Parenteral feeding
Ear tubes	Drain effusion fluid and allow aeration of the middle ear space

6.1.2 Catheter Infection

There is a well recognized association between catheters and infection, and mounting evidence suggesting biofilm involvement (Donlan 2001b). Firstly, the foreign body itself presents a potential surface for bacterial attachment and subsequent biofilm formation. Secondly, externalized catheters present a conduit for bacteria from the skin and external environment to enter the body, along both the inside lumen and the outside surface (Fig. 6.1). Catheters can also become infected from a distant site from the body itself by hematogenous route. Intravenous catheters and urinary catheters are the most commonly used medical devices in USA and a frequent cause of nosocomially acquired bloodstream infections (Trautner and Darouiche 2004). Central venous catheter (CVC) infections have been associated with an in-hospital mortality rate of 12–35%, prolonged hospitalizations, and multiple complications associated with the treatment of these infections (Doshi et al. 2009). With urinary catheters the infection rate is between 10 and 50% of patients undergoing short-term urinary catheterization (7 days) rising to virtually all patients undergoing long-term catheterization (>28 days) (Donlan 2001a, Maki and Tambyah 2001). Patients with a long-term catheter can also develop acute pyelonetritis and sepsis (Gorman and Jones 2003). Encrustation of the urinary catheter is another common complication in patients with long-term indwelling catheterization. It is believed that encrustation is associated with the adhesion and

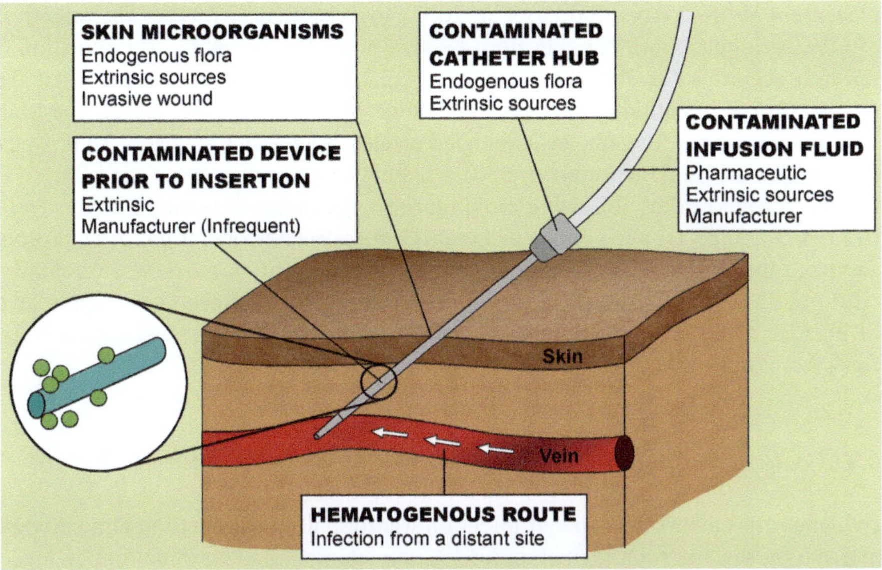

Fig. 6.1 Diagram illustrating some of the various routes of infection associated with a transcutaneous catheter. Skin bacteria can reach the catheter through the insertion site (endogenous flora), or through the hub during manipulation or from contaminated infusion fluid (extrinsic sources). The bacteria can also reach the catheter via hematogenous seeding or from infection from the surrounding environment (Illustration by Henriette Slot)

colonization of urease-producing biofilm bacteria, principally *Proteus mirabilis*, but also with other bacteria, such as *Staphylococcus aureus*, *Pseudomonas aeruginosa*, *Providencia* sp., and *Klebsiella pneumoniae*, which leads to the deposition of rigid crystals, composed primarily of magnesium ammonium phosphate (struvite), calcium phosphate (hydroxyapatite), and calcium oxalate onto the catheter lumen surface (Getliffe 2003, Gray 2001, Park et al. 2002, Stickler 2008). These crystalline deposits may produce pain in the patient due to the occurrence of mechanical trauma to the urethral mucosa upon catheter withdrawal. In addition, the accumulation of deposits contributes to the blockage of the catheter lumen, thereby resulting in retention of urine, ascending infection of the urinary tract, and increased discomfort of the patient (Wilson 2009). Encrustation is common in 40–50% of long-term catheterized patients (Getliffe 2003).

The pathogenesis of infections on catheters and shunts is a multifaceted interaction, and the type of catheter and duration of insertion are important considerations (Trautner and Darouiche 2004). Factors that significantly increased the risk of CVC bloodstream infection were associated with inexperience of the operator and nurse-to-patient ratio in the intensive care unit, catheter insertion with sub-optimal sterile barriers, placement of a CVC in the internal jugular or femoral vein rather than subclavian vein, placement in an old site by guidewire exchange, heavy colonization of the insertion site or contamination of a catheter hub, and duration of CVC placement above 7 days (Safdar et al. 2002). Approximately 65% of the infections of CVC originate from the skin flora (Bouza et al. 2002), while contamination of the hub accounts for 30% of the catheter-related infections and is the most frequent cause of infection in long-term catheters (Bouza et al. 2002, Safdar and Maki 2004). Risk factors for catheter-associated urinary tract infection, based on prospective studies and use of multivariable statistical modeling by Maki and Tambyah (2001), resulted in the following ranking: prolonged catheterization >6 days (relative risk 5.1–6.8), female gender, catheter insertion outside operating room, urology service, other active sites of infection, diabetes, malnutrition, Azotemia (creatinine >2.0 mg/dL), ureteral stent, monitoring of urine output, drainage tube below level of bladder and above collection bag, and antimicrobial-drug therapy (relative risk from 0.1 to 0.4).

6.1.3 Clinical Evidence for Infections Associated with Catheters

Evidence for catheter-associated infection traditionally has relied on clinical signs and symptoms, microbiological culture, Gram staining, and acridine-orange leucocyte cytospin test of blood samples (Bouza et al. 2002, Chatzinikolaou et al. 2004, Cicalini et al. 2002) or on methods that can only be applied following catheter removal (Warwick et al. 2004). The international reference diagnostic method for catheter-related infections is Maki's semi-quantitative "roll plate" method (1973), which is based on the removal of the catheter and rolling the distal tip back and forth on an agar plate (Bouza et al. 2002, Maki et al. 1977). Fourteen colony-forming units (CFU) or more are definitive of an insertion site infection (Maki et al. 1977). There are, however, several problems with these methods for diagnosis. First, it typically takes at least 1 day to obtain a result. Second, since only the catheter tip is rolled back and forth on the agar plate, bacteria further up the catheter and on the internal, luminal site are not included in the analysis, leading to sampling error. Third, it has been shown that many catheters are culture negative after removal in cases where there were distinct signs and symptoms of infections (such as fever and increased number of polymorphonuclear (PMN) leukocytes), suggesting that culture by the Makis method under-represents the infection rate (Cormican 2003, Timsit 2007). Therefore, a high index of suspicion and improved diagnostic methods are indicated for diagnosing, for example, shunt-associated infection (Conen et al. 2008). Recently data have been reported supporting the hypothesis that colonization of CVC is only rarely responsible for catheter-related blood stream infection (Smuszkiewicz et al. 2009) leading to routine microbiological monitoring of CVC tips at the hospital being discontinued to reduce the cost of treatment. However, an alternative conclusion is that culture negative results may represent false negatives. The biofilm paradigm can explain conflicting signs and symptoms of local infection (fever, tenderness, swelling, and redness) in the absence of positive bacterial culture and the recalcitrance to antibiotic treatment (Hall-Stoodley et al. 2004). Emerging evidences, based on culture-free molecular diagnostic techniques and "enhanced" culturing techniques, suggest that conventional microbial culture methods produce some false negatives, particularly with chronic, biofilm-related infections. When non-culture based molecular methods or more sensitive culture procedures are used the positive identification of pathogens is increased (Hall-Stoodley and Stoodley 2009, Hall-Stoodley et al. 2006, Nielsen et al. 2008).

Several reasons might explain the poor sensitivity of conventional culture to detect chronic infections. First, bacteria in biofilms develop a slow (or non-) growing or metabolically dormant phenotype induced by nutrient limitation (Fux et al. 2004, Lewis 2007). Furthermore, bacterial populations in species, such as Staphylococci and *P. aeruginosa*, contain mutant small colony variants (SVC) which are slow growing and might be missed by conventional culture (Neut et al. 2007). SVC have also been shown to be more resistant to antibiotics than the wild type strains and may be selected for in *P. aeruginosa* biofilms (Starkey et al. 2009). Taken together, these observations suggest a link between biofilm formation, chronic infection, and failure to culture.

A second hypothesis for false culture negative results is that clinical sampling might simply "miss" the infecting bacteria. Fluids are typically sampled, rather than foreign bodies or tissues, due to considerations of risk to the patient as well as the convenience of access. However, biofilm bacteria may not be included in such samples. This is suggested by the increased recovery rate of biofilm bacteria by ultrasonication of orthopedic specimens prior to culture (Piper et al. 2009, Trampuz et al. 2007, Tunney et al. 1999). Similarly, Leroy et al. (2007) demonstrated that repeated sampling of effusion from nontypeable *H. influenzae* (NTHi) infected chinchilla ears recovered more NTHi than single samples. However, it is not yet clear whether such "enhanced" culture techniques will translate to catheters. Sonication did not increase the rates of culture-positive results in patients with long-term tunneled catheters (Slobbe et al. 2009) while it was proved to be 20% more sensitive compared to the semi-quantitative roll plate method and > 20% more sensitive than the method of flushing the individual catheter lumens in a study by Sherertz (2004). Finally, residual antibiotics in the clinical specimen might prevent the growth of bacteria which are spread on laboratory agar.

The detection rate of pathogens in clinical specimens has been vastly increased by the direct observation of surfaces by microscopy and the use of polymerase chain reaction (PCR)-based methods. In middle ear biopsies in pediatric patients suffering from otitis media (OM), for example, PCR was used to amplify DNA from genes specific to targeted pathogens associated with OM (Hall-Stoodley et al. 2006, Post et al. 1995). Reverse transcriptase PCR (RT-PCR) was also used to amplify mRNA, to indicate the presence of metabolically active bacteria (Rayner et al. 1998). Fluorescent in-situ hybridization (FISH) staining, based on the principle of fluorescently tagged complementary nucleic acid probes which hybridize with 16S ribosomal RNA (rRNA), has also been increasingly used for microbial identification in clinical samples (Hall-Stoodley et al. 2006, Kirketerp-Moller et al. 2008, Nistico et al. 2009). Unlike PCR, FISH can not only specifically identify pathogenic bacteria in situ, but also is compatible with retaining biofilm architecture and would therefore shed light on biofilm composition and distribution in catheter-associated infections.

6.1.4 Involvement of Biofilms in Catheter Infections

Evidence that biofilms can develop on IV devices, including CVCs, has been well documented, e.g., Murga et al. (2001), Nielsen et al. (2008), and Storti et al. (2005). Colonization of the outer lumen of the catheter by microorganisms is usually attributed to the catheter's proximity to skin flora. Colonization of the catheter inner lumen may be the result of a break in aseptic handling of the device prior to insertion, or, of exposure of the end connectors to contaminated environmental or intravenous fluids. A prospective study by Sabbuba et al. (2003) revealed that a single genotype of *P. mirabilis* persisted in the urinary tract, despite numerous changes of catheters, periods of non-catheterization, and antibiotic-therapy. It was confirmed that strains present in the catheter biofilms were identical to those isolated from the same patient's urine (Sabbuba et al. 2003). Many patients who suffer from recurrent catheter encrustation have been found to have bladder stones, and it is yet unknown whether the stones harbor specific *P. mirabilis* strains, which re-colonize successive catheters or the extent that biofilms might play in the formation of the stones. However electron and confocal microscopy suggest that structured communities of bacterial biofilms are associated with such stones (Marcus et al. 2008, Nickel et al. 1985).

There is indirect evidence to suggest that biofilms were involved in urinary or bloodstream infections caused by *Acinetobacter baumannii* (Rodríguez-Baño and Martí 2008). In this report, biofilm formation was investigated in 92 strains using a microtiter plate assay. Fifty-six (63%) of isolates formed biofilm. All catheter-related urinary or bloodstream infections, and the sole case of cerebral shunt-related meningitis were caused by biofilm-forming strains (Rodríguez-Baño et al. 2008). Another 559 patient study by Diskin et al. (2007) suggests a link between biofilms and infection rates in hemodialysis catheters. Here, 18% of 796 catheters developed sepsis, and the most significant variable was the number of days that the catheter remained in place. The only other two factors which were significant were the administration of intravenous iron or a mid-treatment bolus of heparin. Both iron and heparin have been associated with biofilm development and the authors concluded that these data implicate biofilms as being significant in the infectious process (Diskin et al. 2007).

Donlan (2001a) reviewed that biofilm formation on CVCs was universal, but the extent and location of biofilm formation depended on the duration of catheterization with short-term (<10 days) catheters having greater biofilm formation on the external surface and long-term catheters (30 days or more) having more biofilm formation on the catheter inner lumen. The nature of the fluid administered through CVC may affect microbial growth: Gram-positive organisms (*S. epidermidis* and *S. aureus)* did not grow well in intravenous fluids, whereas Gram-negative aquatic organisms (e.g., *P. aeruginosa*, and *Pantoea* sp. and also *Klebsiella* spp., *Enterobacter* spp. and *Serratia* spp*)* sustained growth (Donlan 2001a). The authors concluded that detection and quantification of biofilms on the inner, as well as outer surface of catheters, can provide the only true picture of biofilm colonization (Donlan 2001a).

6.1.5 Biofilm and Microbial Diversity in Catheters

> "There are known knowns. There are things we know that we know. There are known unknowns. That is to say, there are things that we now know we don't know. But there are also unknown unknowns. There are things we do not know we don't know". US Defense Secretary Donald Rumsfeld on February 12 2002.

A recent European study of IV catheter-related infections found that the microorganisms most often isolated from IV catheters were coagulase-negative staphylococci (CNS) (30–51.5%), followed by *Candida* spp., *S. aureus, Enterococcus* spp., and *Pseudomonas* spp. (Bouza et al. 2002, Bouza et al. 2004, Donlan 2001a). *Acinetobacter* spp. (Bouza et al. 2002, Linde et al. 2002), *Enterobacter* spp. (Bouza et al. 2002), *Klebsiella pneumonae* (Donlan 2001a) (all *Gammaproteobacteria*) and other bacteria have been identified but reported less frequently. In cerebral shunts the most prevalent isolated microorganisms were CNS, *S. aureus*, and *Propionibacterium acnes*. The most common species present in biofilms from urinary catheters were *E. faecalis, P. aeruginosa, E. coli,* and *P. mirabilis* (Stickler 2008).

Furthermore, increasingly polymicrobial and Gram-negative bacilli infections have been observed (Conen et al. 2008, Wang et al. 2004). However, since many biofilm-forming bacteria are poorly detected by traditional culture methods (Fux et al. 2003, Fux et al. 2005), several other potential pathogenic bacteria may be involved in catheter-related infections, increasing the risk of a delayed or incorrect diagnosis and increasing the risk of relapse or systemic infection, pain and discomfort and a suboptimal patient outcome, diagnostic and technical difficulties for the health care provider and increased costs for the health service (Schinabeck and Ghannoum 2003).

Molecular techniques have been used successfully for the identification of bacteria in many other systems (e.g., Juretschko et al. 2002, Neufeld and Mohn 2005, Thomsen et al. 2001), including some medico-related (Ott et al. 2006, Sakamoto et al. 2006, Warwick et al. 2004), but have also been used for the improved diagnosis of microbial infections in a few cases of individual CVCs using broad range 16S rRNA gene sequencing or fingerprinting. This technique is based on the fact that while all bacteria have a 16Sr RNA gene, there is variability in the nucleotide sequence of regions within the gene which can be used to distinguish among bacterial genera and species (see, e.g., Lau et al. 2002, Woo et al. 2000). Nielsen et al. (2008), for example, used a panel of molecular techniques including clone libraries based on broad range 16S rDNA amplification, denaturant gradient gel electrophoresis (DGGE), phylogeny, and FISH to substantially improve the diagnostic outcomes in a study of biofilms removed from CVCs. This report suggests that conventional culture-dependent methods can significantly benefit from supplemental molecular diagnostics. On seven catheters, where no growth was observed with culture, all but one catheter demonstrated bands by DGGE, which was possible to amplify and sequence (Nielsen et al. 2008). Interestingly, there seemed to be increased bacterial diversity in biofilms from catheters with a longer insertion time, suggesting that there might be an ecological succession similar to that seen in dental biofilms which occurs as one species modifies the environment making it conducive for colonization by another species (see also Chapter 4).

6.1.6 New Potentials in Diagnosis

In the previous chapter we briefly discussed some of the molecular methods that are being used to look for biofilms in clinical specimens. At present most of these are for "research purpose only," but it is likely that within a few years many of these techniques will be developed and approved for routine clinical diagnostics. Currently, there are two chief problems with the conventional diagnosis of catheter-related infections in particular and device-related infections in general. First, there are no standard clinical methods which more accurately identify pathogens, independent of culture. A growing body of literature suggests that conventional culture-based methodology is not a reliable predictor of biofilm-related infections and therefore there is a need to get beyond the notion that clinical culture is the "gold standard" by which medical microbiology is judged. In one study, culture positive results accounted for only two-thirds of infections as assessed by blood cultures within the catheter. This was increased by removal of the catheter (DesJardin et al. 1999). The accurate identification of pathogens in catheter-related infections is of direct importance, since it can be immediately used to guide clinical management. Thus better clinical management should offset the increased cost of molecular-based diagnostics such as PCR and FISH. Second, diagnostic methods need to be developed which specifically identify biofilms. How clinical management will be affected at the present time, even if biofilms are definitively diagnosed, depends on the type of device-related infection.

Even though the new diagnostic tools provide valuable possibilities in clinical microbiology, they also provide new challenges, namely how to interpret the findings in a useful and meaningful way. Exactly like Robert Koch experienced more than 100 years ago when he was able to culture and separate different microorganisms from human material and therefore to establish "Kochs postulates" to define when a microbiological finding was a pathogen, one has to recognize that the clinical significance of an unusual microbial species identified by molecular methods remains to be clarified. First of all contamination either by microbes or even by just genetic material has to be considered. Sampling from urinary catheters, for example, is not specifically designed to prevent contamination from the patients, staff flora, or the hospital milieu. Specimens could also be contaminated in the clinical microbiology laboratory, since strict aseptic methods are not used, or, indeed, required. Finally, contamination in the molecular biology laboratory is also a possibility. Because of these issues microscopy has been used to compliment and corroborate PCR-based analysis of clinical specimens (Hall-Stoodley et al. 2006, Stoodley et al 2008, Tunney et al. 1999). The direct visualization of biofilms is strongly suggestive of a growth process and it is unlikely that whole clusters of bacteria could originate from contamination (providing that specimens were properly stored immediately after collection), particularly if the biofilm was firmly attached (i.e., not removed by gentle rinsing).

Concerning the clinical significance of those findings one should implement a modification of Kochs postulates: (1) The microbe has to be identified in the material – best if visualized. (2) The microbe should generate a specific immune response. (3) The microbe must have relevant virulence factors to establish infection. (4) One should ideally be able to establish a relevant animal model. (5) The patient should recover when the microbe is eradicated from the patient. These points remain to be revealed before the "Gold standards" can be changed.

Bouza et al. (2004) investigated the etiology of catheter-related infections in European hospitals by a 1-day laboratory-based, point-prevalence survey based on a questionnaire completed by microbiology laboratories in European Union (EU) and non-EU hospitals. Overall, 19% of catheter tip cultures were polymicrobial and the microbiological diagnosis of intravenous catheter-related infections was increasingly more effective in European Institutions (Bouza et al. 2004). Traditional methods of diagnosis of catheter-associated infection rely on clinical features and quantitative microbiology of blood samples or on methods that can only be applied following CVC removal. Often the patients have been exposed to antibiotics, so even with appropriate use of diagnostic methods, there may remain considerable diagnostic uncertainty (Warwick et al. 2004). These observations suggest that the developments of additional diagnostic methods are necessary in cases where catheter-related infections are suspected. A study by Nielsen et al. (2008) showed that blood cultures, not generally correlated with tip cultures, and the bacteria identified by cultivation must in principle primarily represent bacteria on the external site of the catheters, which however was not necessarily supported by this study. A study by Blot et al. (1999) suggested that measurement of the differential time to positive culture between hub-blood and peripheral-blood cultures was a simple and reliable tool for in-situ diagnosis of catheter-related sepsis in cancer patients. Further studies are needed to confirm these data for short-term catheter placement and whether this technique can detect microorganisms when they are in a biofilm.

A quantitative PCR (qPCR) assay using primers and probe targeted to bacterial 16S ribosomal DNA was used to measure the levels of bacterial DNA in blood samples drawn through the CVC in a population of patients receiving intravenous nutrition (Warwick et al. 2004). Bacterial DNA concentrations were raised in 16 of 16 blood samples taken during episodes of probable bacterial CVC-associated infection compared to only 4 of 29 episodes in which bacterial CVC-associated infection was unlikely. Therefore, the use of this technique, and others like this, has the potential to substantially reduce the removal of CVCs. Using a specific approach for *P. aeruginosa* (Jaffe et al. 2001), qPCR was performed on DNA extracts from the study by Nielsen et al. (2008). Moderate gene copy numbers were detected in three catheters by qPCR where *P. aeruginosa* not was detected using other techniques. High copy numbers of *P. aeruginos* were measured in biofilm from the inner luminal and external side of another catheter, which also was detected by DGGE and cultivation. This clearly illustrates the different sensitivities of the various methods applied. In Nielsen et al. (2008) a clear difference between the bacterial species was found in biofilm on the external (exluminal) and internal (luminal) side of the CVC, which cannot be detected by Maki's method. The clone libraries showed that the bacteria from catheters on the external site were dominated by the *Firmicutes*, especially the staphylococci. The second most abundant group was the "uncultured *Deltaproteobacteria*" and they were only found on the external site. The internal site of the CVC was colonized mainly by the *Gammaproteobacteria*. Polymicrobial infections were observed on most of the catheters and were much more common than the cultivation-dependent methods generally indicated.

However, while PCR-based techniques might improve the detection rate of pathogens, the presence of biofilm can only be inferred from circumstantial factors. Currently microscopic observation is arguably the only direct demonstration of biofilm. PCR-based techniques can identify the presence of bacteria without the need for culture, but currently only microscopy can definitively identify biofilm. In adenoids we have found that conventional hematoxylin and eosin staining and Gram staining of thin sections missed biofilms which were observed by confocal microscopy of the surface of adenoids from pediatric patients suffering from otitis media (Nistico, L., Stoodley, P., and Hall-Stoodley, L. personal communication). Since the biofilms were patchy the thin section was likely to miss transecting the biofilm cell clusters completely or might just contain a few bacterial cells. However, by using the confocal to look at specimens in plain view we were able to gain a more complete perspective of biofilm distribution on the surface. Scanning electron microscopy (SEM) has been used extensively by Professor Bill Costerton to identify biofilms in many clinical specimens, including catheters. However, because confocal and scanning electron microscopy require highly specialized instrumentation and training and are time consuming, in their present form they will remain useful as research tools and are unlikely to become part of a routine diagnostic facility. Therefore, there is a great need for a rapid, high throughput definitive diagnostic for biofilm infection which has minimal operator bias.

Guidelines for the diagnosis of biofilms in infection have been outlined by Parsek (2003), with subsequent suggested modifications by Hall-Stoodley and Stoodley (2009). In brief these are (1) pathogenic bacteria associated with a surface, (2) microscopic examination of infected tissue demonstrates bacterial cell clusters encased in a matrix, which may be of bacterial and host origin (as also mentioned under the modified Kochs postulates), (3) infection is localized, (4) recalcitrance to antibiotic treatment despite demonstrated susceptibility of planktonic bacteria, or (since biofilm infections are often culture negative) recalcitrance to antibiotic therapy inferred from the presence of live bacterial cells in the biofilm from in-situ viability staining or RT-PCR, (5) culture-negative result in spite of clinically documented high suspicion of infection, and (6) ineffective host clearance evidenced by the location of bacterial cell clusters in discrete areas in the host tissue associated with host inflammatory cells.

Scanning and transmitted electron microscopy (SEM and TEM) have been used extensively to detect biofilms in various types of catheters (Machado et al. 2009, Marie and Costerton 1984, Marie et al. 1983); however, electron microscopy cannot determine viability, and generally it is difficult to identify specific pathogens with this technique. In contrast, confocal microscopy has been used to identify microorganisms in biofilms associated with various catheters, e.g., see Davis et al. (2002), Marcus et al. (2008), Nielsen et al. (2008), and Ott et al. (2007) (Fig. 6.2).

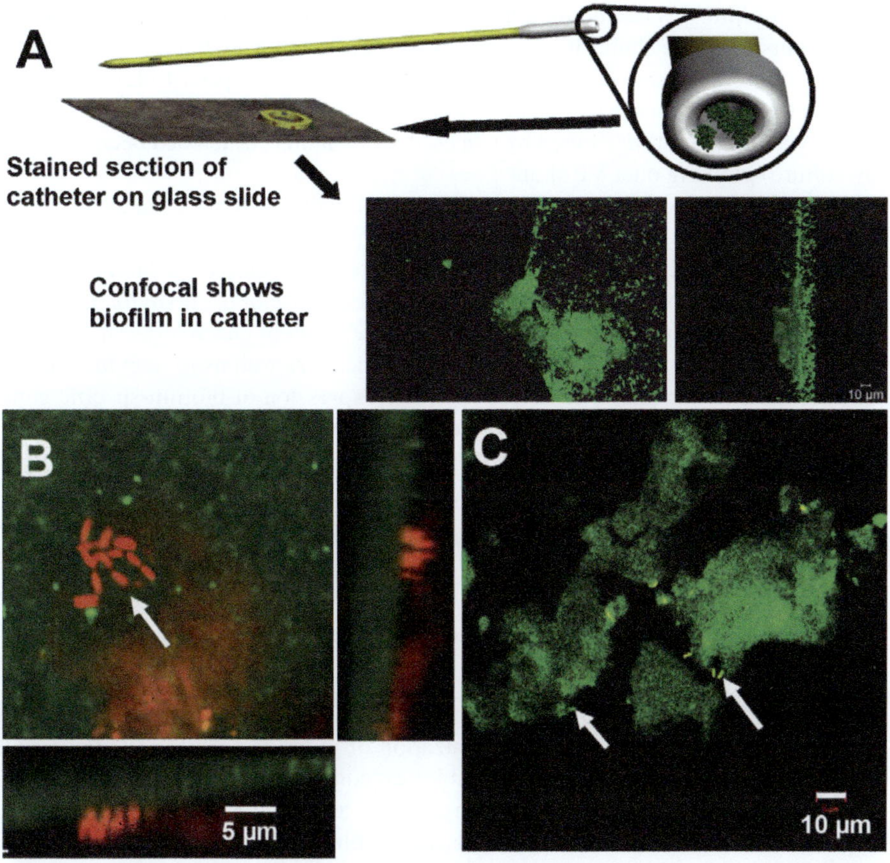

Fig. 6.2 Use of confocal microscopy to detect biofilms on catheters. (**a**) A schematic illustration of biofilm on a catheter. A CVC was cultivated with *E. coli* cells in vitro and hybridized with the probe targeting Bacteria labeled with Fluos (*green*) and the biofilm was observed from different angles using confocal microscopy. Image and illustration by Poul Larsen. (**b**) A small biofilm cluster of bacterial rods (*red*) attached to the tip of a CVC removed from a dialysis patient. The catheter surface was autofluorescent and appeared green. The XZ and YZ saggital sections are shown below and to the right of the main plan view respectively. The bacteria were dividing suggesting that they were viable. *Klebsiella* sp. was identified in this patient's blood culture. The image was taken by Dr. Vikram Gahlot. (**c**) Biofilm from a CVC removed from a patient and hybridized with the probe targeting *Gammaproteobacteria* labeled with Cy3 (*yellow*) and the probe targeting Bacteria labeled with Fluos (*green*). The arrows show *Gammaproteobacteria*-positive cells, which were not identified by cultivation. The image was taken by Trine Rolighed Thomsen

Stoodley et al. (2008) used a combination of PCR, RT-PCR, and confocal microscopy to detect live *S. aureus* in an elbow arthroplasty which had previously been cultured negative on numerous occasions. The same combination was applied to detect a live staphylococcal biofilm occluding an external ventricular drain (EVD) in a patient suffering from Dandy-Walker syndrome and a history of failed ventriculoperitoneal (VP) shunts (Stoodley et al. manuscript in preparation). In the latter case the biofilm was occluding approximately 60% of the lumen, but was not seen on the outside of the shunt. Interestingly CSF fluid taken from the EVD was culture negative for 5 consecutive days, including the sample taken prior to EVD removal. Davis et al. (2002) used live-dead staining and confocal microscopy and SEM to identify a fungal biofilm formed by *Coccidioides immitis* (which had been identified by culture) growing on a VP shunt.

However, while confocal microscopy is arguably the most powerful technique for directly identifying biofilms, in its present format it is not an ideal methodology for transfer to routine diagnostics due to the extensive time required for sample preparation and examination coupled with the requisite expertise of the operator. Currently, the power of confocal microscopy is to more accurately characterize the rates of biofilm infection associated with clinical specimens, as well as to map the location of the infecting bacteria. The detection of the expression of biofilm-specific genes or antigens (Brady et al. 2007, Selan et al. 2002) in fluids, aspirates, or biopsies promises to be a more rapid approach.

6.1.7 Prevention and Treatment

> The dream: "We have to find mechanisms to prevent and control the biofilm and infection and further develop catheter material that resists colonization" (Liedl 2001).

Intervention strategies for biofilm-associated infections include (1) prevention of initial device contamination, (2) minimization of initial microbial cell attachment, (3) use of agents such as high-dose antibiotics, combinations of antibiotics targeted at different niches of the biofilm given as early as possible or anti-biofilm agents in a catheter lock solution to penetrate the biofilm matrix and kill the embedded organisms, or at least suppress the biofilm until it is safe or more feasible to remove the catheter, (4) removal of the infected device, (5) ethanol lock into the lumen of the catheter, and (6) quorum sensing (QS) inhibition. Some anti-bacterials are better than others in treating biofilm-associated bacteria, such as rifampin (in combination with other antibiotics), tigecycline, daptomycin, *N*-acetylysteine (in combination with tigecycline), and ethanol (Aslam 2008). Shanks et al. (2006) demonstrated that heparin alternatives, sodium citrate and sodium EDTA, prevented the formation of *S. aureus* biofilms, suggesting that they may reduce the risk of biofilm-associated complications in indwelling catheters. A novel and potential clinically relevant finding of the study was that citrate at low levels strongly stimulates biofilm formation by *S. aureus*.

The failure of antibiotics to treat most biofilm infections is due to the difficulty of achieving the therapeutically necessary "sterilizing" concentration. The two main issues are delivery and toxicity. These difficulties were addressed in catheter infections in one study by filling the catheter-lumen with pharmacological concentrations of antibiotics in heparin using an "antibiotic lock" technique which was successful in 82.6% of 167 episodes (Mermel et al. 2001). This technique is currently used for uncomplicated infections of surgically implanted catheters involving CNS.

Although impractical for many clinical settings, removal currently remains the best treatment for the definitive cure of catheter-related infections.

However, there are currently many treatments specifically targeted to biofilms that are under development (see also Chapter 14), and the co-development of diagnostics and treatments would appear to be a logical progression. In addition, the development of biofilm diagnostics will allow more precise estimates of the extent of biofilm infection and the various epidemiological factors involved in biofilm catheter infections, with the subsequent design of anti-biofilm devices and procedures.

Catheter related bacteremia can often only be managed by either catheter removal with delayed placement of a new catheter, or an exchange of the infected catheter. However, removal of a biomedical device may be associated with significant problems for the patient. Therefore, strategies to inhibit bacterial attachment are being actively pursued. To prevent colonization two general approaches have been adopted. The first addresses contamination at the site of insertion by the use of prophylactic antibiotics or antiseptics, cutaneous disinfection, pre- and post insertion with various antiseptics, and catheter site care including the increased frequency of dressing care. The second approach focuses on the physical nature of the catheter material with some "anti-infection" strategies including silver impregnated subcutaneous collagen cuffs, antiseptic hubs, antimicrobial surfaces impregnated with antibiotics (e.g., rifampicin and minocycline), silver ions, and chlorhexidine (Spencer 1999).

It is believed that the surface characteristics of the polymers used in today's catheters are, to a large degree, responsible for the risk of infection. In particular the surface roughness, the hydrophobic nature of the surface, and the ability of certain types of common bacteria to attach and metabolize the polymers are at fault (Pourrezaei et al. 1994). Prior to 1950 when venous access required the use of rigid stainless steel needles, catheter infection was a minor problem. This has led material scientists to develop various metal coatings for both inside and outside of catheters. Metallic films have been deposited on Teflon, polymide and latex substrates (Pourrezaei et al. 1994). Specifically, various metal coatings were shown to be very effective in inhibiting the growth of common pathogens (Pourrezaei et al. 1994). However, Ha and Cho (2006) reviewed the role of silver-coated catheters and their effect in preventing infection and concluded that the role of silver in preventing catheter infections has not been supported by qualitative data, and resistance to silver was likely to become a problem with widespread use.

The blocking of QS pathways is another area of promise. QS is a mechanism that bacteria use to co-ordinate their growth phenotype through the production and sensing of secreted signal molecules. It has been demonstrated that biofilm bacteria in which QS is blocked, either by mutation or by administration of QS inhibitory drugs, are sensitive to treatment with tobramycin and H_2O_2, and are more readily phagocytosed by PMNs, in contrast to bacteria with functional QS systems (Bjarnsholt et al. 2005). In a recent study by Kiran et al. (2008) QS inhibitors were tested as potential therapeutics for even the most persistent infections. The animal models reviewed were associated with subcutaneous grafts, CVC, ureteral stent and wound models, and a wound case study. A QS inhibitor, RNAIII inhibiting peptide (RIP) and its non-peptide analog, were shown to prevent or treat infections caused by any staphylococcal strain tested, including antibiotic-resistant strains like (methicillin-resistant *S. aureus*) MRSA (Kiran et al. 2008).

The penultimate goal is a biocompatible material that is completely refractive to biofilm formation and which is active against all pathogens. However, this goal has remained elusive. Over the last 25 years many surfaces or coatings have been proposed based on optimizing various "antifouling" properties including hydrophobicity, surface charge, biocide/antimicrobial/cell-signal leaching surfaces, shedding or degrading surfaces, and lubricious surfaces. While most show some promise in vitro, usually against a challenge under quite narrow conditions, they generally remain in the developmental stage. It is likely that any prevention strategy will require multiple modalities ranging from novel anti-microbial/biofilm materials and designs to improved catheter management and infectious disease control.

6.2 Conclusion

In conclusion the biofilm concept is important in the understanding of catheter-related infections in terms of diagnosis, prevention, and treatment. Different molecular biological techniques have recently been used to improve diagnosis and have increased the understanding of the bacterial diversity and infectious mechanisms. Until recently it was assumed that, outside of natural orifices, medical infections tended to be caused predominantly by a single pathogen. However, emerging evidence from the examination of clinical specimens suggests that polymicrobial infection is more common than previously thought, and it is now of interest to look at the less well-known microorganisms which have been found in biofilm-related infections in order to investigate the question of whether they play a role in the pathology, or represent incidental bystanders. Finally, molecular methods make a more rapid diagnosis possible, even when the patient has received antibiotics and/or when microorganisms are difficult to cultivate. However, they may also indicate whether infectious bacteria are located in a biofilm which may directly result in a better treatment and patient management.

In the future more quantitative assays with a focus on the active fraction of microorganisms in the biofilm will be relevant in clinical microbial departments in hospitals. Additionally, next generation sequencing tool open up for the possibility of an increased understanding of the infectious microorganisms involved in catheter-related infections. For example, Tenover (2007) recently published a pyrosequencing assay which can provide data regarding the presence of multidrug-resistant *M. tuberculosis* directly from positive mycobacterial broth cultures in <1 day. Similarly, Leake identified yeast in chronic wounds using pyrosequencing and determined the relative contribution of the microorganisms present in the sample (Leake et al. 2009). While improved diagnostics and treatment of catheter infections are desirable for more efficient catheter management, the ultimate goal is to design catheters which do not allow sustained microbial colonization or the formation of biofilms in the first place. The development of an infection-free catheter remains an elusive goal, despite efforts reaching back over four decades. However, as the direct examination of clinical specimens using modern techniques raises the awareness of the true extent of the "foreign body–biofilm infection" problem with catheters, an increasing number of academic and industrial researchers are drawn to the problem, bringing ever-imaginative strategies. The awareness that biofilms are often the underlying cause of infections allows researchers to focus on relevant prevention strategies. Increased understanding of basic biofilm developmental biology provides more opportunities for intervention.

Finally, incomplete understanding of the mechanisms of catheter infection has permitted the acceptance of infection as an almost inevitable consequence of catheter use. Now we know that biofilms are the enemy, these high rates of infection should no longer be tolerated.

References

Aslam S (2008) Effect of antibacterials on biofilms. Am J Infect Control 36:S175.e179–S175.e111

Blot F, Nitenberg G, Chachaty E, Raynard B, Germann N, Antoun S, Laplanche A, Brun-Buisson C, Tancrede C (1999) Diagnosis of catheter-related bacteraemia: a prospective comparison of the time to positivity of hub-blood versus peripheral-blood cultures. Lancet 354: 1071–1077

Bouza E, Burillo A, Munoz P (2002) Catheter-related infections: diagnosis and intravascular treatment. Clin Microbiol Infect 8:265–274

Bouza E, San Juan R, Munoz P, Pascau J, Voss A, Desco M (2004) A European perspective on intravascular catheter-related infections report on the microbiology workload, aetiology and antimicrobial susceptibility (ESGNI-005 Study). Clin Microbiol Infect 10:838–842

Brady R, Leid J, Kofonow J, Costerton J, Shirtliff M (2007) Immunoglobulins to surface-associated biofilm immunogens provide a novel means of visualization of methicillin-resistant Staphylococcus aureus biofilms. Appl Environ Microbiol 73:6612–6619

Chatzinikolaou I, Hanna H, Hachem R, Alakech B, Tarrand, J, Raad I (2004) Differential quantitative blood cultures for the diagnosis of catheter-related bloodstream infections associated with short- and long-term catheters: a prospective study. Diagn Microbiol Infect Dis 50:167–172

Cicalini S, Palmieri F, Noto P, Boumis E, Petrosillo N (2002) Diagnosis of intra vascular catheter-related infection. J Vasc Access 3:114–119

Conen A, Walti L, Merlo A, Fluckiger U, Battegay M, Trampuz A (2008) Characteristics and treatment outcome of cerebrospinal fluid shunt-associated infections in adults: a retrospective analysis over an 11-year period. Clin Infect Dis 47:73–82

Cormican M (2003) Device-associated infection: the biofilm-related problem in health care. In: Lens P, Moran AP, Mahony T, Stoodley P, O'Flaherty V (ed) Biofilms in medicine, industry and environmental biotechnology, IWA Publishing, London, UK

Davis L, Cook G, Costerton J (2002) Biofilm on ventriculo-peritoneal shunt tubing as a cause of treatment failure in coccidioidal meningitis. Emerg Infect Dis 8:376–379

DesJardin J, Falagas M, Ruthazer R, Griffith J, Wawrose D, Schenkein D, Miller K, Snydman D (1999) Clinical utility of blood cultures drawn from indwelling central venous catheters in hospitalized patients with cancer. Ann Intern Med 131:641–647

Diskin C, Stokes T, Dansby L, Radcliff L, Carter T (2007) Is systemic heparin a risk factor for catheter-related sepsis in dialysis patients? An evaluation of various biofilm and traditional risk factors. Nephron Clin Pract 107:22

Donlan R (2001a). Biofilms and device-associated infections. Emerg Infect Dis 7:277–281

Donlan R (2001b). Biofilm formation: a clinically relevant microbiological process. Clin Infect Dis 33:1387–1392

Doshi R, Patel G, Mackay R, Wallach F (2009) Healthcare-associated Infections: epidemiology, prevention, and therapy. Mt Sinai J Med 76:84–94
Fux CA, Wilson S, Stoodley P (2004) Detachment characteristics and oxacillin resistance of Staphyloccocus aureus biofilm emboli in an in vitro catheter infection model. J Bacteriol 186:4486–4491
Fux CA, Stoodley P, Hall-Stoodley L, Costerton JW (2003) Bacterial biofilms – a diagnostic and therapeutic challenge. Expert Rev Anti-Infect Ther 1:667–683
Fux CA, Costerton JW, Stewart PS, Stoodley P (2005) Survival strategies of infectious biofilms. Trends Microbiol 13:34–40
Getliffe K (2003) Managing recurrent urinary catheter blockage: Problems, promises, and practicalities. J WOCN 30:146–151
Gorman SP, Jones DS (2003) Biofilm complications of urinary tract devices. In: Wilson M, Devine D (eds) Medical implications of biofilm. Cambridge university press, Cambridge, pp 136–170
Gray M (2001) Managing urinary encrustation in the indwelling catheter. J WOCN 28:226–229
Ha U, Cho Y (2006) Catheter-associated urinary tract infections: new aspects of novel urinary catheters. Int J Antimicrob Agents 11:11
Hall-Stoodley L, Stoodley P (2009) Evolving concepts in biofilm infections. Cell Microbiol 6:6
Hall-Stoodley L, Costerton JW, Stoodley P (2004) Bacterial biofilms: from the natural environment to infectious diseases. Nature Reviews Microbiol 2:95–108
Hall-Stoodley L, Hu FZ, Gieseke A, Nistico L, Nguyen D, Hayes J, Forbes M, Greenberg DP, Dice B, Burrows A, Stoodley P, Post JC, Ehrlich GD, Kerschner J (2006) Direct detection of bacterial biofilms on the middle-ear mucosa of children with chronic otitis media. JAMA 296:202–211
Jaffe R., Lane J, Bates C (2001) Real-time identification of Pseudomonas aeruginosa direct from clinical samples. J Clin Lab Anal 15:131–137
Juretschko S, Loy A, Lehner A, Wagner M (2002) The microbial community composition of a nitrifying-dentrifying activated sludge from an industrial sewage treatment plant analyzed by the full-cycle rRNA approach. J Syst Appl Microbiol 25:84–99
Kiran M, Giacometti A, Cirioni O, Balaban N (2008) Suppression of biofilm related, device-associated infections by staphylococcal quorum sensing inhibitors. Int J Artif Organs 31: 761–770
Kirketerp-Moller K, Jensen P, Fazli M., Madsen K, Pedersen J, Moser C, Tolker-Nielsen T, Givskov M, Bjarnsholt T (2008) The distribution, organization and ecology of bacteria in chronic wounds. J Clin Microbiol 46(8): 2717–22
Lau S, Woo P, Woo G, Yuen K (2002) Catheter-related Microbacterium bacteremia identified by 16S rRNA gene sequencing. J Clin Microbiol 40:2681–2685
Leake J, Dowd S, Wolcott R, Zischkau A (2009) Identification of yeast in chronic wounds using new pathogen-detection technologies. J Wound Care 18:103–108
Leroy M, Cabral H, Figueira M, Bouchet V, Huot H, Ram S, Pelton SI, Goldstein R (2007) Multiple consecutive lavage samplings reveal greater burden of disease and provide direct access to the nontypeable Haemophilus influenzae biofilm in experimental otitis media. Infect Immun 75:4158–72
Lewis K (2007) Persister cells, dormancy and infectious disease. Nat Rev Microbiol 5:48–56
Liedl B (2001) Catheter-associated urinary tract infections. Curr Opin Urol 11:75–79
Linde HJ, Hahn J, Holler E, Reischl U, Lehn N (2002) Septicemia due to *Acinetobacter junii*. J Clin Microbiol 40:2696–2697
Machado J, Suen V, Figueiredo J, Marchini J (2009) Biofilms, Infection, and Parenteral Nutrition Therapy. J Parenter Enteral Nutr 33(4):397–403
Maki D, Tambyah P (2001) Engineering out the Risk of Infection with Urinary Catheters. Emerg Infect Dis 7(2):342–347
Maki D, Weise C, Sarafin H (1977) A semiquantitative culture method for identifying intravenous-catheter-related infection. N Engl J Med 296:1305–1309
Marcus R, Post J, Stoodley P, Hall-Stoodley L, McGill R, Sureshkumar K, Gahlot V (2008) Biofilms in nephrology. Expert Opin Biol Ther 8:1159–1166

Marie T, Costerton J (1984) Scanning and transmission electron microscopy of in situ bacterial colonization of intravenous and intraarterial catheters. SJ Clin Microbiol 19:687–693

Marie T, Noble M, Costerton J (1983) Examination of the morphology of bacteria adhering to peritoneal dialysis catheters by scanning and transmission electron microscopy. J Clin Microbiol 18:1388–1398

Mermel L, Farr B, Sherertz R, Raad I, O'Grady N, Harris J, Craven D (2001) Guidelines for the management of intravascular catheter-related infections. Clin Infect Dis 32: 1249–1272

Murga R, Miller JM, Donlan RM (2001) Biofilm formation by Gram-negative bacteria on central venous catheter connectors: Effect of conditioning films in a laboratory model. J Clin Microbiol 39:2294–2297

Neufeld JD, Mohn WW (2005) Unexpectedly high bacterial diversity in arctic tundra relative to boreal forest soils, revealed by serial analysis of ribosomal sequence tags. Appl Environ Microbiol 71:5710–5718

Neut D, van der Mei H, Bulstra S, Busscher H (2007) The role of small-colony variants in failure to diagnose and treat biofilm infections in orthopedics. Acta Orthop 78:299–308

Nielsen M, Thomsen TR, Moser C, Hoiby N, Nielsen PH (2008). Use of cultivation-dependent and -independent techniques to assess contamination of central venous catheters: a pilot study. BMC Clin Pathol 8

Nickel J, Emtage J, Costerton J (1985) Ultrastructural microbial ecology of infection-induced urinary stones. J Urol 133(4):622–7

Nistico L, Gieseke A, Stoodley P, Hall-Stoodley L, Kerschner JE, Ehrlich GD (2009) Fluorescence in-situ hybridization for the detection of biofilm in the middle ear and upper respiratory tract mucosa. In: Sokolowski B (eds) Auditory and vestibular research methods and protocols. The Humana Press, Totowa, NJ, pp 191–215

Ott S, El Mokhtari N, Rehman A, Rosenstiel P, Hellmig S, Kuhbacher T, Lins M, Simon R, Schreiber S (2007) Fungal rDNA signatures in coronary atherosclerotic plaques. Environ Microbiol 9:3035–3045

Ott S, El Mokhtari N, Musfeldt M, Hellmig S, Freitag S, Rehman A, Kuhbacher T, Nikolaus S, Namsolleck P, Blaut M, Hampe J, Sahly H, Reinecke A, Haake N, Gunther R, Kruger D, Lins M, Herrmann G, Folsch U, Simon R, Schreiber S (2006) Detection of diverse bacterial signatures in atherosclerotic lesions of patients with coronary heart disease. Circulation 113:929–937

Park J, Cho Y, Kwon I, Jeong S, Bae Y (2002) Assessment of PEO/PTMO multiblock copolymer/segmented polyurethane blends as coating materials for urinary catheters: in vitro bacterial adhesion and encrustation behavior. Biomaterials 23:3991–4000

Parsek M, Singh P (2003) Bacterial biofilms: an emerging link to disease pathogenesis. Annu Rev Microbiol 57:677–701

Piper KE, Jacobson MJ, Cofield RH, Sperling JW, Sanchez-Sotelo J, Osmon DR, McDowell A, Patrick S, Steckelberg JM, Mandrekar JN, Sampedro MF, Patel R (2009) Microbiologic diagnosis of prosthetic shoulder infection by use of implant sonication. J Clin Microbiol 47(6):1878–1884

Post J, Preston R, Aul J, Larkins-Pettigrew M, Rydquist-White J, Anderson K, Wadowsky R, Reagan D, Walker E, Kingsley L, Magit A, Ehrlich G (1995) Molecular analysis of bacterial pathogens in otitis media with effusion. JAMA 273:1598–1604

Pourrezaei K, Shvets I, DeLaurentis M, Boxman R, Beard R, Croitoriu N, Mukhtar M, Logan D, Rastogi R (1994) Development of antimicrobial and antithrombogenic coatings for inside and outside of medical catheters. Surface Coastings Technol 68:669–674

Rayner MG, Zhang Y, Gorry MC, Chen Y, Post, JC, Ehrlich GD (1998) Evidence of bacterial metabolic activity in culture-negative otitis media with effusion. JAMA 279:296–299

Rodríguez-Baño J, Martí S, Soto S, Fernández-Cuenca F, Cisneros JM, Pachón J, Pascual A, Martínez-Martínez L, McQueary C, Actis LA, Vila J (2008) Biofilm formation in *Acinetobacter baumannii*: associated features and clinical implications. Clin Microbiol Infec14:276–278

Sabbuba NA, Mahenthiralingam E, Stickler DJ (2003) Molecular Epidemiology of Proteus mirabilis Infections of the Catheterized Urinary Tract. J Clin Microbiol 41:4961–4965

Safdar N, Maki DG (2004) The pathogenesis of catheter-related bloodstream infection with noncuffed short-term central venous catheters. Intensive Care Med 30:62–67

Safdar N, Kluger D, Maki D (2002) A Review of Risk Factors for Catheter-Related Bloodstream Infection Caused by Percutaneously Inserted, Noncuffed Central Venous Catheters: Implications for Preventive Strategies. Medicine 81:466–479

Sakamoto M, Rocas I, Siqueira J Jr, Benno Y (2006) Molecular analysis of bacteria in asymptomatic and symptomatic endodontic infections. Oral Microbiol Immunol 21:112–122

Schinabeck M, Ghannoum M (2003) Catheter-related infections – diagnosis, treatment and prevention. Clin Microbiol Newslett 25:113–118

Selan L, Passariello C, Rizzo L, Varesi P, Speziale F, Renzini G, Thaller M, Fiorani P, Rossolini G (2002) Diagnosis of vascular graft infections with antibodies against staphylococcal slime antigens. Lancet 359:2166–2168

Shanks RMQ, Sargent JL, Martinez RM, Graber ML, O'Toole GA (2006) Catheter lock solutions influence staphylococcal biofilm formation on abiotic surfaces. doi: 10.1093/ndt/gfl170. Nephrol Dial Transplant 21:2247–2255

Sherertz R (2004) Update on vascular catheter infections. Curr Opin Infect Dis 17:303–307

Slobbe L, El Barzouhi A, Boersma E, Rijnders BJ (2009) Comparison of the roll plate and sonication method to diagnose catheter colonisation and bacteraemia in patients with long-term tunnelled catheters. A randomised prospective study. J Clin Microbiol 47(4):885–888

Smuszkiewicz P, Trojanowska I, Tomczak H (2009) Venous catheter microbiological monitoring. Necessity or a habit? Med Sci Monit 15:SC5–8

Spencer R (1999) Novel methods for the prevention of infection of intravascular devices. J Hosp Infect 43:S127–135

Starkey M, Hickman J, Ma L, Zhang N, De Long S, Hinz A, Palacios S, Manoil C, Kirisits M, Starner T, Wozniak D, Harwood, C, Parsek M (2009) *Pseudomonas aeruginosa* rugose small colony variants have adaptations likely to promote persistence in the cystic fibrosis lung. J Bacteriol 191(11):3492–503

Stickler DJ (2008) Bacterial biofilms in patients with indwelling urinary catheters. Nat Clin Pract Urol 5:598–608

Stoodley P, Nistico L, Johnson S, Carabin L-A, Baratz M, Gahlot V, Ehrlich GDE, Kathju S (2008) Direct demonstration of viable *S. aureus* biofilms in an infected total joint arthroplasty. JBJS 90:1751–1758

Storti A, Pizzolitto AC, Pizzolitto EL (2005) Detection of mixed microbial biofilms on central venous catheters removed from Intensive care Unit Patients. Brazilian J Microbiol 36 275–280

Tenover F (2007) Rapid detection and identification of bacterial pathogens using novel molecular technologies: infection control and beyond. Med Microbiol 44:418–423

Thomsen TR, Ramsing NB, Finster K (2001) Biogeochemical and Molecular Signatures of Anaerobic Methane Oxidation in a Marine Sediment. Appl Environ Microbiol 67:1646–1656

Timsit J (2007) Diagnosis and prevention of catheter-related infections. Curr Opin Crit Care 13:563–571

Trampuz A, Piper K, Jacobson M, Hanssen A, Unni K, Osmon D, Mandrekar J, Cockerill F, Steckelberg J, Greenleaf J, Patel R (2007) Sonication of removed hip and knee prostheses for diagnosis of infection. N Engl J Med 357:654–663

Trautner B, Darouiche R (2004) Catheter-associated infections: pathogenesis affects prevention. Arch Intern Med 164:842–850

Tunney M, Patrick S, Curran M, Ramage G, Hanna D, Nixon J, Gorman S, Davis R, Anderson N (1999) Detection of prosthetic hip infection at revision arthroplasty by immunofluorescence microscopy and PCR amplification of the bacterial 16S rRNA gene. J Clin Microbiol 37: 3281–3290

Wang K, Chang W, Shih T, Huang C, Tsai N, Chang C, Chuang Y, Liliang P, Su T, Rau C., Tsai Y, Cheng B, Hung P, Chang C, Lu C (2004) Infection of cerebrospinal fluid shunts: causative pathogens, clinical features, and outcomes. Jpn J Infect Dis 57:44–48

Warwick S, Wilks M, Hennessy E, Powell-Tuck J, Small M, Sharp J, Millar M (2004) Use of quantitative 16S ribosomal DNA detection for diagnosis of central vascular catheter-associated bacterial infection. J Clin Microbiol 42:1402–1408

Wilson M (2009) Biofilm and other causes of pain in catheterization. Br J Community Nurs 14:102–113

Woo PC, Tsoi HW, Leung KW, Lum PN, Leung AS, Ma CH, Kam KM, Yuen KY (2000) Identification of Mycobacterium neoaurum isolated from a neutropenic patient with catheter-related bacteremia by 16S rRNA sequencing. J Clin Microbiol 38:3515–3517

Chapter 7
Osteomyelitis

Graeme A. O'May, Rebecca A. Brady, Ranjani Prabhakara, Jeff G. Leid, Jason H. Calhoun, and Mark E. Shirtliff

7.1 Introduction

Osteomyelitis is defined as an infection of the bone. The pathogenesis of osteomyelitis has been delineated clinically and several types of can be distinguished and classified according to the source of the infecting microorganism (i.e., hematogenous or contiguous focus) and the vascular capability of the infected individual (i.e., with or without generalized vascular insufficiency) (Lew and Waldvogel 2004).

7.1.1 Anatomy and Function of Bone

In order to fully understand the pathogenesis, treatment, and prevention of osteomyelitis, knowledge of the structure and function of bone is necessary. The functions of bone in the body are (i) to support the body's mass against gravity, (ii) act as a shield against blunt or penetrating trauma for certain vital areas of the body (notably the heart (sternum) and brain (skull), and (iii) to provide a solid frame against which muscles can pull in order to provide mobility. These functions have, through the action of natural selection, dictated the structure of bone. Bone must be hard and yet not so much that it is brittle; a little flexibility is necessary before breakage occurs.

The long, straight section of a long bone is called the *diaphysis*; the two ends are termed *epiphyses* (Fig. 7.1). When bones are growing, the junction between the epiphyses and diaphysis contains an actively growing cartilage plate called the *epiphysial plate*; this is where most bone growth and elongation occurs. Such elongation occurs through the generation of additional cartilage, forcing the two ends apart; this new cartilage will be replaced eventually by new bone. The outer surface

M.E. Shirtliff (✉)
Department of Microbial Pathogenesis, Dental School, University of Maryland – Baltimore, 650 West Baltimore Street, Baltimore, MD 21201, USA
e-mail: mshirtliff@umaryland.edu

T. Bjarnsholt et al. (eds.), *Biofilm Infections*, DOI 10.1007/978-1-4419-6084-9_7,

Fig. 7.1 Gross anatomy of a long bone. Taken from http://training.seer.cancer.gov/anatomy/skeletal/classification.html

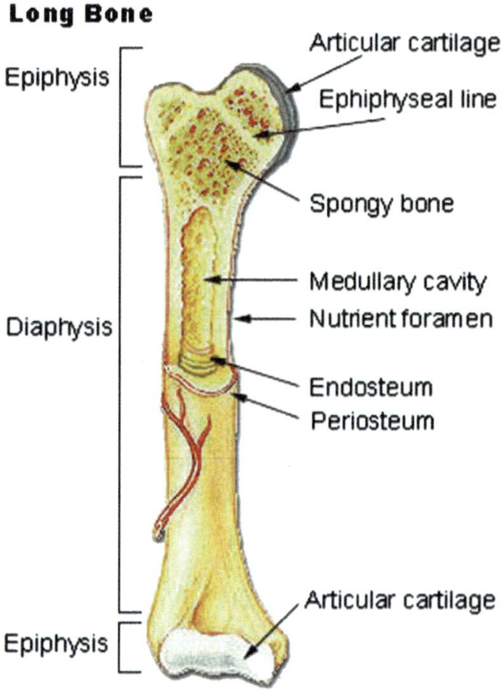

of the epiphyses – where they meet other bones in a joint – are covered with *articular cartilage*, which functions to provide an almost frictionless bearing surface for joint movement.

The diaphysis of a long bone is hollow. Its outer layer is composed of *compact bone*, a hard layer of bone. A hollow cavity within this outer layer is known as the *marrow cavity*. The contents of the marrow cavity alter with age: in children it contains red marrow, a site of red blood cell production, whilst in adults this has been replaced with yellow marrow. Yellow marrow is a fatty tissue which no longer supports production of red blood cells. Epiphyses also are covered with a layer of compact bone, albeit a thinner layer than the diaphysis. Underlying this in the epiphyses is *spongy bone*, a network of strengthening crossbeam-like bony plates and rods called *trabeculae*. Within and formed by this network are a multitude of small spaces which in some bones contain red marrow. The interior of many irregular (short or flat) shaped bones also contains spongy marrow.

Microscopically, compact bone has a layered structure consisting of, directly beneath the outer surface, several rings known as the *circumferential lamellae*. These extend around the entire circumference of the bone. Deeper into the lumen of the bone are located cylinder-shaped structural units known as the *Haversian systems* (Fig. 7.2). Each of these systems is centered around a *Haversian canal*, within which are located nerves and blood vessels. Running perpendicular to the Haversian canal are *Volkmann's canals*; these provide a conduit for nerves and blood vessels

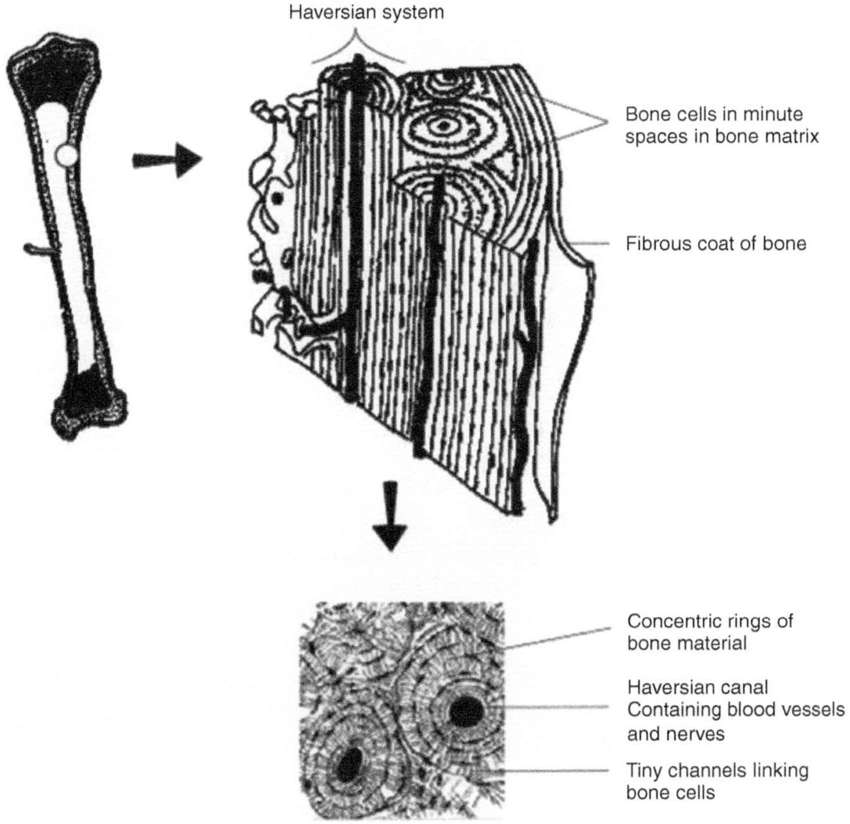

Fig. 7.2 Haversian systems of the compact bone in long bones. http://en.wikibooks.org/wiki/Anatomy_and_Physiology_of_Animals/The_Skeleton

going to and from the periosteum to the Haversian canal. The *Haversian lamellae* lie around each Haversian canal; between each lamellum lies the lacunae within which are located the *osteocytes*, inactive bone-producing cells trapped after they laid down bone. *Canaliculi* link each lacunae to its neighbors and function to transport nutrients and waste materials. The cylindrical shape of the Haversian systems renders the gaps between them triangular; these spaces are filled with *interstitial lamellae*, composed of material that previously formed Haversian systems. This material is continually being destroyed and rebuilt, giving rise to the interstitial lamellae.

Surrounding the bone is a sheet of connective tissue known as the *periosteum*. Its inner osteogenic layer gives rise to osteoblasts; these cells, after being trapped by the bone they have created are the source of the osteocytes. Osteoblasts reside on the surface of the bone where they manufacture the protein and mineral matrix of new circumferential lamellae. Other cell types within bone include osteoclasts; these large multinucleated cells, formed from monocytes, function to dissolve and resorb bone. A balance between the action of osteoblasts and osteoclasts is vital for the continuing structural integrity of bone.

7.2 Types and Pathogenesis of Osteomyelitis

Two basic types are recognized, these being hematogenous and contiguous focus osteomyelitis. The primary difference between these is the source of the infecting microorganisms; in hematogenous osteomyelitis the infective agent originates in the bloodstream. Contiguous focus osteomyelitis can result from either direct introduction of the infective agent to the bone (as in traumatic injuries) or from an adjacent soft tissue infection (Brady et al. 2008).

7.2.1 Hematogenous Osteomyelitis

Hematogenous osteomyelitis accounts for *circa* 20% of the total cases of osteomyelitis. Primary hematogenous osteomyelitis is caused by direct seeding of bone from an infective agent present in the vasculature. This type of osteomyelitis is more predominant in infants and children; however, it is not unknown in the adult population (Lew and Waldvogel 2004). In adult individuals, hematogenous osteomyelitis is more usually caused by secondary infection; bacteria gain access to the bloodstream and colonize distal bone. Reactivation of a dormant focus of hematogenous osteomyelitis that an individual suffered during infancy or childhood and "arrested" can also be the source of a hematogenous osteomyelitis. Hematogenous osteomyelitis is most common in the distal tibia; the lesion is usually close to the metaphysis. This type of osteomyelitis is usually located in either a long bone (i.e., tibia, ulna, radius, etc.) or in a vertebra and is most often caused by a single etiologic agent. Symptoms at presentation are usually one or more of malaise, lethargy, fever, tenderness (at or above the site of infection) and a decreased range of motion in the affected limb (Carek et al. 2001).

The anatomy of the metaphyseal region, where the blood flow is sluggish and disordered, explains why the long bones (tibia, femur) are most frequently involved in osteomyelitis (Shirtliff et al. 1999). This slowing of blood flow allows bacteria to settle and initiate colonization with a resultant inflammatory response. Minor trauma likely predisposes the infant or child to infection by producing a small hematoma, vascular obstruction, and a subsequent bone necrosis that is susceptible to inoculation from a transient bacteremia (Morrissy and Haynes 1989). Generally, an acute infection will develop within 2 weeks of disease onset (Carek et al. 2001); typically this results in local cellulitis and a breakdown of leukocytes, increased bone pressure, decreased pH, and decreased oxygen tension. The end result of the action of these physiologic factors is compromise of the circulation within the bone and further spread of infection. In infants, infection may spread to the joint surfaces through the vascularized growth plate (Jackson and Nelson 1982). However, in children greater than 1 year old, the growth plate lacks capillaries and the infection tends to be confined to the metaphysis and diaphysis. The joint is thus usually spared unless the metaphysis is intracapsular. Infants and children suffering from hematogenous osteomyelitis usually have normal soft tissue enveloping the infected bone and are able to mount an efficient metabolic response to the infection. They also have the potential to absorb large sequestra and generate a significant response to the infection in the periosteal region. Because of this resorbing ability, if appropriate antimicrobial therapy is begun before the onset of significant bone necrosis, the younger patient has an excellent probability of halting or resolving the infection without surgical intervention (Berendt and Byren 2004).

Chronic hematogenous osteomyelitis can be said to occur beginning several weeks to months after onset of the disease (Carek et al. 2001). The existing cortex is usually viable. The *involucrum* – an area of live, encasing bone surrounding infected dead bone within a compromised soft tissue envelope – is the hallmark sign of chronic osteomyelitis (Mader et al. 1980). The involucrum contains necrotic marrow and endosteal bone. In normal bone, necrosis is a vital part of the cycle as it signals granulation tissue to resorb dead bone at the junction of living and dead tissue. Some of the dead cortex will usually detach from the living bone and form a sequestrum. After complete separation, or sequestration, the dead bone is eroded by granulation tissue and destroyed. However, in some cases the area of dead bone is too large to be resorbed, or the host response is compromised. This can lead the process of resorption to be inadequate, and may cause the formation of an involucrum. The involucrum affords mechanical continuity and assists in maintenance of function during healing. Involucra have an irregular surface and often have holes through which pus may move into the surrounding soft tissues and eventually drain to the skin surface, forming a draining sinus tract (Mader et al. 1996). The purpose of the involucrum is to isolate the infection from the remaining healthy bone. Involucrum development occurs upon establishment of the infection, after fibrous tissue and chronic inflammatory cells surround granulations and dead bone. New bone forms, as a result of the vascular reaction to the infection, from the periosteum, endosteum, and cortex. Involucra may continue to increase thickening for weeks or months and eventually form a portion of, or in some cases, all of a new bone shaft.

Though the involucrum functions to contain the infection, decreases in vascularity and low oxygen tension due to the presence of this structure can lead to decreased effectiveness of the host response; chronic disease can then ensue. Dead bone functions as an inert surface for the attachment of bacteria and the formation of biofilm. This form of infection, coupled with the host's inability to resorb the dead bone, results in a very complicated disease to treat because bacteria in a biofilm are 50–500 times more resistant to antimicrobial agents than their planktonic, free-floating counterparts. Therefore, debridement (surgical removal of infected bone and/or surrounding soft tissue) is often necessary for these infections to resolve.

Between 2 and 7% of total hematogenous osteomyelitis cases are located within a vertebra (Tyrrell et al. 1999). Incidences of this type of osteomyelitis are increasing due to the rising proportion of the population composed of aging adults, who possess both risk factors for bacteremia and deteriorating spinal pathology (Berendt and Byren 2004). In vertebral osteomyelitis (as well as all other locations), polymorphonuclear leukocytes (PMNs) are present due to the acute inflammatory response (see also Chapter 12). Degradative enzymes released from disintegrating PMNs, together with vascular ischemia and release of bacterial products, can cause an extension of the infection into the cartilaginous end-plate, disc, and/or proximal regions. Posterior extension of the infection is an especial difficulty as it can in some cases lead to abscesses in either the epidural or subdural spaces; in particularly serious cases, meningitis can result. Extension of the focus of infection anteriorly or laterally can lead to paravertebral, retropharyngeal, mediastinal, subphrenic, or retroperitoneal abscesses. Additionally, the rich venous networks within the bones of the spinal column can lead to efficient and rapid spread to adjacent vertebrae.

7.2.2 Contiguous Focus Osteomyelitis

In the past several years there has been a marked decline in hematogenous osteomyelitis with a concurrent rise in contiguous disease (Espersen et al. 1991). Although the term "contiguous focus" implies that the infection stems from an adjacent soft tissue infection, chronic contiguous focus osteomyelitis can also begin as an acute infection, with the microbes being directly inoculated into the bone at the time of trauma (Healy and Freedman 2006). Infection can also be spread by nosocomial contamination during preoperative or intraoperative procedures. The age distribution of contiguous focus osteomyelitis prevalence peaks in both the young and the elderly; infections occurring in younger individuals are usually a result of trauma and related surgery whilst in older individuals it is secondary for surgical procedures and decubitus ulcers. Also, if the osteomyelitis is secondary to a penetrating trauma then multiple microorganisms may be involved; they are termed polymicrobic infections.

Trauma contributes to osteomyelitis infections in several ways besides the obvious direct inoculation of bacteria through the skin barrier and into the soft tissues and bone beneath. Damage of any sort to tissue tends to cause a decrease in blood supply to the affected area, which itself can cause formation of necrotic areas of inert tissue. Bacteria are then able to bind to this essentially inert tissue surface and infection can be the unfortunate end result. Indeed, trauma has been shown to depress the immune and inflammatory responses to bacterial invasion. Degree of severity of tissue injury is thought to be correlated with risk of infection; the presence of bacteria within the tissues in a wound is not always sufficient in and as of itself for establishment of osteomyelitis (Ziran 2007).

7.3 Etiology of Bacterial Osteomyelitis

A number of bacteria of diverse genera capable of causing – or more correctly – have been recovered from cases of osteomyelitis. *Staphylococcus* spp. cause the majority of cases and are fully capable of causing osteomyelitis in individuals of any age and with functioning immune systems. Of course, other pathogenic microorganisms

osteomyelitis; these include *Enterococcus* spp., *Streptococcus* spp., *Pseudomonas aeruginosa, Enterobacter* spp., *Mycobacterium* spp., and various anaerobic and mycoidal species (specifically *Candida* spp.). Each of these pathogenic genera individually represents a very small minority of infections when compared to that represented by *Staphylococcus* spp. The immature or compromised immune status of the host is the primary cause of initial infection and development into a persistent and chronic osteomyelitis.

Hematogenous osteomyelitis is generally monomicrobiotic in nature, i.e., a single bacterial taxon is isolated from the infected region. Polymicrobial hematogenous osteomyelitis is rare (Lew and Waldvogel 2004). In younger individuals, aged under 1 year, *Staphylococcus aureus*, *Streptococcus agalactiae*, and *Escherichia coli* are most frequently recovered from infected bone, while in the child (ages between 1 and 18), *S. aureus, Streptococcus pyogenes*, and *Haemophilus influenzae* are the most common organisms isolated. After the age of four, the incidence of osteomyelitis from which *H. influenzae* is recovered decreases. However, the overall incidence of *H. influenzae* as a cause of osteomyelitis is decreasing because of the *H. influenzae* vaccine now given to children (De Jonghe and Glaesener 1995). In adults, *S. aureus* is the most common organism isolated (Shirtliff et al. 1999). Other pathogenic microorganisms associated with osteomyelitis include *Enterococcus* spp., *Streptococcus* spp., *Pseudomonas aeruginosa, Enterobacter* spp., *Mycobacterium* spp., as well as anaerobic and mycoidal species (specifically *Candida* spp.). Each of these, individually represents a small minority of infections. The immature or compromised immune status of the host is the primary cause of both initial infection and development into a persistent and chronic osteomyelitis infection by these other species. In hematogenous vertebral osteomyelitis, aerobic Gram-negative rods are sometimes found, with the urinary tract or intravenous drug use as the source of infection (Berendt and Byren 2004). *P. aeruginosa* and *Serratia marcescens* have a high incidence in intravenous drug users (Holzman and Bishko 1971, Sapico 1996). It should be stressed, however, that while these varied species have been known to cause the disease, *S. aureus* produces the vast majority of osteomyelitis infections in all age groups.

Contiguous focus osteomyelitis located within a vertebra is usually a polymicrobial infection from which anaerobic or facultative anaerobic are often isolated. Alternative sources of infection include the genitourinary tract, adjacent skin and soft tissue, respiratory tract, an infected intravenous line site, endocarditis, dental infection (see also Chapter 4), as well as sources not known (Sapico and Montgomerie 1979, Berendt and Byren 2004). Positive cultures are at present thought to be very important for diagnosis, since other conditions such as trauma and vertebral collapse may simulate infection. Typically multiple organisms are isolated from individuals suffering osteomyelitis secondary to a diabetic foot infection. These are typically two or more of: *S. aureus*, coagulase-negative *Staphylococcus spp., Streptococcus spp., Enterococcus spp.*, Gram-negative bacilli, and various anaerobes (Calhoun et al. 1988b, Berendt and Byren 2004, Rao and Lipsky 2007). Aerobic Gram-negative bacilli are commonly present in a mixed infection (Calhoun et al. 1988b).

In contrast to hematogenous osteomyelitis, multiple pathogenic species are usually isolated from the infected bone in cases of contiguous focus osteomyelitis. Once more, staphylococci are involved in a majority of cases, with *S. aureus* and coagulase-negative staphylococci accounting for 75% of bacteria recovered from such infections (Mader et al. 1996). These data further reinforce the critical importance of the genus *Staphylococcus* in the pathogenesis of osteomyelitis. However, Gram-negative bacilli and anaerobic bacteria of various genera are also found, albeit at a lower prevalence, in these situations. The infection usually manifests within 1 month after inoculation of the organisms from trauma, surgery, or a soft tissue infection. Patients usually present with a low-grade fever, pain local to the site of infection, and sinus tract drainage.

7.3.1 Staphylococcus *spp.*

Staphylococci are by far the most common etiologic agent recovered from cases of osteomyelitis (Shirtliff et al. 1999). The most important pathogen of this genus is without doubt *S. aureus*.

S. aureus is a Gram-positive, ubiquitous bacterial species. *S. aureus* is a normal commensal of the human nostrils; *ca* 20% of the population are permanently colonized with this bacterium, while a further 60% are transient carriers (Kluytmans et al. 1997). The presence of *S. aureus* alone will not usually lead to illness; however, if the mucosal or skin surfaces are breached and the microorganism gains access to the tissues beneath, serious infection can ensue (Fitzpatrick et al. 2005b). Due to the increasing participation of *S. aureus* in osteomyelitis and other types of infection (see below), its swift development of multiple-antibiotic resistance, and its predilection to move from an acute infection to one that is biofilm-mediated, persistent, chronic and recurrent, this pathogen continues to receive considerable attention. The virulence mechanisms by which this pathogen colonizes the host, evades and destroys the immune response, and persists are outlined below.

7.3.1.1 Virulence

Staphylococcus spp. have been shown to be the causative agent of a plethora of infections [e.g., tropical pyomyositis, lower respiratory infections (pneumonia), superficial skin infections (boils, sties and carbuncles), localized abscesses, endocarditis, toxic shock syndrome, serious skin infections (furunculosis), food poisoning, bacteremia, empyema, pyopneumothorax, and exfoliative diseases] and are by far the etiologic agent isolated most commonly from cases of osteomyelitis. Therefore, it is unsurprising that the most important pathogen of the genus, *S. aureus*, has evolved a wide variety of virulence products and mechanisms in order to cause disease. The pathogenesis of staphylococcal infections is multifactorial and it is difficult to determine the precise role of any given factor in infection. Most of the virulence factors whose function is known appear specifically adapted to persistence, immune evasion, and infection within the host. Staphylococcal products with a role in infection can be categorized as those responsible for (i) adherence, (ii) direct host damage, or (iii) immunoavoidance. There exist also a number of enzymes and extracellular proteins whose role in virulence is at present unclear. Staphylococcal virulence factors have a specific role in the colonization and infection process in osteomyelitis; their expression is coordinated throughout the various stages of infection. Therefore, the differential regulation of these virulence factors due to staphylococcal population levels and environmental factors is vital for successful colonization and establishment of infection.

S. aureus produces a large number of extracellular and cell-associated products that contribute to virulence and persistent infection. Most of these seem to be specifically adapted to survival and infection within the host. During early exponential growth when cell density is low, proteins that promote adherence and colonization (such as fibronectin binding protein, protein A, staphylokinase, and coagulase) are expressed. When cell growth reaches high densities, production of the adherence and colonization factors is suppressed, while secreted toxins and enzymes are expressed [such as enterotoxins B, C and D, epidermolytic (exfoliative) toxin A, α, β, and δ hemolysins, serine protease, nuclease, type 5 capsular polysaccharide, clumping factor, leukocidin, phosphatidyl-specific phospholipase C, fatty acid modifying enzyme, lipase, hyaluronate lyase (hyaluronidase), and toxic shock syndrome toxin (TSST) 1]. These proteins are produced after exponential growth in planktonic, batch culture has ceased (i.e., the culture has entered stationary phase), and are known to cause damage to host tissues, thus obtaining nutrients for pathogen growth and dissemination.

The expression of most of these staphylococcal products is under partial or complete control of the staphylococcal accessory regulator (*sar*) and the accessory gene regulator (*agr*) system. During early logarithmic growth, a protein encoded by *rot* (repressor of toxins) inhibits the expression of *agr*-activated virulence factors (McNamara et al. 2000). Once activation of the *agr* and *sar* regulatory loci occurs during late exponential phase, there is an increased transcription of an *agr* regulatory RNA molecule known as RNAIII (Balaban and Novick 1995). RNAIII blocks transcription of surface protein genes and upregulates transcription of genes encoding extracellular pathogenicity factors. In this way, *S. aureus* is able to sense when its population density has increased to the point where colonization has been successful. One of the major mechanisms by which *S. aureus* evades clearance by effector cells and molecules of the immune system is by formation of *biofilm*.

7.3.1.2 Adherence

For successful initiation of biofilm formation and infection, any pathogen must colonize the target tissue; the first step in this process is adherence. *Staphylococcus* spp. possesses a large number of adhesins for host proteins that allow adherence to the extracellular matrix in bone. These are known as "microbial surface components recognizing adhesive matrix molecules" (MSCRAMMS) (Herrmann et al. 1988, Yacoub et al. 1994, Ryden et al. 1997). Some host matrix proteins and their functions are fibronectin and laminin (adherence proteins), elastin (imparts elastic properties), collagen (structural support), and hyaluronic acid (a glycosaminoglycan that is rich in the joints and the matrix and provides cushioning through hydration of its polysaccharides). A number of bone or joint-specific matrix proteins are recognized by MSCRAMMS. These include osteopontin (a soluble phosphoprotein that acts as a cytokine and osteoclast attachment protein and is needed for bone injury repair and remodeling), bone sialoprotein (interacts with osteoblasts and acts as a nucleator for calcium hydroxyapatite formation), and vitronectin (an adhesive glycoprotein involved in adhesion regulation and the coagulation, fibrinolytic, and complement cascades; also allows for bone resorption when bound to osteoclasts). Eight adhesin genes have been determined and include genes encoding fibrinogen binding proteins (*fib*, *cflA*, and *fbpA*) (Boden and Flock 1994, McDevitt et al. 1994, Cheung et al. 1995), fibronectin binding proteins (*fnbA* and *fnbB*) (Jonsson et al. 1991), a collagen receptor (*cna*) (Patti et al. 1992), an elastin binding protein (*ebpS*) (Park et al. 1996), and a broad specificity adhesin (*map*) that mediates low level binding of several proteins including osteopontin, collagen, bone sialoprotein, vitronectin, fibronectin, and fibrinogen (McGavin et al. 1993). Also, this microorganism has been shown to possess a number of other host protein-binding receptors in which the genes have not yet been determined. These include a laminin (52 kDa) (Lopes et al. 1988), a lactoferrin (450 kDa) (Naidu et al. 1992), and a transferrin (42 kDa) (Modun et al. 1994) binding protein. The staphylococcal receptor that binds laminin may be used in extravasation (Lopes et al. 1985). These receptors were found in *S. aureus* but were absent from the noninvasive pathogen *S. epidermidis* (Lopes et al. 1985). The lactoferrin and transferrin receptors bind to these host iron acquisition proteins and may be used as adhesins and/or as iron acquisition mechanisms. In addition, *S. aureus* expresses a 42-kDa protein, Protein A, which is bound covalently to the outer peptidoglycan layer of their cell walls. This adherence protein binds to the host platelet gC1qR (a multifunctional, ubiquitously distributed cellular protein, initially described as a binding site for the globular heads of the complement complex C1q) (Nguyen et al. 2000). Therefore, Protein A may be able to promote adhesion to sites of vascular injury and thrombosis and has been implicated as an important colonization factor. Protein A production is repressed by the *sar* locus via both RNAIII-dependent and independent mechanisms during post-exponential phase growth (Cheung et al. 1997). This protein is also associated with *S. aureus* immunoavoidance (see below). Many of these and other staphylococcal cell wall proteins must be exported out of the bacterial cell in order to interact with the extracellular environment. This export can be either a targeting process (the protein is exported and has binding domains for cell wall secondary polymers such as teichoic acids) or a sorting process (a C-terminal conserved amino acid sequence, LPXTG, that directs the export and covalent attachment to the peptidoglycan) (Navarre et al. 1996).

Increasing evidence supports the importance of staphylococcal surface components as virulence determinants by enabling initial colonization. In a number of studies, mutants in these receptors strongly reduced the ability of staphylococci to produce infection. In addition, there was significant binding of *S. aureus* to bone sialoprotein, fibronectin, and collagen type 1 in a mouse model, indicating that adherence remains a key phase in the early stages of infection (Bremell and Tarkowski 1995). Expression of adhesins permits the attachment of the pathogen to cartilage. Inoculation of mice with mutants of the collagen adhesin gene showed that septic arthritis occurred 43% less often than in the corresponding wild type (Switalski et al. 1993). Collagen adhesin positive strains were also associated with the production of high levels of IgG and interleukin-6 (Switalski et al. 1993). In a murine septic arthritis model, inoculation of mice with mutants of the collagen adhesin gene showed that septic arthritis occurred 43% less often than in the corresponding wild type (Switalski et al. 1993). Also, vaccination with a recombinant fragment of the *S. aureus* collagen adhesin was able to reduce the sepsis-induced mortality rate to 13%, compared with 87% in the control group (Nilsson et al. 1998). However, the role of collagen adhesion of *S. aureus* as a major virulence factor has been recently questioned since approximately 30–60% of clinical isolates do not display collagen binding in vitro or the *cna*-encoded collagen adhesin (Thomas et al. 1999). Staphylococcal fibronectin-binding proteins (FbpA and FbpB) may have a major role in colonization during musculoskeletal infections. In a recent study, all of the tested clinical isolates ($n = 163$) contained one or both of the coding regions for these binding proteins and 95% of these strains had a comparable fibronectin binding capacity to that seen in a staphylococcal reference strain known to efficiently bind fibronectin (Peacock et al. 2000). In addition, an in vivo study of endocarditis in a rat model showed that mutants deficient for fibronectin-binding protein were 250-fold less adherent to traumatized heart valves (Kuypers and Proctor 1989). Also, *S. aureus* adherence to miniplates from iliac bones of guinea pigs was three times higher than the adhesin-defective mutant strain (Fischer et al. 1996). It is likely that fibronectin-binding proteins play an important role in bone and joint infections, especially those associated with initial trauma or implanted medical devices (Patel et al. 1987).

7.3.1.3 Staphylococcal Biofilm

Staphylococcus spp. within the host produces a multilayered "biofilm" community (Fig. 7.3). A biofilm is a modular community of microbes embedded within a host- and/or microbe-derived hydrated matrix of exopolymeric substances that exists at a phase or density interface. This interface is, in most cases, between a solid or semi-solid support [e.g., soft tissue or bone (Fig. 7.4) and a liquid medium (e.g., extracellular fluid, blood, mucin, etc.)] (Wimpenny 2000). Biofilm thickness can vary from a single cell layer to (more commonly in infections) a thick community of cells embedded within a thick polymeric matrix. Structural analyses have demonstrated that in vitro biofilms possess a sophisticated architecture in which microcolonies exist in discrete pillar or mushroom-shaped structures (Costerton et al. 1995). Between these structures, an intricate channel network provides access to environmental nutrients. It has been hypothesized that the development and maintenance of this phenotype may be mediated through the action of quorum sensing systems in biofilm-producing microbes (McLean et al. 1997, Stickler et al. 1998, Parkins et al. 2000, Singh et al. 2000).

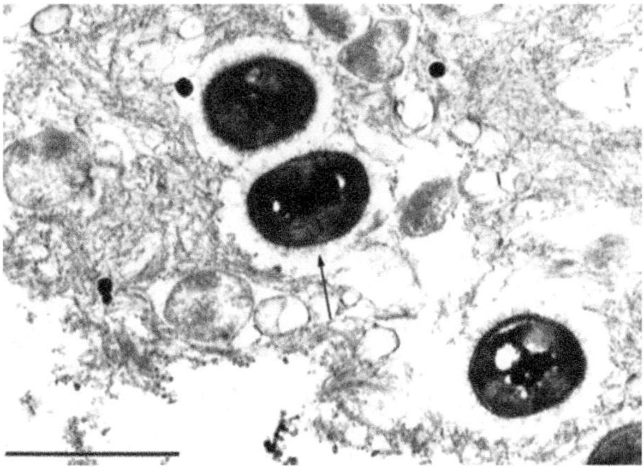

Fig. 7.3 Scanning electron micrograph showing *S. aureus* microcolony (*dark arrow*) growing within bone. Scale bar represents 2 μm

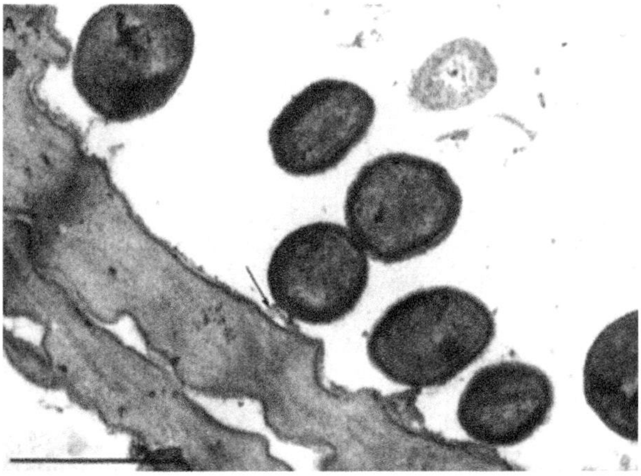

Fig. 7.4 Scanning electron micrograph showing *S. aureus* attaching to host bone, as indicated by the *dark arrow*. Scale bar represents 3 μm

By adopting this sessile mode of life, biofilm-embedded microbes benefit from a number of advantages over their planktonic counterparts. One such advantage is the capability of the extracellular matrix, or glycocalyx, to entrap and thus concentrate a number of environmentally derived nutrients, such as carbon, nitrogen, and phosphate (Beveridge et al. 1997). Another benefit to growing as a biofilm is facilitation of resistance to a number of removal mechanisms, for example, elimination by antimicrobial and/or antifouling agents, shear stress from fluid flow, host phagocytic clearance and opsonization by antibodies, and host oxygen radical and protease defenses (see also Chapter 13). This innate resistance to antimicrobial factors is mediated through a number of mechanisms including the very low metabolic levels and radically down regulated rates of cell division of the deeply entrenched microbes. While low metabolic rates may explain a great deal of the antimicrobial resistance properties of biofilms, other factors could add to the community's resistance capacity, adding to the cumulative effect. One such may be the capability of the biofilm matrix to act as a "diffusion barrier" to slow down the penetration of some types of antimicrobial agent (Xu et al. 2000). For example, the reactive chlorine species (such as hypochlorite, chloramines, or chlorine dioxide) present in a number of antimicrobial/antifouling agents may be deactivated by reaction with the topmost parts of the biofilm before disseminating into the deeper community (De Beer et al. 1994). In another study, alginate (a component of *P. aeruginosa* exopolysaccharide) was shown to be able to induce α-helical conformation in antimicrobial peptides and likely entraps these peptides, preventing their diffusion into the biofilm (Chan et al. 2004).

Persistence in the host in the face of attack by the components of the immune system by clinical strains of *Staphylococcus* spp. is assisted through a number of properties of the glycocalyx. Since the normal phagocytic processes are devoted towards the removal of the glycocalyx and the implant, local immune deficiency and damage to adjacent host tissue can occur through an accumulation of immune effector molecules. Therefore, the energy and resources of the immune system that would normally be used to fight infection are subverted. Third, the glycocalyx may activate monocyte production of prostaglandin E2 to indirectly inhibit T-cell proliferation (Stout et al. 1992). Lastly, this glycocalyx has been shown to directly inhibit polymorphonuclear leukocytes (Ferguson et al. 1992).

An additional advantage conferred upon bacteria by the biofilm manner of growth is the potential for dispersion to distal sites via detachment from the community. Microcolonies, or parts of microcolonies, detach under the influence of mechanical fluid shear or through a genetically programmed response that mediates the detachment process (Boyd and Chakrabarty 1994). In the direction of fluid flow this detached population travels to distal regions of the host or water system to attach and begin anew biofilm formation in previously non-colonized areas. In addition, detachment and seeding of virgin surfaces may be accomplished by the migration of single, motile cells from the cores of attached microcolonies (Sauer et al. 2002). Therefore, this advantage allows an enduring bacterial source population that is resilient against antimicrobial agents and the host immune response, while simultaneously enabling continuous shedding to encourage bacterial spread. Thus a bacterial biofilm present within the body can act as a reservoir of pathogens that survive antibiotic administration and re-establish infection upon cessation of treatment.

The multilayered *S. aureus* biofilm is embedded within a polysaccharide glycocalyx (Gristina et al. 1985). This glycocalyx develops on devitalized bone (such as the involucrum) or medically implanted devices (Akiyama et al. 1993). The presence of implants are a predisposing factor in the development of infection since they are coated in host proteins soon after implantation, and this provides an excellent source of attachment for any bacteria remaining after debridement surgery (Herrmann et al. 1988, McDevitt et al. 1994, Gracia et al. 1997, Francois et al. 1998). Once attached, the bacteria can form the glycocalyx, or slime layer, which protects the bacteria from normal host defenses and systemic antibiotics (Gristina and Costerton 1984, Duguid et al. 1992, Darouiche et al. 1994, Oie et al. 1996). This pathogen usually grows in coherent microcolonies in the adherent biofilm that is often so extensive; the underlying infected bone or implant surface is obscured. This layer slows the inward diffusion of a number of antimicrobials (Gristina and Costerton 1984, Duguid et al. 1992, Darouiche et al. 1994, Oie et al. 1996). In addition, the anaerobic nature of the deep layers of the biofilm results in a dramatically reduced growth rate and metabolic activity which is one of the main mechanisms allowing bacterial escape from the bactericidal and bacteriostatic effects of antimicrobial therapy (Anderl et al. 2003). The presence of persister cells also contribute to the biofilm mediated resistance (Lewis 2005). These cells are a metabolically inactive portion of the total population making this subset of the population resistant to antimicrobial agents. Upon cessation of antimicrobial therapy, the dormant state of peristers is reversible and the infection can be reactivated. Furthermore, those bacteria that survive antibiotic clearance often develop resistance to the impregnated antibiotic and regrow. This resistance has been clinically demonstrated by the isolation of small colony variants of *S. aureus* resistant to gentamicin from the wounds of patients treated with gentamicin-impregnated PMMA beads (von Eiff et al. 1997). Also, the glycocalyx displays antiphagocytic properties, thereby allowing the bacteria to evade clearance by the host's immune system (Ferguson et al. 1992, Dasgupta 1996, Shiau and Wu 1998), although more recent evidence has been contradictory (Leid et al. 2002). The glycocalyx is mainly composed of teichoic acids (80%) and staphylococcal and host proteins (Hussain et al. 1993). Host proteins such as fibrin are derived from the conversion of fibrinogen by the staphylococcal coagulase–prothrombin complex (see below) (Akiyama et al. 1997).

Another important component of biofilm produced by staphylococci is extracellular DNA (eDNA). eDNA as a component of the extracellular matrix of biofilms was noted first in *P. aeruginosa* (Whitchurch et al. 2002, Steinberger and Holden 2005, Allesen-Holm et al. 2006). Such is the importance of eDNA in *P. aeruginosa* infections that application of DNAse concurrently with antibiotic therapy, is being used for treatment of cystic fibrosis patients (Shah et al. 1995, Gibson et al. 2003) (see also Chapter 14). The finding that DNAse present on skin cells can lessen biofilm formation (Eckhart et al. 2007) reinforces the importance of eDNA to biofilm viability. Rice and co-workers have recently demonstrated that eDNA is important for biofilm formation and adherence in *S. aureus*, and that release of eDNA appears, at least in part, to be mediated by the function of *cidA*-encoded murein hydrolase (Rice et al. 2007). This gene product is a holin homologue and has been shown to play a role in cell lysis; thus it is believed that this gene allows lysis of *S. aureus* biofilm cells and release of their DNA into the extracellular milieu. Other cellular factors that may be involved in release of DNA include autolysins such as Atl or induction of prophages that lead to lysis (Webb et al. 2003). In *S. epidermidis*, the autolysin AtlE was shown to be important in release of chromosomal DNA and subsequent initial attachment during early biofilm formation, as an *atlE* mutant did not release DNA and had lower biofilm-forming capacity (Qin et al. 2007).

Biofilm produced by *S. epidermidis* contains both the capsular polysaccharide/adhesin (PS/A) that mediates cell adherence to biomaterials, and a polysaccharide intercellular adhesin (PIA) that may mediate bacterial accumulation into cellular aggregates (Heilmann et al. 1996, McKenney et al. 1998). PS/A is a high-molecular-mass (>250 kDa) molecule composed of acid-stable polymers of β-1,6-linked glucosamine. PIA is a polymer of β-1,6-linked *N*-acetyl glucosamine residues with a molecular mass of less than 30 kDa that is synthesized through genes present on the intercellular adhesion locus (*ica*) (McKenney et al. 1998, Miyazaki et al. 1999). *S. aureus* and other *Staphylococcus* spp. also contain an *ica* locus and its deletion results in the loss of biofilm-forming ability (Miyazaki et al. 1999). The presence of glycocalyx was noted in 76% of *S. aureus*, 57% of *Staphylococcus epidermidis*, 75% of *Escherichia coli*, and 50% of *Pseudomonas aeruginosa* clinical osteomyelitis isolates (Alam et al. 1990).

In more recent work, however, the importance of PIA in biofilm formation and relevance to infection has been called into question. The *ica* gene cluster is not present in all *S. epidermidis* strains (Ziebuhr et al. 1997). Moreover, up to 30% of *S. epidermidis* biofilms have been found to be PIA-negative (Rohde et al. 2007); another study that focused on prosthetic hip infection found that only about one third of patients infected with *S. epidermidis* carried an *ica*-positive strain (Nilsdotter-Augustinsson et al. 2007). While the *ica* cluster is commonly found in *S. aureus* isolates, the importance of PIA production to virulence is uncertain. For example, in a guinea pig model of biofilm infection, deletion of *ica* and, thus, lack of PIA production caused no decrease in virulence (Francois et al. 2003), and deletion of *ica* in the clinical isolate UAMS-1 did not lead to lesser biofilm formation either in vitro or in an in vivo mouse model of catheter infection (Beenken et al. 2004). As well, several clinical isolates of MRSA (that were *ica*-positive) have been identified in which biofilm production is independent of *ica,* as increased transcription of the operon was not seen during glucose-mediated biofilm growth. As well, under NaCl-induced biofilm growth, though *ica* transcription was increased, levels of biofilm production were not similarly heightened (Fitzpatrick et al. 2005a). Deletion of the *ica* locus in one of these isolates did not lead to a lessened ability to form a biofilm; however, the same deletion in a laboratory strain of *S. aureus* did abrogate biofilm formation (Fitzpatrick et al. 2005a). Other studies support this idea and show that, in 114 MRSA clinical isolates, PIA production did not correlate with biofilm production, and deletion of the *ica* locus in six of these isolates did not lead to lessened biofilm formation (O'Neill et al. 2007). However, in methicillin-sensitive *S. aureus* (MSSA), there was a correlation between PIA production and biofilm formation, and deletion of *ica* abolished biofilm formation (O'Neill et al. 2007). Thus, it seems likely that the *ica* locus' contribution to biofilm development is strain- and environment-dependent, and that there are different mechanisms of biofilm development in MRSA vs. MSSA. In those *S. epidermidis* and *S. aureus* strains in which biofilm formation is not dependent on PIA expression, protein adhesin(s) seem to be the most important factors. For example, in *S. epidermidis*, the accumulation-associated protein (Aap) is found at elevated levels in the proteinaceous biofilm (Hennig et al. 2007, Rohde et al. 2007). Clearly, more work should be done to fully elucidate the alternative mechanisms that contribute to biofilm formation by these microorganisms.

A variety of other genes and their products have been demonstrated to be involved in the development of staphylococcal biofilms. There exists evidence that attachment of bacterial cells to a polymer surface – of course necessary for biofilm formation – may be promoted by a *S. epidermidis* autolysin (Heilmann et al. 1996); *S. aureus* posesses a homologue of this gene (*atl*) which may have a similar function, perhaps through DNA release as discussed above. Teichoic acid structure is also crucial in the development of biofilms. Specifically, the addition of D-alanine esters to teichoic acids via the *dltA* gene product may be an important factor in imparting the proper charge balance on the Gram-positive cell surface, assisting in the physicochemical elements of initial attachment and biofilm formation. Another *S. aureus* gene product, known as the "biofilm associated protein" (Bap), was discovered via transposon mutagenesis to be required for biofilm formation on inert surfaces. However, the significance of this protein remains debatable since the *bap* gene was detected in only 5% of bovine mastitis isolates and none of the 75 clinical isolates evaluated. Gene expression in planktonic (shaken) *versus* biofilm (static) *S. aureus* cultures was evaluated by Becker and co-workers; five genes whose expression were increased in biofilms were identified (Becker et al. 2001). These included the genes encoding threonyl-tRNA synthetase (upregulated by amino acid starvation), three oxygen starvation response genes and a gene that encodes the ATPase ClpC. Microarrays are a powerful tool to investigate global gene expression. This technique, when used to study differential gene expression between these conditions, suggested 48 genes the transcription of which was increased at least twofold in the biofilm compared to planktonic conditions (Beenken et al. 2004). Taken together, these data suggest that genes involved in cell wall synthesis and pH balance are important in biofilms, whereas toxin and protease production is higher during planktonic growth (Beenken et al. 2004, Resch et al. 2006).

7.4 Properties of the Host Immune Response in the Development of Osteomyelitis

Biofilm formation by *S. aureus* makes eradication of the pathogen extremely difficult. Devitalized tissue induced by the staphylococcal toxins and the early inflammatory response not only present a suitable substrate for further bacterial adherence, but also interfere with the host's ability to mount an effective immune response against *S. aureus*, potentially leading to the development of a chronic infection (see also Chapter 11). One reason for this involves the functional impairment of phagocytic cells that are important during the early innate response to *S. aureus* infection. For example, a reduction in the amount of superoxide (a mediator of bacterial killing) produced within professional phagocytic blood cells of the infected host may occur (Roisman et al. 1983). Another mechanism by which dead bone can produce locally compromised immunity is through frustrated phagocytosis (Roisman et al. 1983), during which professional phagocytes undergo apoptosis when encountering a substrate of a size that is beyond its phagocytic capacity. The resulting release of reactive products may cause accidental host tissue damage and local vascular insufficiency, thereby increasing the predisposition to chronic infection development (Leid et al. 2002).

One theory behind the ineffectual phagocytosis of *S. aureus* growing in a biofilm has been that leukocytes are unable to penetrate into the depths of a biofilm to where the bacteria reside, thus leading to insufficient clearance and persistence. A study by Leid et al. (Leid et al. 2002), however, showed via time-lapse video microscopy that leukocytes do, in fact, attach to and enter a biofilm. Once inside the biofilm, however, these cells were unable to phagocytose the bacteria found there, but did produce inflammatory cytokines. This study supports the idea that there are other mechanisms of immune evasion at work besides frustrated phagocytosis, which prevent the proper engulfment of bacteria in a biofilm. One clue may be that the leukocytes that penetrated the biofilm in these studies were permeable to the large, intercalating dye, propidium iodide. Therefore, their phagocytic activity may have been disabled by the staphylococcal production of pore forming toxins, including α-hemolysin, γ-hemolysin, leukocidin, and the recently described phenol-soluble modulin-like peptides (Wang et al. 2007).

Besides recruitment of phagocytic cells to the site of infection, the innate immune system responds to peptidoglycan (via *N*-formyl methionine proteins and teichoic acids) by producing proinflammatory cytokines, such as IL-1, IL-6, and TNF-α, as well as C reactive protein. These factors enable the host to mount a protective inflammatory response that often contains and may resolve the infection. However, when the infection is not cleared by the innate immune system, *S. aureus* is well equipped to persist by a number of strategies. One such is elicitation of inadequate cell mediated (Th1) and humoral (Th2) adaptive immune responses (see also Chapter 12).

The timed expression of *S. aureus* virulence factors by its quorum-sensing system promotes host $CD4^+$ helper T cells to release Th1 cytokines, including IL-12, IFN-γ, and TNF-α, resulting in a shift of the adaptive immune system to an ineffective Th1 cell-mediated immune response (Leid et al. 2002). Because this Th1 immune response is often inadequate for clearing the early biofilm form, it enables *S. aureus* to form a fully mature biofilm and, hence, a persistent infection. In a study using a murine model of acute *S. aureus* biofilm infection, the increase in central cytokines of cell mediated immunity (IL-2 and IFN-γ) appeared to be only transient, while inflammatory cytokines remained at elevated levels around biofilm-infected tissue (Yoon et al. 1999). This cytokine profile resulted in the initial expansion and activation of T-cell subsets followed by apoptosis. In this way, *S. aureus* seemed to interfere with the antibacterial immune response by down regulating T cell-mediated immunity and cytokine production. In addition, the Th1 response produced by staphylococcal infection is ineffective in the low oxygen partial pressures found in biofilms and in infected tissues where immune cell function is inhibited. A study performed in mice also found that high levels of IFN-γ (a Th1 cytokine) play a detrimental role in staphylococcal infection, and IL-4 and IL-10 (Th2 cytokines) are involved in host resistance to infection through regulation of IFN-γ (Sasaki et al. 2000).

The Th2 antibody-mediated response is also ineffective against a mature bacterial community. This Th2 response has, however, been previously shown to be readily effective at clearing a biofilm infection in the early phase of formation (Nayak et al.

2004, Shkreta et al. 2004, Sun et al. 2005). Unfortunately, this antibody-mediated response is down regulated by both the host cytokines associated with the initial response to *S. aureus* infection, most notably IFN-γ, as well as by *S. aureus* production of superantigens, capsule, and other toxins. Although the antibody-mediated immune response does eventually recover and is again able to mount a response against the biofilm by then the fully mature biofilm is resistant to antibody-mediated clearance.

7.5 Diagnosis, Treatment and Prevention of Osteomyelitis

7.5.1 Diagnosis

During acute infection, if the proper antibiotic is started early, the infection will usually clear after 2–4 weeks of treatment (Berendt and Byren 2004). However, diagnosing these infections during this early, clearable state is difficult. Radiographic changes during acute infections are usually not discernible until 1–1.5 weeks after inception of the disease. Magnetic resonance imaging (MRI) is effective in diagnosing acute infections in the absence of metal implants, but there is a lag time after previous surgery or infection (Berendt and Byren 2004).

In chronic osteomyelitis there exist large areas of devitalized cortical and cancellous bone within the wound. Because antibiotics do not penetrate well into devitalized bone (Healy and Freedman 2006), dead areas must be completely debrided, including devitalized scar tissue, marrow, and cortex. The soft tissue covering the area of bone trauma must heal; if this does not occur, the infection will persist and a new infection could form. Compromise of local soft tissue is a major reason for continued drainage. Diagnosis of chronic infection can often be made by radiography. Other techniques include radionuclide scans, though these lack specificity (Berendt and Byren 2004).

The presence of general vascular insufficiency makes suitable therapy and management of chronic contiguous osteomyelitis complicated. Most patients fitting this description have diabetes mellitus (Calhoun et al. 1988a), and range from 35 to 70 years of age. Due to the large increase in the diabetic population, osteomyelitis in the diabetic foot is now considered the most common bone infection (Berendt and Byren 2004). The small bones of the feet, as well as the talus, calcaneus, distal fibula, and tibia are commonly involved in this category of infection. Often, the infection is commenced by minor trauma to the feet, such as infected nail beds, cellulitis, or trophic skin ulceration. Neuropathy in these patients impairs the proper functioning of the foot as well as protective pain responses, leading to progression of soft tissue infections into underlying bone (Berendt and Byren 2004).

Osteomyelitis in those individuals with compromised vasculature can be difficult to diagnose. The patient may present with any of a large number of complaints, including ingrown toenails, a perforating foot ulcer, cellulitis, or a deep space infection. Examination shows decreased dorsal pedis and posterior tibia pulses, poor capillary refill, and decreased sensation; however, fever and systemic toxicity are often absent. Although arrest of the infection is desirable, a more achievable treatment goal is to contain the infection and preserve the functional integrity of the involved limb. Debridement and ablation are often essential. The intractable character of this type of infection often leads to recurrent bone infections, even after suitable therapy. Partial removal of the infected bone is almost always necessary.

7.5.2 Antimicrobial Chemotherapy

Because staphylococci are by far the infectious agent most commonly recovered from cases of osteomyelitis, it is advisable to commence empirical therapy upon patient presentation with a regimen that includes an anti-staphylococcal agent (Berendt and Byren 2004). Following a definitive diagnosis by laboratory testing, the appropriate, more specialized antibiotic can be applied if it is found that the susceptibility profile of the infecting microorganism renders inappropriate the empirically selected antibiotic. Treatment usually lasts at least 4 weeks and is administered intravenously, but duration does vary markedly with age; length of treatment tends to be shorter in children (Jaberi et al. 2002). In adult individuals, *S. aureus* is generally treated with nafcillin or with cefazolin, clindamycin, vancomycin, ciprofloxacin, or levofloxacin being given as alternatives should treatment with the former drug fail (Lew and Waldvogel 2004).

7.5.3 Novel Treatments

Given the difficulties in treating osteomyelitis with conventional antimicrobial agents and the tendency of biofilm infections to resist clearance by such agents, it is apparent that novel treatments are necessary (see also Chapter 14). Recently the use of anti-PIA antibodies to prevent attachment or the formation of PIA in general has been investigated by various authors (McKenney et al. 1999, Maira-Litran et al. 2005, Kelly-Quintos et al. 2006). Another option is to coat medical devices prior to implantation. An enzyme produced by *Actinobacillus actinomycetemcomitans* known as "dispersin B" (DspB) is capable of cleaving PIA (Itoh et al. 2005). However, as mentioned above, many clinical isolates do not appear to express the PIA polysaccharide. Finally, Balaban and co-workers have advocated the approach of using the RIP heptapeptide, which is proposed to inhibit RNAIII-activated virulence factors, in the treatment of biofilm-associated infections (Balaban et al. 2003, Balaban et al. 2003, Giacometti et al. 2003, Balaban et al. 2005, Balaban et al. 2007). The suggested mechanism is inhibition of quorum sensing in *S. aureus* leading to reduced biofilm formation that is less recalcitrant to the action of antibiotics. However, the validity of this claim remains unclear since other authors have shown that the *agr* system works to increase levels of biofilm detachment and that disruption of the QS system leads to increased biofilm formation (Vuong et al. 2000, Vuong et al. 2003, Otto 2004, Vuong et al. 2004, Kong et al. 2006). Therefore, whether or not RIP will truly be an effective anti-biofilm agent is open to question. Thus it is apparent that the number of novel therapies that are under development and could be effective against all clinical isolates of *S. aureus* is limited. To date, surgical intervention and debridement remains the most effective method of treatment of biofilm-associated infections. In osteomyelitis infections, this means debridement of the infected bone and, on occasion, the surrounding soft tissue.

7.5.4 Prevention

That prevention of a disease is always better than waiting for the disease to occur and then attempting a cure has been known since the time of Hippocrates. Of course, the best and most effective way of preventing an infectious disease is through vaccination; this is no less true in the case of osteomyelitis. Since vaccine development tends to focus on one microorganism at a time and *S. aureus* is one of the main causative agents of osteomyelitis, the following text will focus on anti-*S. aureus* vaccine development efforts to-date and particularly on those directed against biofilm infections.

A recent review of such efforts was published in late 2008 by Schaffer and Lee. A summary of recent vaccine efforts, adapted from (Schaffer and Lee 2008), is provided in Table 7.1 below. The majority of such efforts have focused upon identifying and testing staphylococcal antigens expressed during planktonic growth. Whilst a proportion of staphylococcal infections are undoubtedly caused by bacterial cells living planktonically within body fluids (probably the best example being septicemia), a significant proportion of staphylococcal infections are mediated by biofilms. Given the knowledge that bacteria residing within biofilms express markedly different proteomes than the same cells living planktonically, it seems likely that vaccines directed against planktonic antigens will be ineffective against biofilm-type infections. Therefore, selection of antigens for inclusion in a vaccine which targets biofilm infections must take into account the biofilm phenotype during antigen selection. Indeed, it is possible that multiple antigens will be required in order to protect against planktonic and biofilm-type *S. aureus* infections.

Efforts to identify *S. aureus* antigens expressed during biofilm growth have been undertaken recently. Brady et al. reported that a number of surface antigens are expressed uniquely by *S. aureus* during biofilm growth (Brady et al. 2006). In this study, rabbits were infected with experimental *S. aureus* osteomyelitis using

Table 7.1 Summary of anti-*S. aureus* vaccines currently in development

Name	Component details	Current status
StaphVax™	Capsular polysaccharides types 5 and 8	Phase III failed
V710	IsdB; iron-regulated surface determinant	Phase II
PNAG	Poly-*N*-acetyl glucosamine	Experimental
ETI-211	Anti-protein A mAb linked to anti-CR1 mAb	Experimental
Alpha hemolysin	Secreted *S. aureus* protein; important virulence determinant	Experimental
SEB	Staphylococcal enterotoxin B toxoid; proteasome	Experimental

strain MRSA-M2 (isolated from a patient with osteomyelitis) (Mader and Shirtliff 1999) and sera drawn at early and late time-points post-infection. This sera was used to immunoblot two-dimensional SDS-PAGE gels upon which had been separated whole cell proteins of *S. aureus* grown in an in vitro biofilm model. In this way, the authors identified antigens that were both present during biofilm growth and immunogenic during osteomyelitis infection. The authors also fractionated cells and detected a number of cell-surface proteins expressed during biofilm growth and immunogenic in the rabbit model of osteomyelitis. Such proteins would be ideal candidates for inclusion in an anti-*S. aureus* biofilm infection vaccine.

The same group verified further the appropriateness of a number of these cell-wall antigens by direct immunovisualization of their presence and spatial distribution within intact *S. aureus* biofilm (Brady et al. 2007). Polyclonal antisera to each of five of the protein antigens discovered previously to be immunogenic in the animal model of osteomyelitis and expressed during biofilm growth (Brady et al. 2006) was raised in rabbits. Antibodies were then added to S. aureus biofilm in an in vitro flow model. A secondary goat anti-rabbit F(ab')$_2$ conjugated to Alexafluor-633 was added, followed by the DNA-interlocating stain SYTO9. Data suggested that all of the proteins investigated were present within the biofilm. Interestingly, however, expression of the proteins was not homogenous; each of the proteins was detected only in a proportion of the total community. The authors suggest that this spatially heterogeneous distribution of antigen expression within biofilm means that using only one biofilm-specific antigen in a vaccine will be ineffective as only a part of the whole community will be targeted by immune effectors. Therefore, the untargeted community will be unaffected and the infection able to persist.

To this end, the same group investigated the efficacy of a quadrivalent vaccine preparation containing four protein antigens (75 μg each) found to be both present in a biofilm in vitro and immunogenic in the rabbit model of osteomyelitis together with TiterMax^{TM} adjuvant (Brady et al. 2008, O'May et al. 2008). Four groups of animals were used: an untreated control group, one treated with vancomycin (40 mg/kg; 2/day) post-infection, one vaccinated pre-infection and one that received both pre-infection vaccination and post-infection vancomycin (40 mg/kg; 2/day) treatment. Animals who received both vaccination and vancomycin treatments showed significant differences in radiological, clinical (limping), and bacteriological signs of osteomyelitis when compared to the untreated controls, whilst those which received either vaccination or vancomycin alone did not. The authors postulate that this is because whilst vaccination is able to prevent establishment of a biofilm within the bone, planktonic (and, therefore, not targeted by the biofilm-specific vaccine) cells survive and are able to cause infection. Conversely, when only vancomycin was used, biofilm was able to form but the planktonic (and, therefore, vancomycin-sensitive) cells were killed. When the two were combined infection was prevented since both biofilm and planktonic cells were targeted. Work is now focusing upon both testing the efficacy of the vaccine in other models of biofilm infection and identifying an antigen expressed by planktonic *S. aureus* cells which will remove the need for vancomycin treatment.

7.6 Conclusion

The properties of bacterial biofilm are critical to the pathogenesis of osteomyelitis and, therefore, also to the development of novel methods of prevention and treatment. Bone provides a stable, non-ablative surface for biofilm formation, allowing invading bacteria to take refuge from the action of the immune system and any antimicrobial chemotherapy used in an attempt to clear the infection. The etiological agent responsible for the vast majority of cases of osteomyelitis, in both children and adults, *Staphylococcus aureus*, is a potent biofilm-forming bacterium, lending further weight to the assertion that biofilm is critical in the pathogenesis of osteomyelitis. Despite this, however, the majority of efforts in development of treatments and preventative therapies remain directed towards planktonic microorganisms. It seems likely that only a resolution of this dichotomy will lead to effective therapies for this highly debilitating infection.

References

Akiyama H, Torigoe R, Arata J (1993) Interaction of *Staphylococcus aureus* cells and silk threads *in vitro* and in mouse skin. J Dermatol Sci 6(3):247–257

Akiyama H, Ueda M, Kanzaki H et al (1997) Biofilm formation of *Staphylococcus aureus* strains isolated from impetigo and furuncle: role of fibrinogen and fibrin. J Dermatol Sci 16(1):2–10

Alam SI, Khan KA, Ahmad, A (1990) Glycocalyx positive bacteria isolated from chronic osteomyelitis and septic arthritis. Ceylon Med J 35(1):21–23

Allesen-Holm M, Barken KB, Yang L et al (2006) A characterization of DNA release in Pseudomonas aeruginosa cultures and biofilms. Mol Microbiol 59(4):1114–1128

Anderl JN, Zahller J, Roe F et al (2003) Role of nutrient limitation and stationary-phase existence in *Klebsiella pneumoniae* biofilm resistance to ampicillin and ciprofloxacin. Antimicrob Agents Chemother 47(4):1251–1256

Balaban N, Cirioni O, Giacometti A et al (2007) Treatment of *Staphylococcus aureus* biofilm infection by the quorum-sensing inhibitor RIP. Antimicrob Agents Chemother 51(6): 2226–2229

Balaban N, Giacometti A, Cirioni O et al (2003) Use of the quorum-sensing inhibitor RNAIII-inhibiting peptide to prevent biofilm formation in vivo by drug-resistant *Staphylococcus epidermidis*. J Infect Dis 187(4):625–630

Balaban N, Gov Y, Bitler A et al (2003) Prevention of *Staphylococcus aureus* biofilm on dialysis catheters and adherence to human cells. Kidney Int 63(1):340–345

Balaban N, Novick R P (1995) Autocrine regulation of toxin synthesis by *Staphylococcus aureus*. Proc Natl Acad Sci USA 92(5):1619–1623

Balaban N, Stoodley P, Fux C A et al (2005) Prevention of staphylococcal biofilm-associated infections by the quorum sensing inhibitor RIP. Clin Orthop Relat Res (437):48–54

Becker P, Hufnagle W, Peters G et al (2001) Detection of differential gene expression in biofilm-forming versus planktonic populations of *Staphylococcus aureus* using micro-representational-difference analysis. Appl Environ Microbiol 67(7):2958–2965

Beenken KE, Dunman PM, McAleese F et al (2004) Global gene expression in *Staphylococcus aureus* biofilms. J Bacteriol 186(14):4665–4684

Berendt T, Byren I (2004) Bone and joint infection. Clin Med 4(6):510–518

Beveridge TJ, Makin SA, Kadurugamuwa JL et al (1997) Interactions between biofilms and the environment. FEMS Microbiol Rev 20(3–4):291–303

Boden MK, Flock JI (1994) Cloning and characterization of a gene for a 19 kDa fibrinogen-binding protein from Staphylococcus aureus. Mol Microbiol 12(4):599–606

Boyd A, Chakrabarty AM (1994) Role of alginate lyase in cell detachment of *Pseudomonas aeruginosa*. Appl Environ Microbiol 60(7):2355–2359

Brady RA, Leid JG, Calhoun JH et al (2008) Osteomyelitis and the role of biofilms in chronic infection. FEMS Immunol Med Microbiol 52(1):13–22

Brady RA, Leid JG, Camper AK et al (2006) Identification of *Staphylococcus aureus* proteins recognized by the antibody-mediated immune response to a biofilm infection. Infect Immun 74(6):3415–3426

Brady RA, Leid JG, Kofonow J et al (2007) Immunoglobulins to surface-associated biofilm immunogens provide a novel means of visualization of methicillin-resistant *Staphylococcus aureus* biofilms. Appl Environ Microbiol 73(20):6612–6619

Brady RA, O'May GA, Leid JG et al (2008). Protective vaccine against chronic infections due to *Staphylococcus aureus*. ASM General Meeting, Boston, MA

Bremell T, Tarkowski, A (1995) Preferential induction of septic arthritis and mortality by superantigen-producing staphylococci. Infect Immun 63(10):4185–4187

Calhoun JH, Cantrell J, Cobos J et al (1988a) Treatment of diabetic foot infections: Wagner classification, therapy, and outcome. Foot Ankle 9(3):101–106

Calhoun KH, Shapiro RD, Stiernberg CM et al (1988b) Osteomyelitis of the mandible. Arch Otolaryngol Head Neck Surg 114(10):1157–1162

Carek PJ, Dickerson LM, Sack JL (2001) Diagnosis and management of osteomyelitis. Am Fam Physician 63(12):2413–2420

Chan C, Burrows LL, Deber CM (2004) Helix induction in antimicrobial peptides by alginate in biofilms. J Biol Chem 279(37):38749–38754

Cheung AI, Projan SJ, Edelstein RE et al (1995) Cloning, expression, and nucleotide sequence of a Staphylococcus aureus gene (fbpA) encoding a fibrinogen-binding protein. Infect Immun 63(5):1914–1920

Cheung AL, Bayer MG, Heinrichs JH (1997) sar Genetic determinants necessary for transcription of RNAII and RNAIII in the agr locus of Staphylococcus aureus. J Bacteriol 179(12): 3963–3971

Costerton JW, Lewandowski Z, Caldwell DE et al (1995) Microbial biofilms. Annu Rev Microbiol 49:711–745

Darouiche RO, Dhir A, Miller AJ et al (1994) Vancomycin penetration into biofilm covering infected prostheses and effect on bacteria. J Infect Dis 170(3):720–723

Dasgupta MK (1996) Biofilm causes decreased production of interferon-gamma. J Am Soc Nephrol 7(6):877–882

De Beer D, Srinivasan R, Stewart PS (1994) Direct measurement of chlorine penetration into biofilms during disinfection. Appl Environ Microbiol 60(12):4339–4344

De Jonghe M, Glaesener G (1995) Type B *Haemophilus influenzae* infections. Experience at the Pediatric Hospital of Luxembourg. Bull Soc Sci Med Grand Duche Luxemb 132(2): 17–20

Duguid IG, Evans E, Brown MR et al (1992) Effect of biofilm culture upon the susceptibility of Staphylococcus epidermidis to tobramycin. J Antimicrob Chemother 30(6):803–810

Eckhart L, Fischer H, Barken KB et al (2007) DNase1L2 suppresses biofilm formation by *Pseudomonas aeruginosa* and *Staphylococcus aureus*. Br J Dermatol 156(6):1342–1345

Ferguson DA Jr, Veringa EM, Mayberry WR et al (1992) *Bacteroides* and *Staphylococcus* glycocalyx: chemical analysis, and the effects on chemiluminescence and chemotaxis of human polymorphonuclear leucocytes. Microbios 69(278):53–65

Fischer B, Vaudaux P, Magnin M et al (1996) Novel animal model for studying the molecular mechanisms of bacterial adhesion to bone-implanted metallic devices: role of fibronectin in *Staphylococcus aureus* adhesion. J Orthop Res 14(6):914–920

Fitzpatrick F, Humphreys H, O'Gara JP (2005a) Evidence for *icaADBC*-independent biofilm development mechanism in methicillin-resistant *Staphylococcus aureus* clinical isolates. J Clin Microbiol 43(4):1973–1976

Fitzpatrick F, Humphreys H, O'Gara JP (2005b) The genetics of staphylococcal biofilm formation–will a greater understanding of pathogenesis lead to better management of device-related infection? Clin Microbiol Infect 11(12):967–973

Francois P, Tu Quoc PH, Bisognano C et al (2003) Lack of biofilm contribution to bacterial colonisation in an experimental model of foreign body infection by *Staphylococcus aureus* and *Staphylococcus epidermidis*. FEMS Immunol Med Microbiol 35(2):135–140

Francois P, Vaudaux P, Lew PD (1998) Role of plasma and extracellular matrix proteins in the physiopathology of foreign body infections. Ann Vasc Surg 12(1):34–40

Giacometti A, Cirioni O, Gov Y et al (2003) RNA III inhibiting peptide inhibits in vivo biofilm formation by drug-resistant Staphylococcus aureus. Antimicrob Agents Chemother 47(6):1979–1983

Gibson RL, Burns JL, Ramsey BW (2003) Pathophysiology and management of pulmonary infections in cystic fibrosis. Am J Respir Crit Care Med 168(8):918–951

Gracia E, Fernandez A, Conchello P et al (1997) Adherence of *Staphylococcus aureus* slime-producing strain variants to biomaterials used in orthopaedic surgery. Int Orthop 21(1):46–51

Gristina AG, Costerton JW (1984) Bacterial adherence and the glycocalyx and their role in musculoskeletal infection. Orthop Clin North Am 15(3):517–535

Gristina AG, Oga M, Webb LX et al (1985) Adherent bacterial colonization in the pathogenesis of osteomyelitis. Science 228(4702):990–993

Healy B, Freedman A (2006) Infections. BMJ 332(7545):838–841

Heilmann C, Schweitzer O, Gerke C et al (1996) Molecular basis of intercellular adhesion in the biofilm-forming *Staphylococcus epidermidis*. Mol Microbiol 20(5):1083–1091

Hennig S, Nyunt Wai S, Ziebuhr W (2007) Spontaneous switch to PIA-independent biofilm formation in an ica-positive Staphylococcus epidermidis isolate. Int J Med Microbiol 297(2):117–122

Herrmann M, Vaudaux PE, Pittet D et al (1988) Fibronectin, fibrinogen, and laminin act as mediators of adherence of clinical staphylococcal isolates to foreign material. J Infect Dis 158(4):693–701

Holzman RS, Bishko F (1971) Osteomyelitis in heroin addicts. Ann Intern Med 75(5):693–696

Hussain M, Wilcox MH, White PJ (1993) The slime of coagulase-negative staphylococci: biochemistry and relation to adherence. FEMS Microbiol Rev 10(3–4):191–207

Itoh Y, Wang X, Hinnebusch BJ et al (2005) Depolymerization of b-1,6-N-acetyl-D-glucosamine disrupts the integrity of diverse bacterial biofilms. J Bacteriol 187(1):382–387

Jaberi FM, Shahcheraghi GH, Ahadzadeh M (2002) Short-term intravenous antibiotic treatment of acute hematogenous bone and joint infection in children: a prospective randomized trial. J Pediatr Orthop 22(3):317–320

Jackson MA, Nelson JD (1982) Etiology and medical management of acute suppurative bone and joint infections in pediatric patients. J Pediatr Orthop 2(3):313–323

Jonsson K, Signas C, Muller HP et al (1991) Two different genes encode fibronectin binding proteins in *Staphylococcus aureus*. The complete nucleotide sequence and characterization of the second gene. Eur J Biochem 202(3):1041–1048

Kelly-Quintos C, Cavacini LA, Posner MR et al (2006) Characterization of the opsonic and protective activity against *Staphylococcus aureus* of fully human monoclonal antibodies specific for the bacterial surface polysaccharide poly-N-acetylglucosamine. Infect Immun 74(5):2742–2750

Kluytmans J, van Belkum A, Verbrugh H (1997) Nasal carriage of *Staphylococcus aureus*: epidemiology, underlying mechanisms, and associated risks. Clin Microbiol Rev 10(3):505–520

Kong KF, Vuong C, Otto M (2006) *Staphylococcus* quorum sensing in biofilm formation and infection. Int J Med Microbiol 296(2–3):133–139

Kuypers JM, Proctor RA (1989) Reduced adherence to traumatized rat heart valves by a low-fibronectin-binding mutant of *Staphylococcus aureus*. Infect Immun 57(8):2306–2312

Leid JG, Shirtliff ME, Costerton JW et al (2002) Human leukocytes adhere to, penetrate, and respond to *Staphylococcus aureus* biofilms. Infect Immun 70(11):6339–6345

Lew DP, Waldvogel FA (2004) Osteomyelitis. Lancet 364(9431):369–379

Lewis K (2005) Persister cells and the riddle of biofilm survival. Biochemistry (Mosc) 70(2): 267–274

Lopes JD, Da-Mota GF, Carneiro CR et al (1988) Evolutionary conservation of laminin-binding proteins. Braz J Med Biol Res 21(6):1269–1273

Lopes JD, dos Reis M, Brentani RR (1985) Presence of laminin receptors in *Staphylococcus aureus*. Science 229(4710):275–277

Mader JT, Ortiz M, Calhoun JH (1996) Update on the diagnosis and management of osteomyelitis. Clin Podiatr Med Surg 13(4):701–724

Mader JT, Shirtliff M (1999) The rabbit model of bacterial osteomyelitis of the tibia. In: Zak O, Sande MA (eds) Handbook of animal models of infection. Academic, London

Maira-Litran T, Kropec A, Goldmann D A et al (2005) Comparative opsonic and protective activities of Staphylococcus aureus conjugate vaccines containing native or deacetylated Staphylococcal Poly-N-acetyl-beta-(1-6)-glucosamine. Infect Immun 73(10):6752–6762

McDevitt D, Francois P, Vaudaux P et al (1994) Molecular characterization of the clumping factor (fibrinogen receptor) of *Staphylococcus aureus*. Mol Microbiol 11(2):237–248

McGavin MH, Krajewska-Pietrasik D, Ryden C et al (1993) Identification of a *Staphylococcus aureus* extracellular matrix-binding protein with broad specificity. Infect Immun 61(6): 2479–2485

McKenney D, Hubner J, Muller E et al (1998) The *ica* locus of *Staphylococcus epidermidis* encodes production of the capsular polysaccharide/adhesin. Infect Immun 66(10): 4711–4720

McKenney D, Pouliot KL, Wang Y et al (1999) Broadly protective vaccine for *Staphylococcus aureus* based on an in vivo-expressed antigen. Science 284(5419):1523–1527

McLean RJ, Whiteley M, Stickler DJ et al (1997) Evidence of autoinducer activity in naturally occurring biofilms. FEMS Microbiol Lett 154(2):259–263

McNamara PJ, Milligan-Monroe KC, Khalili S et al (2000) Identification, cloning, and initial characterization of rot, a locus encoding a regulator of virulence factor expression in *Staphylococcus aureus*. J Bacteriol 182(11):3197–3203

Miyazaki E, Chen JM, Ko C et al (1999) The *Staphylococcus aureus rsbW* (orf159) gene encodes an anti-sigma factor of SigB. J Bacteriol 181(9):2846–2851

Modun B, Kendall D, Williams P (1994) Staphylococci express a receptor for human transferrin: identification of a 42-kilodalton cell wall transferrin-binding protein. Infect Immun 62(9):3850–3858

Morrissy RT, Haynes DW (1989) Acute hematogenous osteomyelitis: a model with trauma as an etiology. J Pediatr Orthop 9(4):447–456

Naidu AS, Andersson M, Forsgren A (1992) Identification of a human lactoferrin-binding protein in Staphylococcus aureus. J Med Microbiol 36(3):177–183

Navarre WW, Daefler S, Schneewind O (1996) Cell wall sorting of lipoproteins in *Staphylococcus aureus*. J Bacteriol 178(2):441–446

Nayak DK, Asha A, Shankar KM et al (2004) Evaluation of biofilm of *Aeromonas hydrophila* for oral vaccination of *Clarias batrachus* – a carnivore model. Fish Shellfish Immunol 16(5): 613–619

Nguyen T, Ghebrehiwet B, Peerschke EI (2000) *Staphylococcus aureus* protein A recognizes platelet gC1qR/p33: a novel mechanism for staphylococcal interactions with platelets. Infect Immun 68(4):2061–2068

Nilsdotter-Augustinsson A, Koskela A, Ohman L et al (2007) Characterization of coagulase-negative staphylococci isolated from patients with infected hip prostheses: use of phenotypic and genotypic analyses, including tests for the presence of the ica operon. Eur J Clin Microbiol Infect Dis 26(4):255–265

Nilsson IM, Patti JM, Bremell T et al (1998) Vaccination with a recombinant fragment of collagen adhesin provides protection against *Staphylococcus aureus*-mediated septic death. J Clin Invest 101(12):2640–2649

O'May GA, Brady RA, Leid JG et al (2008). A quadrivalent vaccine directed against biofilm-upregulated immunogens is protective against chronic osteomyelitis caused by methicillin-resistant *Staphylococcus aureus*. ICAAC 2008, October 25–28, Washington, DC

O'Neill E, Pozzi C, Houston P et al (2007) Association between methicillin susceptibility and biofilm regulation in Staphylococcus aureus isolates from device-related infections. J Clin Microbiol 45(5):1379–1388

Oie S, Huang Y, Kamiya A et al (1996) Efficacy of disinfectants against biofilm cells of methicillin-resistant *Staphylococcus aureus*. Microbios 85(345):223–230

Otto M (2004) Quorum-sensing control in Staphylococci – a target for antimicrobial drug therapy? FEMS Microbiol Lett 241(2):135–141

Park PW, Rosenbloom J, Abrams WR et al (1996) Molecular cloning and expression of the gene for elastin-binding protein (ebpS) in Staphylococcus aureus. J Biol Chem 271(26): 15803–15809

Parkins MD, Ceri H, Storey DG (2000). Differential virulence factor expression in *Pseudomonas aeruginosa* biofilms. American Society for Microbiology Biofilms, Big Sky, MT

Patel AH, Nowlan P, Weavers ED et al (1987) Virulence of protein A-deficient and a-toxin-deficient mutants of *Staphylococcus aureus* isolated by allele replacement. Infect Immun 55(12): 3103–3110

Patti JM, Jonsson H, Guss B et al (1992) Molecular characterization and expression of a gene encoding a *Staphylococcus aureus* collagen adhesin. J Biol Chem 267(7):4766–4772

Peacock SJ, Day NP, Thomas MG et al (2000) Clinical isolates of *Staphylococcus aureus* exhibit diversity in fnb genes and adhesion to human fibronectin. J Infect 41(1):23–31

Qin Z, Ou Y, Yang L et al (2007) Role of autolysin-mediated DNA release in biofilm formation of Staphylococcus epidermidis. Microbiology 153(Pt 7):2083–2092

Rao N, Lipsky BA (2007) Optimising antimicrobial therapy in diabetic foot infections. Drugs 67:195–214

Resch A, Leicht S, Saric M et al (2006) Comparative proteome analysis of *Staphylococcus aureus* biofilm and planktonic cells and correlation with transcriptome profiling. Proteomics 6(6):1867–1877

Rice KC, Mann EE, Endres JL et al (2007) The *cidA* murein hydrolase regulator contributes to DNA release and biofilm development in *Staphylococcus aureus*. Proc Natl Acad Sci USA 104(19):8113–8118

Rohde H, Burandt EC, Siemssen N et al (2007) Polysaccharide intercellular adhesin or protein factors in biofilm accumulation of Staphylococcus epidermidis and Staphylococcus aureus isolated from prosthetic hip and knee joint infections. Biomaterials 28(9):1711–1720

Roisman FR, Walz DT, Finkelstein AE (1983) Superoxide radical production by human leukocytes exposed to immune complexes: inhibitory action of gold compounds. Inflammation 7(4): 355–362

Ryden C, Tung HS, Nikolaev V et al (1997) Staphylococcus aureus causing osteomyelitis binds to a nonapeptide sequence in bone sialoprotein. Biochem J 327(Pt 3):825–829

Sapico FL (1996) Microbiology and antimicrobial therapy of spinal infections. Orthop Clin North Am 27(1):9–13

Sapico FL, Montgomerie JZ (1979) Pyogenic vertebral osteomyelitis: report of nine cases and review of the literature. Rev Infect Dis 1(5):754–776

Sasaki S, Nishikawa S, Miura T et al (2000) Interleukin-4 and interleukin-10 are involved in host resistance to *Staphylococcus aureus* infection through regulation of gamma interferon. Infect Immun 68(5):2424–2430

Sauer K, Camper AK, Ehrlich GD et al (2002) *Pseudomonas aeruginosa* displays multiple phenotypes during development as a biofilm. J Bacteriol 184(4):1140–1154

Schaffer AC, Lee JC (2008) Vaccination and passive immunisation against *Staphylococcus aureus*. Int J Antimicrob Agents 32 Suppl 1:S71–78

Shah PL, Scott SF, Geddes DM et al (1995) Two years experience with recombinant human DNase I in the treatment of pulmonary disease in cystic fibrosis. Respir Med 89(7):499–502

Shiau AL, Wu CL (1998) The inhibitory effect of *Staphylococcus epidermidis* slime on the phagocytosis of murine peritoneal macrophages is interferon-independent. Microbiol Immunol 42(1):33–40

Shirtliff ME, Cripps MW, Mader JT (1999). Retrospective review of 728 patients with long bone osteomyelitis. American Society for Microbiology 99th General Meeting. Chicago, IL:c410

Shkreta L, Talbot BG, Diarra MS et al (2004) Immune responses to a DNA/protein vaccination strategy against *Staphylococcus aureus* induced mastitis in dairy cows. Vaccine 23(1):114–126

Singh PK, Schaefer AL, Parsek MR et al (2000) Quorum-sensing signals indicate that cystic fibrosis lungs are infected with bacterial biofilms. Nature 407(6805):762–764

Steinberger RE, Holden PA (2005) Extracellular DNA in single- and multiple-species unsaturated biofilms. Appl Environ Microbiol 71(9):5404–5410

Stickler DJ, Morris NS, McLean RJ et al (1998) Biofilms on indwelling urethral catheters produce quorum-sensing signal molecules *in situ* and *in vitro*. Appl Environ Microbiol 64(9):3486–3490

Stout RD, Ferguson KP, Li YN et al (1992) Staphylococcal exopolysaccharides inhibit lymphocyte proliferative responses by activation of monocyte prostaglandin production. Infect Immun 60(3):922–927

Sun D, Accavitti MA, Bryers JD (2005) Inhibition of biofilm formation by monoclonal antibodies against *Staphylococcus epidermidis* RP62A accumulation-associated protein. Clin Diagn Lab Immunol 12(1):93–100

Switalski LM, Patti JM, Butcher W et al (1993) A collagen receptor on *Staphylococcus aureus* strains isolated from patients with septic arthritis mediates adhesion to cartilage. Mol Microbiol 7(1):99–107

Thomas MG, Peacock S, Daenke S et al (1999) Adhesion of *Staphylococcus aureus* to collagen is not a major virulence determinant for septic arthritis, osteomyelitis, or endocarditis. J Infect Dis 179(1):291–293

von Eiff C, Bettin D, Proctor RA et al (1997) Recovery of small colony variants of *Staphylococcus aureus* following gentamicin bead placement for osteomyelitis. Clin Infect Dis 25(5): 1250–1251

Vuong C, Gerke C, Somerville GA et al (2003) Quorum-sensing control of biofilm factors in *Staphylococcus epidermidis*. J Infect Dis 188(5):706–718

Vuong C, Kocianova S, Voyich JM et al (2004) A crucial role for exopolysaccharide modification in bacterial biofilm formation, immune evasion, and virulence. J Biol Chem 279(52): 54881–54886

Vuong C, Saenz HL, Gotz F et al (2000) Impact of the *agr* quorum-sensing system on adherence to polystyrene in *Staphylococcus aureus*. J Infect Dis 182(6):1688–1693

Wang R, Braughton KR, Kretschmer D et al (2007) Identification of novel cytolytic peptides as key virulence determinants for community-associated MRSA. Nat Med 13(12):1510–1514

Webb JS, Thompson LS, James S et al (2003) Cell death in *Pseudomonas aeruginosa* biofilm development. J Bacteriol 185(15):4585–4592

Whitchurch CB, Tolker-Nielsen T, Ragas PC et al (2002) Extracellular DNA required for bacterial biofilm formation. Science 295(5559):1487

Wimpenny J (2000) An overview of biofilms as functional communities. In: Allison DG, Gilbert P, Lappin-Scott HM, Wilson M (eds) Community structure and co-operation in biofilms. Cambridge University Press, Cambridge, UK

Xu KD, McFeters GA, Stewart PS (2000) Biofilm resistance to antimicrobial agents. Microbiology 146(Pt 3):547–549

Yacoub A, Lindahl P, Rubin K et al (1994) Purification of a bone sialoprotein-binding protein from Staphylococcus aureus. Eur J Biochem 222(3):919–925

Yoon KS, Fitzgerald RH Jr, Sud S et al (1999) Experimental acute hematogenous osteomyelitis in mice. II. Influence of *Staphylococcus aureus* infection on T-cell immunity. J Orthop Res 17(3):382–391

Ziebuhr W, Heilmann C, Gotz F et al (1997) Detection of the intercellular adhesion gene cluster (*ica*) and phase variation in *Staphylococcus epidermidis* blood culture strains and mucosal isolates. Infect Immun 65(3):890–896

Chapter 8
The Importance of Biofilms in Chronic Rhinosinusitis

Jeff G. Leid, Emily K. Cope, Stacy Parmenter, Mark E. Shirtliff, Scot Dowd, Randall Wolcott, Randall Basaraba DVM, Darrell Hunsaker, James Palmer, and Noam Cohen

8.1 Introduction

There is mounting evidence that bacterial and possibly fungal biofilms play an important role in the etiology and persistence of Chronic Rhinosinusitis (CRS). CRS affects nearly 16–25% of the US population each year, with billions of dollars of annual healthcare expenditures dedicated to its treatment (Gliklich and Metson 1995). Unfortunately, the recalcitrant nature of the disease, which often exhibits a chronic relapsing course, significantly contributes to these healthcare costs. The reasons for the persistent nature of the disease are likely secondary to a number of underlying pathophysiologic mechanisms. Asthma, allergic rhinitis, Gram-positive and Gram-negative infections, aspirin-sensitive asthma, fungus, osteitis, nasal polyposis, superantigens, and other factors have been implicated as etiologies contributing to the development of CRS. The chronic inflammation that develops as a fundamental hallmark of the disease can both cause and be a consequence of dysfunctional mucociliary clearance. Ultimately, stasis of sinonasal secretions will lead to subsequent infection and/or persistent inflammation. In some cases, persistent and recurrent infections occur despite multiple therapeutic interventions for CRS. These chronic infections often involve a particularly resistant form of microbial growth that is manifested by communities of bacteria called biofilm.

8.1.1 Background

Medical scientists have worked under the premise espoused more than 160 years ago by Robert Koch that bacterial infections were caused by individual bacteria floating in purulent fluid or invading animal tissues. Methods of culture were developed to grow these bacteria from swabs or needles which withdrew the fluid containing free floating (planktonic) bacteria. Regrowth of these bacteria on media with

J.G. Leid (✉)
Center for Microbial Genetics and Genomics, Northern Arizona University, Flagstaff, AZ 86011, USA
e-mail: jeff.leid@nau.edu

T. Bjarnsholt et al. (eds.), *Biofilm Infections*, DOI 10.1007/978-1-4419-6084-9_8,

discs imbedded with antibiotics identified drugs to use to eliminate the infections. Unfortunately, some infections did not respond to the identified antibiotics. These "chronic" infections included osteomyelitis, cystitis, cardiac valvulitis, prostatitis, mastoiditis, dental plaque, sinusitis, etc. Based on over 30 years of research by environmental, industrial, and medical scientists, a massive bank of knowledge is now available to medical researchers highlighting the importance of biofilms in the treatment of chronic infectious diseases in humans.

This includes the upper respiratory tract where the presence of biofilms has been demonstrated in chronic otitis media, cholesteatoma, and chronic adenoiditis among other important diseases (Post, Stoodley et al. 2004). Colonization and infection with *Pseudomonas aeruginosa*, a common microorganism isolated from the sinus mucosa, has been linked to CRS as these bacteria are particularly resistant to antibiotic therapy and have the potential to drive chronic disease. It is well accepted that *P. aeruginosa*, in a biofilm state, plays important roles in bacterial persistence and antibiotic resistance in chronic infections, such as otitis media (see also Chapter 3) and cystic fibrosis lung disease (see also Chapter 10) (Hoiby et al. 2001, Ha et al. 2007). If CRS develops from acute bacterial sinusitis, then this progression into a chronic disease parallels other biofilm-related diseases. However, since the etiology of CRS remains to be defined, it is unclear whether *P. aeruginosa* is a frank pathogen of CRS or whether other communities of bacteria are responsible for antibiotic resistance and chronic inflammation. Clearly the etiology of CRS must be defined so that therapeutic efforts can be targeted to the organisms responsible for the disease.

8.1.2 Biofilms Defined

A biofilm is an assemblage of microbial cells that is irreversibly associated with a surface, although free floating cell clusters can occur, and these microbial communities are often enclosed in a matrix of primary polysaccharide material (Costerton et al. 1999, Webb et al. 2003). Biofilm-associated organisms differ from their planktonic counterparts in the genes that are transcribed, resulting in an altered resistance to antibiotics and the human immune system (Donlan 2000, Becker et al. 2001, Bollinger et al. 2001, Stewart and Costerton 2001, Vallet et al. 2001, Whiteley et al. 2001, Donlan 2002, Donlan and Costerton 2002, Parsek and Singh 2003, Smith and Iglewski 2003, Head and Yu 2004, Ren et al. 2004, Leid et al. 2002, 2005, 2009). These communities form on a wide variety of surfaces including living tissues, indwelling medical devices, industrial or potable water systems, and natural aquatic systems. The interface between a surface and an aqueous medium (e.g., water and blood) provides an ideal environment for the attachment and growth of microorganisms. The nature of the substratum, conditioning films that form on that substratum, hydrodynamics of the aqueous medium, characteristics of that medium, and various properties of the cell surface all play a role in determining the rate and extent of microbial attachment and biofilm formation (Donlan 2000, Becker et al. 2001, Bollinger et al. 2001, Stewart and Costerton 2001, Vallet et al. 2001, Whiteley et al. 2001, Donlan 2002, Donlan and Costerton 2002, Parsek and Singh 2003, Smith and Iglewski 2003, Head and Yu 2004, Ren et al. 2004).

Though every microbial biofilm is unique, some structural attributes are considered universal. Biofilms contain microcolonies of bacterial cells that are separated from other microcolonies by interstitial voids. Easily observed in vitro, liquid flow occurs in these voids (also called water channels) allowing diffusion of nutrients, gasses, and antimicrobial agents. Biofilm architecture is heterogeneous in space and time, constantly changing as a result of internal and external processes. Proximity of cells within or between microcolonies provides an ideal niche for the exchange of extrachromosomal plasmids encoding resistance to multiple antibiotics, exchange of nutrients through gradients, and communication by quorum sensing molecules (Costerton et al. 1999, Donlan 2000, O'Toole et al. 2000, Hoiby et al. 2001, Mah and O'Toole 2001, Leid et al. 2002b, Mah et al. 2003, Parsek and Singh 2003, Webb et al. 2003, Leid et al. 2005). Biofilm-associated organisms have dramatically reduced susceptibility to various types of antimicrobial agents, either because the biofilm structure impedes transport of the agent to the cell surface or because the cells within the biofilm exhibit an altered physiology (Donlan 2002). The clinical relevance of these biofilm communities is the existence of a group or groups of organisms that completely tolerate antibiotic challenge and resist host immunity. These microbes may then serve as a reservoir for further development of antibiotic resistance in the population. A paper in the journal Nature correlated the importance of biofilm-specific gene expression and resistance to antibiotics (Mah et al. 2003). For the first time, it was clear that a genetic program specific to the biofilm mode of growth was essential for broad spectrum antibiotic resistance. We have followed up on these studies and demonstrated that a biofilm-specific program also exists that dictates susceptibility to challenge from the host's defenses (Leid et al. 2005, Leid et al. 2009).

8.1.3 Biofilms and Disease

Biofilms have now been implicated in many infectious diseases, including dental caries, periodontitis, otitis media, musculoskeletal infections, necrotizing fasciitis, biliary tract infection, osteomyelitis, bacterial prostatitis, native and prosthetic valve endocarditis, chronic urinary tract infections, cystic fibrosis pneumonia, and now chronic rhinosinusitis (see also the other clinical chapters of this book). Furthermore, nosocomial-type infections are caused by bacterial biofilms. These include ICU pneumonia, sutures, AV shunts, scleral buckles, contact lenses, urinary catheter cystitis, endotracheal tubes, Hickman catheters, central venous catheters, and pressure equalization tubes (Mah and O'Toole 2001, Ehrlich et al. 2002, Post et al. 2004). Although the idea of biofilm communities as the cause of diseases, especially chronic diseases, is still not widely accepted in the practice of medicine, the Centers for Disease Control and Prevention estimates that ~60–70% of all infections are biofilm related. As more focus is directed to microbes growing in the biofilm lifestyle, it is likely that other diseases will be indentified that clearly have biofilm links.

The widely accepted conceptual paradigm of a biofilm comes from a multitude of studies that have either looked at bacterial attachment to plastic or glass, under static or shear settings. In many instances, these communities have been described as having large towers or mushrooms extending away from the substratum with community populations ranging between 10^9 and 10^{12} organisms. However, documentation of such extensive communities in vivo from examination of infected tissue is less pictorially and numerically dramatic. Indeed, in the original description of *P. aeruginosa* biofilms in CF sputum by Costerton, the electron micrographs demonstrated clumps of *P. aeruginosa* bacteria containing between 20 and 50 organisms, not extensive communities that are often described from in vitro assays (Costerton et al. 1999, Klausen et al. 2006). When more physiological substrates are employed for in vitro characterization of biofilms, the communities are often much less extensive and mimic the clumps originally described by Costerton (Landry et al. 2006). Parsek's group demonstrated that *P. aeruginosa* growing on either bovine or human mucin under shear forces formed antibiotic resistant communities that phenotypically represented small clusters of organisms. We have demonstrated similar in vivo findings in a mouse model of endophthalmitis (Fig. 8.1, Leid et al. 2002a). These more physiological biofilms were quite different from the extensive phenotypic data of *P. aeruginosa* biofilms that collectively created the original conceptual paradigm of biofilm communities. Nonetheless, these extensive in vitro studies have pushed biofilm research to the forefront of medicine and have defined many important characteristics that are associated with microorganisms growing as heterogeneous communities.

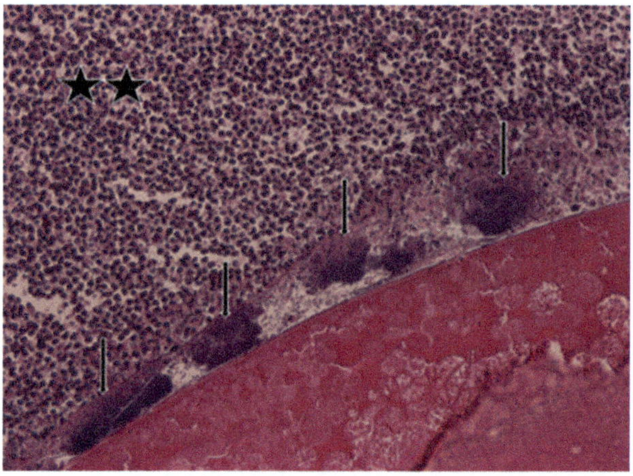

Fig. 8.1 Histograph of biofilm-mediated endophthlamitis. The *arrows* point to clusters of *Staphylococcus aureus* that are attached to the lens of the eye. The stars represent the inflammatory response to presence of the biofilm organisms

8.1.4 Head and Neck Biofilm-Related Diseases

A comprehensive introduction to biofilms in the Head and Neck was most recently reviewed in Current Opinion in Otolaryngology (Post et al. 2004). Because of that, we will not detail all of the biofilm diseases here but will highlight a few head and neck diseases that either exhibit a polymicrobial nature or inherent antimicrobial tolerance. Otitis media with effusion has been definitively shown to be associated with mucosal biofilms of known middle ear pathogens, such as *Haemophilus influenzae*, *Streptococcus pneumonia*, and *Moraxella catarrhalis* (Hall-Stoodley et al. 2006) (see also Chapter 3). Polymicrobial adenoid biofilms have been identified and associated sinusitis has been shown to be reduced in patients after undergoing adenoidectomy (Zuliani et al. 2006). Tonsillitis may at times be a biofilm process, mediated by the presence of multiple organisms, as are all infections involving tracheotomy and tympanostomy tubes and other implanted materials (Vlastarakos et al. 2007). Antibiotic otic drops, in an in vitro study of tympanostomy tubes containing *P. aeruginosa* biofilms, reduced colony forming units by day 5 through day 21. However the antibiotics did not halt progression of the biofilms (Oxley et al. 2007). All of these are examples of biofilm infections that are common to the head and neck, yet are extremely difficult to treat because of the inherent resistance that is associated with biofilm communities.

8.1.5 Factors Contributing to Biofilm Antibiotic Resistance

The inherent resistance of these communities to antibiotics and the host immune system is still being elucidated, although much more is known about antimicrobial resistance mechanisms (Mah and O'Toole 2001, Gilbert et al. 2002, Mah et al. 2003, Parsek and Singh 2003, Bagge et al. 2004, Leid et al. 2005, Ooi et al. 2008) (see also Chapter 13). As mentioned, biofilm-associated organisms have dramatically reduced susceptibility to various types of antimicrobial agents, either because the biofilm structure impedes transport of the agent to the cell surface or because the cells within the biofilm exhibit an altered physiology. A paper in the journal Nature correlated the importance of biofilm-specific gene expression and resistance to antibiotics (Mah et al. 2003). This paper was important because it was the first description of biofilm-specific genetic resistance mechanisms against a fairly broad antimicrobial spectrum. Our group has followed up on this study and demonstrated that biofilm-specific genes are also responsible for resistance to the host's immune defenses (Leid et al. 2005). Additionally, communication between the bacteria within the community dictate resistance to the host, and intricate studies have demonstrated that communication of the biofilm organisms can lead directly to leukocidal activity and bacterial persistence (Christensen et al. 2007, Jensen et al. 2007).

8.1.5.1 Antibiotic Resistance

Biofilms evade host defenses and demonstrate decreased susceptibility to systemic and local antibiotic therapy (Davies et al. 1998, Mah and O'Toole 2001). The exopolysaccharide alginate in *P. aeruginosa* could lead to decreased penetration of antibiotics into the biofilm. However, studies showing that antibiotics can diffuse efficiently into biofilms contradict this theory (Mah and O'Toole 2001). Because water comprises a large portion of the biofilm mass, this allows for diffusion of antibiotics down water channels into the core regions of the biofilm. Resistance could then be conferred by deactivating or neutralizing positively charged antibiotics interacting with the negatively charged polymers of the biofilm matrix. A third theory suggests that bacteria could lie in a non-growing state of suspended animation in the basal layers of the biofilm due to the accrual of waste products and depletion of needed substrates. This could confer relative resistance to antibiotics as most antibiotics work only on dividing bacteria. Finally, decreased diffusion of antibiotics into the bacterial cytoplasm due to fewer porins in the bacterial cell wall is another possible method of resistance. Fewer porins could develop as a stress response due to osmotic forces changing nutrient gradients. In reality, antibiotic resistance is probably a result of a combination of these mechanisms.

Aminoglycoside antibiotics may potentially induce biofilm formation in some bacteria at subtherapeutic doses. Hoffman and colleagues demonstrated the induction of biofilm formation in *P. aeruginosa* and *Escherichia coli* when exposed to subtherapeutic concentrations of these antibiotics (Hoffman et al. 2005). Certain Pseudomonads have a gene named the aminoglycoside response regulator (arr) that confers this biofilm-specific aminoglycoside resistance. The ubiquitous nature of this response in Pseudomonads as well as other bacteria is currently being studied (Hoffman et al. 2005, da Fonseca et al. 2008). As topical sinus irrigations with gentamicin or tobramycin are often prescribed in patients with CRS, this could be a potential source of bacterial biofilm development for *P. aeruginosa* and other microorganisms, especially at subtherapeutic concentrations. A recent paper by O'Toole and colleagues demonstrated that tobramycin, even at high doses, enhanced *P. aeruginosa* biofilm formation on cultured human cells (Anderson et al. 2008). Along these lines, Walker and colleagues demonstrated that live human neutrophils could enhance biofilm formation by serving as necrotic debris that *P. aeruginosa* utilized as scaffolding for biofilm formation (Walker et al. 2005, Parks et al. 2009). This idea of necrotic debris from the host response serving as a microniche for biofilm formation, combined with the antibiotic resistance seen in these communities, could explain the initial pathogenesis of many biofilm diseases, including chronic rhinosinusitis.

Biofilm development is a cyclical process involving initial attachment and mature biofilm formation followed by detachment and potential reseeding of other parts of an implanted medical device or perhaps other organs and tissues in the human body (Costerton et al. 1999). As bacteria transition from planktonic organisms to attached, multicellular communities, specific genes are differentially expressed that aid in biofilm formation and establishment of a community of microorganisms resistant to antibiotics and the human immune system (Sauer et al. 2002, Leid et al. 2002, 2005, 2009). The first step in development is initial, reversible attachment. This step is typically augmented by some conditioning of the attachment surface. During initial attachment, which occurs over seconds and minutes, the bacteria transition from reversible attachment to irreversible attachment and many biofilm specific genes and gene products are up-regulated within 6 h post attachment. Quorum sensing genes are turned on and result in further maturation of the biofilm under specific growth conditions or secretion of virulence factors that aid in developing the infectious nidus (Parsek and Singh 2003). As the biofilm transitions into a mature community with extensive three-dimensional architecture, exopolymeric substances are (matrix) produced (Hentzer et al. 2001, Nivens et al. 2001, Shirtliff et al. 2002). All of these steps, from initial attachment and biofilm maturation to detachment and reinfection, serve as distinct mechanisms against both antibiotic and human leukocyte killing. Studies of 14 strains of *S. pneumoniae*, a CRS-relevant pathogen like *P. aeruginosa*, clearly demonstrated the complexity of the mature biofilm structure. These strains were shown to differ in architecture, protein mass, and cell counts (Allegrucci et al. 2006), and document the important difference between clinical strains that cause disease and many of the laboratory strains that have lost both genetic diversity and pathogenic potential through serial passage (Jelsbak et al. 2007).

8.2 Biofilms and Chronic Rhinosinusitis: What Is the Evidence?

Based on the problems arising from recalcitrant infections in chronic rhinosinusitis and emerging research on biofilms in other chronic disease, multiple studies were undertaken to examine the role that biofilms might play in chronic rhinosinusitis (Post et al. 1996). In the first study of biofilms and CRS, Perloff and Palmer examined frontal sinus stents, placed during surgery, by scanning electron microscopy to demonstrate biofilms on the stents of six patients. Morphologic structures characteristic of biofilm growth including water channels, glycocalyx coatings, and a three-dimensional microcolonies on all six of the stents were evident by SEM, and all sinus cultures were positive for *P. aeruginosa.* Additionally, they examined sterile stents in vitro placed in inoculated media for 48 h, which showed the presence of biofilms (Perloff and Palmer 2004). Following this, cultures were taken from 16 consecutive patients and were assessed for biofilm presence on excised mucosa using SEM. All 16 demonstrated signs of infection including cilia loss, and 25% (4/16) demonstrated near total coverage of the ciliary surface and morpholic appearance of a biofilm (Cryer et al. 2004). These studies, including the ability of *P. aeruginosa* to form biofilms on living tissue, were confirmed in an animal model of infection using New Zealand White Rabbits (Perloff and Palmer 2005).

Additional studies have documented biofilm formation in samples taken from patients after functional endoscopic sinus surgery (FESS) by SEM. Ramadan et al. obtained intra-operative samples from the ethmoid bullae from five patients undergoing FESS, all of which demonstrated morphologic criteria of biofilms on SEM (Ramadan et al. 2005). A follow-up study by Sanclement et al. observed biofilms in 80% (24/30) of patients compared to 0/4 controls (Sanclement et al. 2005). In addition, six patient samples were examined by transmission electron microscopy and demonstrated bacterial structures on the mucosal surfaces, which correlated with biofilm structures on SEM in these patients. The limit of all of these studies was the lack of identification of the organisms that were seen as biofilms on the human sinus tissue.

Sanderson et al. used confocal laser scanning microscopy (CLSM) in combination with Fluorescence In Situ Hybridization (FISH) analysis to examine intra-operative samples taken from 18 patients with CRS and five controls undergoing septoplasty (Sanderson et al. 2006). The analysis found 78% (14/18) of patients with detectable bacteria in a biofilm matrix contained *H. influenzae.* They also found *S. pneumoniae*, and *S. aureus*, whereas *P. aeruginosa* was notably absent. Ferguson and Stolz utilized TEM in conjunction with bacterial cultures to demonstrate biofilms on 50% (2/4) of patient samples taken intra-operatively in presumed CRS, both of which grew out *P. aeruginosa.* The other two patients were discovered to have a non-bacterial etiology to the CRS (Ferguson and Stolz 2005). However, the limited patient sample size for this study makes it hard to determine the significance of these findings. Healy and colleagues followed up the initial FISH studies on CRS samples by demonstrating that fungal elements intermixed with *H. influenzae* biofilm communities in patients with allergic rhinosinusitis (Healy et al. 2008). Finally, Psaltis et al. used CLSM to demonstrate biofilms on mucosal biopsies from the middle meatus/ethmoid area in 45% (17/38) patients undergoing FESS for CRS, compared with 0% (0/9) controls undergoing endoscopic transsphenoidal hypophysectomy (Psaltis et al. 2007).

In the largest patient sample study to date, Prince et al. demonstrated that clinical isolates from CRS patients grew as biofilms in vitro using the Calgary Biofilm Assay. Of 157 samples, they noted a biofilm formation rate of 28.6% (Prince et al. 2008). However, it is unclear whether the in vitro CBA approach is an appropriate mimic for the in vivo environment that likely enhances biofilm formation for bacterial species. Speciation of the cultures demonstrated that *S. aureus* was the most commonly isolated organism (33%), but that 20% of patients had either *Pseudomonal* infection or polymicrobial infections containing *P. aeruginosa*. Additionally, this study linked the number of prior surgeries to an increased likelihood of harboring biofilm-forming strains of bacteria (Prince et al. 2008). Collectively, these studies outlined the role of biofilms in CRS and have led to many groups studying a specific organism in the context of CRS pathogenesis.

There is a complete lack of concordance between the studies above, which may be explained by a number of factors in this emerging field. These include differences in sites considered as specimen versus control, lack of universal definition of biofilm from mucosal tissue biopsies, inconsistencies in individual hospital microbiologic laboratory protocols, differences in technique for assessing biofilms (SEM vs. CSLM), and perhaps epidemiologic differences in the endemic microbial flora between regions. Nonetheless, the common theme is that biofilms are present, likely even in "healthy" sinus tissue, and that these biofilm communities are composed of different groups of microorganisms that either promote health or disease. Many examples exist in the literature demonstrating that biofilms of specific microorganisms have the ability to modulate the host immune response, including inflammation. It is quite possible that this situation exists in CRS and results in chronic inflammation. The importance of the polymicrobial nature of the disease is discussed below.

In an examination of the role of biofilms in disease progression, Coticchia and colleagues compared the mucosal surface area in pediatric patients undergoing adenoidectomy for chronic rhinosinusitis versus those undergoing surgery for obstructive sleep apnea (OSA). In the seven patients with CRS, SEM demonstrated biofilm covering greater than 90% of the adenoid mucosa, as opposed to a mean coverage of 1.9% in those children with OSA (Coticchia et al. 2007). Additional studies have shown that the benefit of adenoidectomy appears unrelated to adenoid size (Maw 1985, Gates 1994), but instead are derived from removal of the biofilm as a source of potential re-infection of the sinus mucosa. This fits with the model of biofilm-mediated chronic infection as described earlier.

With these finding in mind, additional reports have examined the post-operative outcomes for patients with biofilms and continue to suggest clinical relevance for biofilms in disease pathogenesis and persistence. Bendouah et al. followed 19 patients for 1 year post-FESS and patients were assessed for evolution of disease by presence or absence of symptoms, as well as endoscopic examination of the nasal cavities (Bendouah et al. 2006). Only 5/19 patients were deemed to have favorable evolution of symptoms following surgery. None of these patients had *P. aeruginosa* or *S. aureus* isolates that formed biofilms by in vitro culture assays, although 3/5 had biofilm-forming isolates of coagulase-negative staphylococci. Poor post-operative evolution was associated with *S. aureus* or *P. aeruginosa* biofilms; 13 of 14 patients with poor outcomes demonstrated either *S. aureus* or *P. aeruginosa* biofilms, or both, and the authors suggested that biofilms may play a role in the chronicity of the disease. However, no mucosal samples were taken to demonstrate the presence or absence of biofilms in situ in these patients.

In a pre-operative study by Psaltis et al., 40 patients undergoing FESS for CRS had mucosal samples taken intra-operatively. These samples were examined using confocal laser scanning microscopy and 50% were found to have biofilms, but the authors were unable to speciate these biofilms to determine which organisms were present. Although there were no objective differences in pre-operative symptom scores, patients with biofilms had significantly higher Lund–McKay scores then those without. However, after 8 months of follow-up, the biofilm-positive patients were more likely to have ongoing post-operative symptoms and endoscopic findings of continuing mucosal inflammation when compared to the patients in whom no biofilms were noted (Psaltis et al. 2008). These studies provide evidence that biofilms indeed play an active role in perpetuating inflammation in CRS patients and may explain the recurrent and resistant nature of this disease. Although further investigations are needed, persistent infectious biofilms are a likely contributing factor to medically recalcitrant CRS. A greater understanding of biofilm-associated CRS is required to develop novel therapies directed at prevention and eradication.

8.3 Etiology of CRS

CRS is most likely the manifestation of the interaction of multiple host and environmental factors suggesting that there may be genetic or epigenetic influences that predispose to disease. Environmental factors that have been proposed include viral, bacterial, and/or fungal colonization as well as exposure to inhaled substances, such as cigarette smoke or allergens. The most reported CRS-associated bacteria

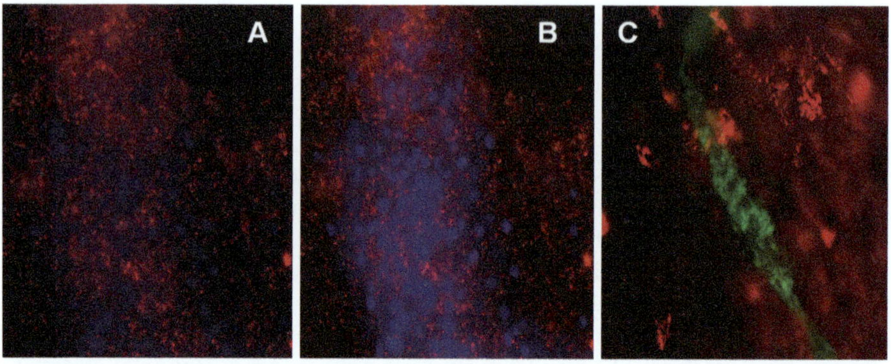

Fig. 8.2 Fluorescent micrographs of explanted sinonasal tissue from human chronic rhinosinusitis patients undergoing functional endoscopic sinus surgery. The *red spheres* are *Haemophilus influenzae* (FISH stain), the *blue* are eukaryotic nuclei (DAPI), and the *green* is a pan-fungal stain (**C**)

in the literature are *S. aureus*, *P. aeruginosa*, coagulase-negative Staphylococci, *S. pneumoniae*, and *Moraxella catarrhalis*. Using species-specific DNA probes, we recently reported the presence of *H. influenzae* in ~80% of CRS patient lesions (Sanderson et al. 2006, Fig. 8.2a,b). This report was the first to demonstrate the prevalence of *H. influenzae* in CRS-associated nasal tissues. Within the last half decade, much attention has also been directed toward the contribution of fungi (Hamilos and Lund 2004, Gosepath and Mann 2005), specifically Alternaria (Kennedy 2004) and Bipolaris (Buzina et al. 2003) or toxigenic *S. aureus* (Bernstein et al. 2003) in the sinonasal mucosa during development of polypoid disease. Subsequently, we have demonstrated the presence of fungi and *H. influenzae* in CRS lesions associated with allergic rhinitis (Healy et al. 2008, Fig. 8.2c). These initial studies began a paradigm shift in CRS.

8.4 Evidence that Chronic Rhinosinusitis Is a Polymicrobial Disease

As medical technology has advanced, the idea of a single organism causing a disease has become outdated. Sophisticated techniques have identified new groups of microorganisms in every tissue of the human body. The same is certainly true for the human sinus cavity. The use of molecular tools has clearly shown that standard microbial culture techniques only identify 10–40% of the microorganisms present in many diseases. This is confounded by the inherent attachment of these communities making them hard targets for standard clinical diagnostics (Veeh et al. 2003). Although not all of these microbes are likely responsible for the disease state, it is reasonable to assume that combinations of these microorganisms are. Yet, besides studies of dental microorganisms or the organisms in chronic wounds, very little knowledge of community-directed disease has evolved. Chronic Rhinosinusitis is the perfect opportunity to innovate the world of medicine by introducing, demonstrating, and resolving the importance of microbial communities in a disease that reduces the quality of life in a quarter of the human population. By understanding the importance of the community composition, and not just the presence of frank pathogens (e.g. *S. aureus*, *S. pneumonia*, and *H. influenzae*), we may be able to innovate the treatment strategies for disease by targeting specific pathogens for elimination while promoting the growth of others. We may even be able to turn the microbes against one another by modulating their communication systems. However, we must first identify the pathogens responsible for CRS. At its worst, this line of research will positively benefit the quality of life of millions of patients by resolving the etiology of CRS. At its best, it will revolutionize the practice of medicine by laying the groundwork for disease-specific, patient-specific treatment depending upon the microbes that are present and causing disease.

8.5 Biofilms and CRS – Making the Case for a Paradigm Shift

Biofilms have now been implicated in many infectious diseases, as outlined above (see the other clinical chapters of this book). Many of these diseases are polymicrobial in nature. As mentioned earlier, studies have identified biofilms in the sinus mucosa of patients with CRS (Cryer et al. 2004, Sanderson et al. 2006, Healy et al. 2008). Additionally, CRS has recently been classified into CRS with nasal polyps (CRSwNP) and CRSsNP (without). One possible explanation for the different clinical presentations of the disease may be found in the microbiological members of the offending biofilm community; i.e., heterogeneous biofilms containing fungal elements and/or multiple species could produce a polypoid inflammatory response (Fig. 8.2c above), while biofilms devoid of these microbes produce a non-polypoid mucosal inflammation.

Because medicine has relied on standard clinical microbiological techniques such as bacterial culture, there are currently only a handful of pathogens that have been characterized in CRS. However, many reports have demonstrated that biofilms, simply by their attached nature, are recalcitrant to standard, clinical diagnostic techniques (Veeh et al. 2003). Thus, patients with CRS offer a unique opportunity to investigate the relationship between microbial communities, the in vivo ecology, and the resultant disease state(s).

Although we, and others, have demonstrated the importance of biofilms in CRS, the collective scientific literature would also suggest that communities of microorganisms exist in healthy tissue throughout the human body. Therefore, the presence of a biofilm is likely not unique to any polymicrobial disease state. What *is* unique to the development of pathology is the *composition* of the microbial community, interacting through communication systems such as quorum sensing, that *collectively* cause disease. Communication between two common CRS pathogens has already been shown to enhance the virulence of one organism and alter the host response (Ratner et al. 2005). One of the best arguments for this community theory of disease in CRS, besides the preliminary data presented here, comes from recent publications in the Otorhinolaryngology literature. In the 2008 Jan–Feb issue of *American Journal of Rhinology*, as well as in recent issues of *Laryngoscope*, many authors suggested the importance of single, distinct microorganisms in CRS. Collectively, however, these reports challenge the current CRS paradigm and support the community theory we are proposing here. Most importantly, data from our group support this paradigm shift in the study of CRS as we clearly show distinct microbial communities in CRS vs. non-CRS tissue. Confirmation of these data will elucidate the critical targets for novel therapeutic strategies in the management of this common, costly, and debilitating disease. It is clear to our group that a major point being missed in current CRS studies is the critical polymicrobial combination(s) driving persistent mucosal inflammation leading to the initiation and progression to CRS.

In support of our paradigm-shifting hypothesis, we have utilized diverse molecular approaches to resolve the polymicrobial nature of CRS. The first technique we employed was Terminal Restriction Fragment Length Polymorphism (TRFLP). This molecular protocol has been utilized for some time in environmental microbiology, and we have successfully adapted this technique to CRS. For these studies, total DNA was isolated from the samples, 16S rDNA was amplified by universal primers, and the amplicons run on an Applied Biosystems ABI sequencer. The resultant TRFLP patterns were compared using relative fragment peak heights. NMDS analysis of TRFLP patterns was performed using PC-ORD4 software (MjM Software, Gleneden Beach OR) and cluster analysis was conducted with the same software using the Jaccard distance measure and Ward's method for linking groups. When CRS samples were compared to non-CRS samples by cluster analysis, the CRS microbial communities clustered together (Fig. 8.3, arrow) whereas the control tissue clustered in distinct clades (Fig. 8.3).

To further characterize the microbial communities in CRS, we employed Denaturing High Pressure Liquid Chromotography (DHPLC). In principle, the DHPLC technique provides a "signature" or fingerprint consisting of dozens of absorbance peaks, each of which correspond to a single bacterial species. The area under each peak corresponds to the amount of DNA from that species that is present in the population. By separating and quantifying these mixed samples, as well as through downstream collection and DNA sequencing, it is possible to quickly identify the bacteria in a given population as well as semi-quantitatively provide their relative abundance by peak height. For these studies, 16S rDNA was amplified from total DNA as above and the fragments from the heterogeneous population were injected into a heated HPLC column. These fragments were partially heat denatured. While the double stranded sections of the partially denatured PCR fragments stick tightly to the column, the single strands of the fragments reduce this binding strength. Elution buffer was then passed through the column at increasing concentrations, eluting the 16S rDNA amplicons off of the column depending not only upon the percent GC but also the relative nucleotide sequence order. This allowed the 16S fragments from each species to elute at distinct times and peaks. Once eluted off of the column, the species were detected by spectrometry and peaks were collected for immediate sequencing (see Fig. 8.4). Strikingly, the two

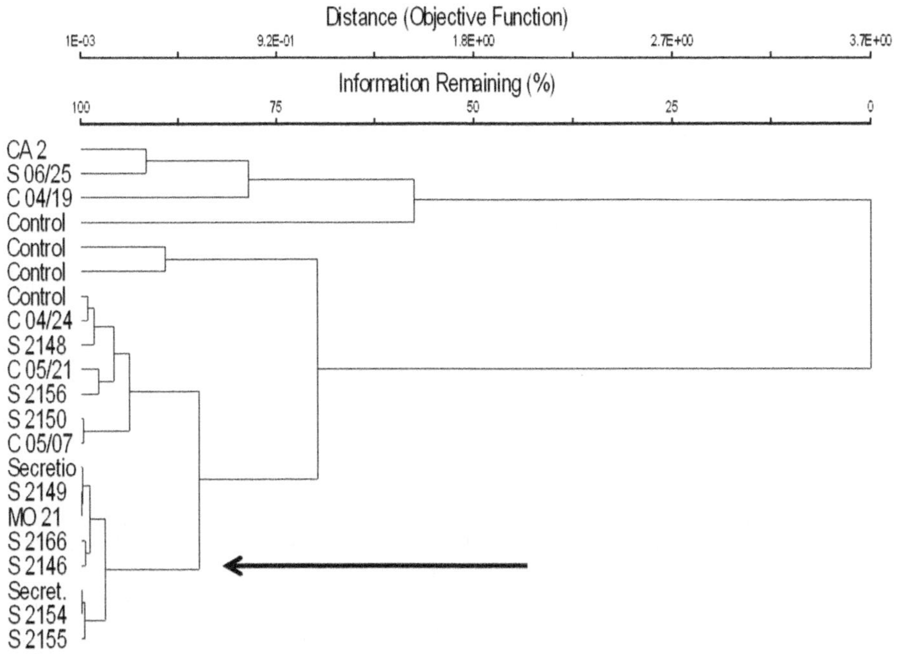

Fig. 8.3 TRFLP data demonstrating the clustering of control or CRS (*arrow*) microbial populations as characterized from explanted human sinonasal tissue

control patients shown in Fig. 8.4 had almost identical microbial profiles. In contrast, the microbial diversity in the CRS patients was large. Upon sequencing, we have identified microbes never before reported in CRS, including two species of Bordetella. These data confirm the TRFLP results and, along with 454 sequencing of the microbes in the CRS tissue which is ongoing in our group, will definitively resolve the microbial populations in CRS for the first time. Moreover, these data provide strong evidence for a new polymicrobial community driven paradigm for CRS with resultant and appropriate changes in patient care.

Many studies have shown multiple bacterial species on culture of CRS patients. In a recent Current Opinion in Microbiology, Wargo discusses the evidence for and the implications of the polymicrobial nature of these infections (Wargo and Hogan 2006). Using Fluorescent In Situ Hybridization (FISH), Confocal Scanning Laser Microscopy (CSLM), and explanted human tissues to examine sinus mucosa in patients undergoing Endoscopic Sinus Surgery (ESS) for CRS, the authors found no correlation between the bacteria in the biofilms in the tissue and the bacteria isolated on culture (Leid et al., unpublished). These same patterns have held no matter what molecular approaches we have utilized, including 454 pyrosequencing of human sinus tissue; that is, the molecular data very rarely match the clinical microbiology data obtained from culture. We are presently analyzing the microbial communities within healthy and CRS-associated sinus mucosa through a variety of molecular techniques. The results not only demonstrate a difference in microbial community

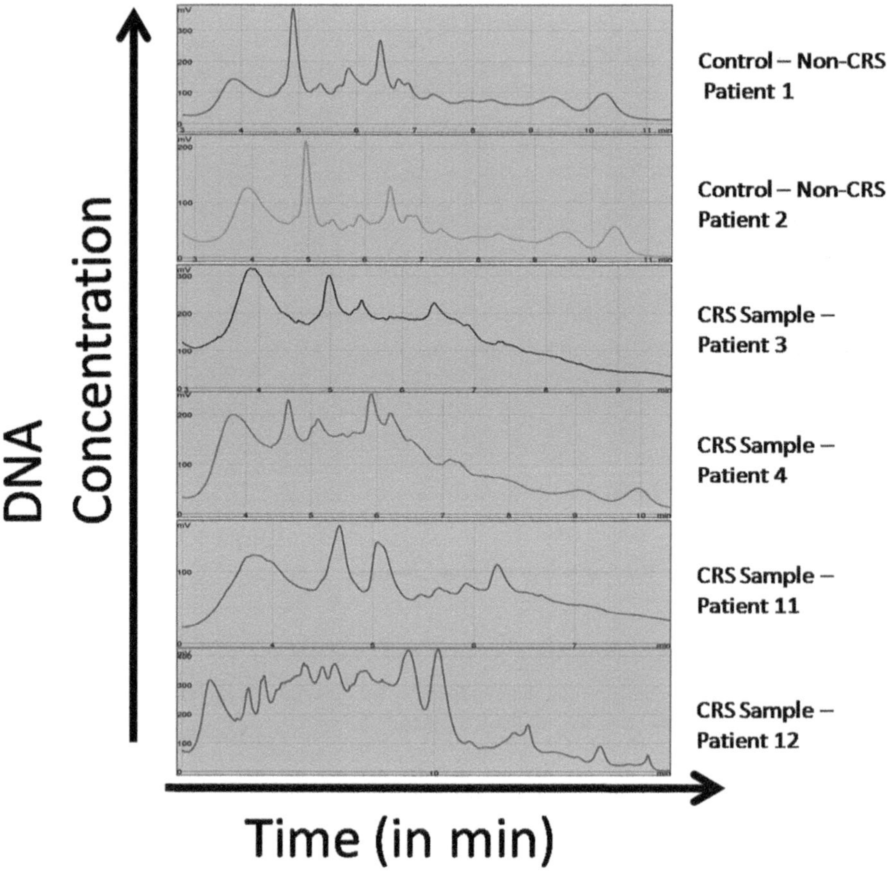

Fig. 8.4 DHPLC profiles of control and CRS patients. Each peak represents a unique bacterial species. Note that the bacterial species are fairly consistent among the control tissues, whereas the CRS patient microbial profiles are diverse

structure, but also solidify the growing evidence of the limitations of data that culture provides because it rarely correlates with the true microbial population in this disease (Psaltis et al. 2007)

8.6 Other Factors Contributing to Development of Biofilm on Rhinosinus Mucosa

It is well recognized that damaged mucosa, dry tissue, and foreign bodies provide excellent substrates for bacterial attachment, leading to biofilm formation. Recent studies by our group have demonstrated biofilm formation on what has been considered to be normal sinonasal mucosa. Environmental factors contributing to CRS were discussed above. Host factors implicated in the development of CRS include ciliary impairment (primary ciliary dyskinesia), mucus homeostasis dysfunction (cystic fibrosis), atopy, asthma, immuno compromised states (immunoglobulin deficiency or HIV), and paranasal sinus anatomic variations (Cohen and Kennedy 2005). The combination of environmental (microorganisms) and host factors results in the common histopathology finding of mucosal inflammation with the physiologic consequence of marked decrease in sinonasal mucociliary clearance. It is interesting to consider that the majority of these factors also contribute to ciliary damage or dysfunction. Is this relationship due to the contribution of these factors to adherence of bacteria (biofilm communities) to the sinus respiratory epithelium, thereby facilitating biofilm formation? There is still much to be learned from the study of explanted CRS tissue from humans.

8.7 Pathophysiology of the Biofilm Communities

Biofilms have a heterogeneous morphology, because the biofilm phenotype is highly dependent on the surrounding environment. An example of this heterogeneity is demonstrated with bacterial biofilms that form on mucosal surfaces, often referred to as a *mucosal biofilm* (Post et al. 2004). These bacterial biofilms exhibit unique cascades of gene expression when compared to biofilms that form on inert surfaces. Since they form in the special environment of ciliated mucosa, which has known antimicrobial characteristics, these results are interesting because ciliated mucosa is expected to have some protection from biofilm formation. Mucosal biofilms are modified by the host inflammatory response and may incorporate some of the host proteins, waste products, and cellular debris in their composition.

S. aureus and *P. aeruginosa* are notorious pathogens in both lower and upper airway disease. Both organisms are able to produce biofilms. *S. aureus* can produce exotoxins that are active as superantigens to specific immunity. Some believe the superantigens play a role in the development of CRS in certain individuals (Bachert et al. 2002, Tripathi et al. 2004). On the other hand, *P. aeruginosa* is a Gram-negative bacterium that is frequently associated with long-term respiratory diseases. Gram-negative bacterial CRS is particularly recalcitrant in nature. Gram-negative sinusitis, specifically *Pseudomonas*, has been studied extensively in the past and has been noted to cause an intense transmucosal injury far greater than experimental sinusitis using other bacteria associated with sinusitis, such as *S. pneumoniae* (Bolger et al. 1997). However, these results may have been more linked to the choice of the animal model than the role of these pathogens in CRS. Persistent CRS disease with either of these infectious organisms that is recalcitrant to antibiotic therapy is nicely explained by the presence of bacterial biofilms.

8.8 Current Paradigms in the Medical and Surgical Management of CRS

The definition of maximal medical therapy for the management of chronic rhinosinusitis has not been standardized. The medications used, timing, and doses vary from physician to physician and across the specialties treating the disease. There are no published guidelines or FDA approved medications for CRS, most likely a result of the undefined disease etiology. What is uniformly accepted is the general principle of maximal medical therapy to promote ventilation of obstructed sinus and drainage of stagnant secretions, by decreasing mucosal inflammatory responses and eliminating inciting bacterial and/or fungal infection. A common treatment regimen consists of a prolonged course of a broad spectrum antibiotic, oral steroid, nasal saline irrigations, and nasal steroid sprays +/- nasal antihistamine sprays based on evidence of atopy. The choice of antibiotic is ideally guided by endoscopic culture of the middle meatus. When one is not attainable, a broad-spectrum antibiotic that covers Gram-positive, Gram-negative, and anaerobic microorganisms is chosen. The exact duration of therapy varies from physician to physician, but patients are commonly treated with an initial three-week course, augmented by an additional three weeks of therapy for sub-optimal response. The course of antibiotics is accompanied by a three-week tapering course of oral steroids. After the initial three weeks, a repeat CT scan and nasal endoscopy is performed. If there is symptomatic improvement, but still significant findings of rhinosinusitis radiographically or endoscopically, an additional three weeks of antibiotics and/or oral steroids may be prescribed. If there is no symptomatic or clinical improvement, the patient is given the option of a longer course of medical therapy or surgical intervention.

The surgical management of CRS and nasal polyposis has evolved over the years. External facial incisions, extensive nasal packing, and prolonged hospital stays have been replaced by outpatient surgery, endoscopic techniques, minimal packing, and computer aided technology that have stretched the limits of sinus surgery. As a result, functional endoscopic sinus surgery (FESS) has become a popular adjunct to the medical management of chronic rhinosinusitis. Utilizing mucosa sparing techniques, the extirpation of sinus cells and the creation and maintenance of patient sinuses have demonstrated excellent long term results through subjective and objective measures (Senior et al. 1998). By employing these therapeutic approaches, many patients with CRS demonstrate improvement following appropriate medical and surgical therapies. However, a subpopulation of symptomatic patients exists with recalcitrant CRS. With its poorly defined etiology, the current standard of care for CRS is being administered in an imperfect fashion. Resolving the microbial communities in CRS, and linking those communities to patient treatment outcomes, will have a real and dramatic impact on patient health.

8.9 Treatments

A number of techniques have been evaluated for their capability to manage and control biofilms in environmental science. Materials and coatings to help reduce initial cell adhesion to surfaces and a variety of treatments aimed at decreasing or destroying already formed biofilms have been evaluated. These include heat, chemical treatments, antibiotics, sonication, quorum-sensing analogs, cleaning regimens, low-power laser therapy, and lectins (Oulahal-Lagsir et al. 2000, Hammer and Bassler 2003, Nandakumar et al. 2004, Parkar et al. 2004, Sheehan et al. 2004, Woodworth et al. 2007). Furthermore, new investigations into biological control agents, such as bacteriophages and protozoa, have shown promise although they are still years away from clinical application (Kadouri and O'Toole 2005). Many of these treatments are prohibitive in humans due to detrimental effects on host cells. The pursuit of a reliable method for the elimination of human biofilm infections is ongoing and likely the holy grail of biofilm research (see also Chapter 14).

Although CRS may have many independent inciting factors, including bacterial infection (whether planktonic or biofilm-mediated), genetics, reactive airways, anatomic abnormality, fungal infection, and allergy, the mainstays of therapy remain the same: anti-inflammatory and antimicrobial agents combined with surgical ventilation. Even though FESS has a high success rate, there are large population of patients that progress back to CRS. Other treatments for biofilms include novel methods of antibiotic therapy. Investigators have demonstrated that low dose macrolide therapy at levels far below the established minimal inhibitory concentration for *Pseudomonas* can decrease biofilm formation (Gillis and Iglewski 2004, Wozniak and Keyser 2004). However, the underlying mechanism behind this decrease has yet to be elucidated. Additionally, studies also demonstrate that antibiotic challenge can actually *increase* biofilm formation in *P. aeruginosa*, thus bringing into question the effective use of antibiotics in therapy (Anderson et al. 2008). Indeed, recent publications suggest that antimicrobial therapy of CRS was unwarranted because of the lack of success of treatment.

Topical saline irrigations are often utilized as a mechanical debridement of the mucosal surface following sinus surgery. Chemical surfactants, such as detergents, have anti-microbial activity by disruption of bacterial cell walls. Baby shampoo is an inexpensive, commercially available solution containing multiple chemical surfactants. Our prior studies demonstrated an in vitro anti-biofilm effect on *Pseudomonas* biofilm formation using a 1% solution of baby shampoo (Chiu et al. 2008). We subsequently studied its effects in a prospective study of symptomatic post-functional endoscopic sinus surgery (FESS) patients who irrigated twice a day for 4 weeks. Sixty percent of patients noted improvement in specific symptoms of thickened mucus and postnasal drainage with the formulation. Other surfactant-containing agents are currently under investigation including a combination of citric acid and zwitterionic surfactant. One study demonstrated significant reductions in biofilms on human chronic sinusitis mucosal specimens with hydrodynamic administration of this solution (Desrosiers et al. 2007).

Future directions for biofilm-associated CRS include investigations into the nature of the biofilm at the molecular and cellular levels. Molecular targets of specific aspects of the biofilm lifecycle continue to show promise. Disrupting the type IV *pili* attachment phases of *Pseudomonas* is one potential target of ongoing research (Gallant et al. 2005). Disrupting quorum sensing could be the most specific and unique target for biofilm therapeutics. A variety of novel mechanisms, including the substitution of furanones and the enzymatic cleavage of acyl-homoserine lactones (one of the quorum sensing signals), can interfere with quorum sensing (Koch et al. 2005). Targeting quorum-sensing signals at the molecular level is an area of continuing research and has potential for biofilm interventions and eradication (Dong et al. 2001). However, multiple lines of investigation should be considered in order to maximize the development of biofilm-specific treatments that show promise in animal models of chronic disease.

8.10 Scientific Challenge/Importance of CRS – Challenge to an Existing Paradigm

While multiple etiologies, including host and environmental factors, contribute to the development of CRS, a universal histopathologic finding is uncharacterized mucosal inflammation. Regardless of the respective causes of CRS, clinicians are left with a significant patient population with persistent and recurrent inflammation and infections that are recalcitrant to antimicrobial or surgical therapy. These patients have chronic sinonasal irritation with a dramatic loss in their quality of life. Many studies have investigated CRS-associated local (sinonasal) and systemic inflammation in the context of specific bacterial and fungal pathogens. The current paradigm of CRS is that the presence of a single pathogen, such as *Staphylococcus aureus*, *P. aeruginosa*, or *Streptococcus pneumonia*, leads to inflammatory events which over time result in chronic disease. To innovate the approach to clinical treatment of CRS, scientists must begin to focus on the polymicrobial nature of the disease. A unique approach to understand this disease may be based on the microbial ideals of ecology, versus the standard approach of studying virulence factors or single genes of single organisms. This new direction in CRS research, and the role of polymicrobial biofilms, will bring fundamental changes in patient treatment and improved patient health.

8.11 Conclusions

Since their initial characterization over four decades ago, biofilms have become a major focus of study for many chronic diseases. Because of their inherent resistance, both to antimicrobial agents and to components of the host's immune system, they often initiate inflammatory responses that ultimately destroy healthy tissue. Although these communities have been associated with a number of diseases in humans, only recently this concept has been applied to chronic rhinosinusitis. As with most biofilm-mediated diseases, the original observations were of microbial communities attached to human cells, visualized by electron microscopy. These early studies were paramount to the study of CRS because the biofilm phenotype explained much of the disease pathology. However, the etiology of the disease remains undefined, and because of this, patient treatment regimes are incomplete at best. Functional endoscopic sinus surgery, where biofilm-containing tissue is removed, has provided a great deal of clinical relief to patients, but this therapy is not effective for approximately 30% of CRS patients. In order to improve patient treatment, the microorganisms, living as biofilm communities, must be defined. Utilizing microbial ecological approaches, as well as state-of-the-art molecular techniques, it is clear that CRS is a polymicrobial disease. Once the offending microbes are elucidated, it may be possible to either drive the selection and expansion of specific microbes through quorum sensing analogs (for example) or develop selective means of bacterial-specific killing (for example, bacteriophage therapy). As the data supporting the contribution of biofilms to the persistence of CRS build, it becomes more evident that novel anti-biofilm therapies must be developed.

References

Allegrucci M, Hu FZ et al (2006) Phenotypic characterization of Streptococcus pneumoniae biofilm development. J Bacteriol 188(7):2325–2235

Anderson GG, Moreau-Marquis S et al (2008) In vitro analysis of tobramycin-treated Pseudomonas aeruginosa biofilms on cystic fibrosis-derived airway epithelial cells. Infect Immun 76(4):1423–1433

Bachert C, Gevaert P et al (2002) Staphylococcus aureus superantigens and airway disease. Curr Allergy Asthma Rep 2(3):252–258

Bagge N, Hentzer M et al (2004) Dynamics and spatial distribution of beta-lactamase expression in Pseudomonas aeruginosa biofilms. Antimicrob Agents Chemother 48(4): 1168–1174

Becker P, Hufnagle W et al (2001) Detection of differential gene expression in biofilm-forming versus planktonic populations of Staphylococcus aureus using micro-representational-difference analysis. Appl Environ Microbiol 67(7):2958–2965

Bendouah Z, Barbeau J et al (2006) Biofilm formation by Staphylococcus aureus and Pseudomonas aeruginosa is associated with an unfavorable evolution after surgery for chronic sinusitis and nasal polyposis. Otolaryngol Head Neck Surg 134(6):991–996

Bernstein JM, Ballow M et al (2003) A superantigen hypothesis for the pathogenesis of chronic hyperplastic sinusitis with massive nasal polyposis. Am J Rhinol 17(6):321–326

Bolger WE, Leonard D et al (1997) Gram negative sinusitis: a bacteriologic and histologic study in rabbits. Am J Rhinol 11(1):15–25

Bollinger N, Hassett DJ et al (2001) Gene expression in Pseudomonas aeruginosa: evidence of iron override effects on quorum sensing and biofilm-specific gene regulation. J Bacteriol 183(6):1990–1996

Buzina W, Braun H et al (2003) Bipolaris spicifera causes fungus balls of the sinuses and triggers polypoid chronic rhinosinusitis in an immunocompetent patient. J Clin Microbiol 41(10): 4885–4887

Chiu AG, Palmer JN et al (2008) Baby shampoo nasal irrigations for the symptomatic post-functional endoscopic sinus surgery patient. Am J Rhinol 22(1):34–37

Christensen LD, Moser C et al (2007) Impact of Pseudomonas aeruginosa quorum sensing on biofilm persistence in an in vivo intraperitoneal foreign-body infection model. Microbiology 153(Pt 7):2312–2320

Cohen NA, Kennedy DW (2005) Endoscopic sinus surgery: where we are-and where we're going. Curr Opin Otolaryngol Head Neck Surg 13(1):32–38

Costerton JW, Stewart PS et al (1999) Bacterial biofilms: a common cause of persistent infections. Science 284(5418):1318–1322

Coticchia J, Zuliani G et al (2007) Biofilm surface area in the pediatric nasopharynx: Chronic rhinosinusitis vs obstructive sleep apnea. Arch Otolaryngol Head Neck Surg 133(2): 110–114

Cryer J, Schipor I et al (2004) Evidence of bacterial biofilms in human chronic sinusitis. ORL J Otorhinolaryngol Relat Spec 66(3):155–158

da Fonseca EL, Freitas Fdos S et al (2008) Detection of new arr-4 and arr-5 gene cassettes in clinical Pseudomonas aeruginosa and Klebsiella pneumoniae strains from Brazil. Antimicrob Agents Chemother 52(5):1865–1867
Davies DG, Parsek MR et al (1998) The involvement of cell-to-cell signals in the development of a bacterial biofilm. Science 280(5361):295–298
Desrosiers M, Myntti M et al (2007) Methods for removing bacterial biofilms: in vitro study using clinical chronic rhinosinusitis specimens. Am J Rhinol 21(5):527–532
Dong YH, Wang LH et al (2001) Quenching quorum-sensing-dependent bacterial infection by an N-acyl homoserine lactonase. Nature 411(6839):813–817
Donlan RM (2000) Role of biofilms in antimicrobial resistance. ASAIO J 46(6):S47–52
Donlan RM (2002) Biofilms: microbial life on surfaces. Emerg Infect Dis 8(9):881–890
Donlan RM, Costerton JW (2002) Biofilms: survival mechanisms of clinically relevant microorganisms. Clin Microbiol Rev 15(2):167–193
Ehrlich GD, Veeh R et al (2002) Mucosal biofilm formation on middle-ear mucosa in the chinchilla model of otitis media. JAMA 287(13):1710–1715
Ferguson BJ, Stolz DB (2005) Demonstration of biofilm in human bacterial chronic rhinosinusitis. Am J Rhinol 19(5):452–457
Gallant CV, Daniels C et al (2005) Common beta-lactamases inhibit bacterial biofilm formation. Mol Microbiol 58(4):1012–1024
Gates GA (1994) Adenoidectomy for otitis media with effusion. Ann Otol Rhinol Laryngol Suppl 163:54–58
Gilbert P, Maira-Litran T et al (2002) The physiology and collective recalcitrance of microbial biofilm communities. Adv Microb Physiol 46:202–256
Gillis RJ, Iglewski BH (2004) Azithromycin retards Pseudomonas aeruginosa biofilm formation. J Clin Microbiol 42(12):5842–5845
Gliklich RE, Metson R (1995) The health impact of chronic sinusitis in patients seeking otolaryngologic care. Otolaryngol Head Neck Surg 113(1):104–109
Gosepath J, Mann WJ (2005) Role of fungus in eosinophilic sinusitis. Curr Opin Otolaryngol Head Neck Surg 13(1):9–13
Ha KR, Psaltis AJ et al (2007) A sheep model for the study of biofilms in rhinosinusitis. Am J Rhinol 21(3):339–345
Hall-Stoodley L, Hu FZ et al (2006) Direct detection of bacterial biofilms on the middle-ear mucosa of children with chronic otitis media. JAMA 296(2):202–211
Hamilos DL, Lund VJ (2004) Etiology of chronic rhinosinusitis: the role of fungus. Ann Otol Rhinol Laryngol Suppl 193:27–31
Hammer BK, Bassler BL (2003) Quorum sensing controls biofilm formation in Vibrio cholerae. Mol Microbiol 50(1):101–104
Head NE, Yu H (2004) Cross-sectional analysis of clinical and environmental isolates of Pseudomonas aeruginosa: biofilm formation, virulence, and genome diversity. Infect Immun 72(1):133–144
Healy DY, Leid JG et al (2008) Biofilms with fungi in chronic rhinosinusitis. Otolaryngol Head Neck Surg 138(5):641–647
Hentzer M, Teitzel GM et al (2001) Alginate overproduction affects Pseudomonas aeruginosa biofilm structure and function. J Bacteriol 183(18):5395–5401
Hoffman LR, D'Argenio DA et al (2005) Aminoglycoside antibiotics induce bacterial biofilm formation. Nature 436(7054):1171–1175
Hoiby N, Krogh Johansen H et al (2001) Pseudomonas aeruginosa and the in vitro and in vivo biofilm mode of growth. Microbes Infect 3(1):23–35
Jelsbak L, Johansen HK et al (2007) Molecular epidemiology and dynamics of Pseudomonas aeruginosa populations in lungs of cystic fibrosis patients. Infect Immun 75(5):2214–2224
Jensen PO, Bjarnsholt T et al (2007) Rapid necrotic killing of polymorphonuclear leukocytes is caused by quorum-sensing-controlled production of rhamnolipid by Pseudomonas aeruginosa. Microbiology 153(Pt 5):1329–1338

Kadouri D, O'Toole GA (2005) Susceptibility of biofilms to Bdellovibrio bacteriovorus attack. Appl Environ Microbiol 71(7):4044–4051

Kennedy DW (2004) Pathogenesis of chronic rhinosinusitis. Ann Otol Rhinol Laryngol Suppl 193:6–9

Klausen M, Gjermansen M et al (2006) Dynamics of development and dispersal in sessile microbial communities: examples from Pseudomonas aeruginosa and Pseudomonas putida model biofilms. FEMS Microbiol Lett 261(1):1–11

Koch B, Liljefors T et al (2005) The LuxR receptor: the sites of interaction with quorum-sensing signals and inhibitors. Microbiology 151(Pt 11):3589–3602

Landry RM, An D et al (2006) Mucin-Pseudomonas aeruginosa interactions promote biofilm formation and antibiotic resistance. Mol Microbiol 59(1):142–151

Leid JG, Costerton JW et al (2002) Immunology of Staphylococcal biofilm infections in the eye: new tools to study biofilm endophthalmitis. DNA Cell Biol 21(5–6):405–413

Leid JG, Shirtliff ME et al (2002) Human leukocytes adhere to, penetrate, and respond to Staphylococcus aureus biofilms. Infect Immun 70(11):6339–6345

Leid JG, Willson CJ et al (2005) The exopolysaccharide alginate protects Pseudomonas aeruginosa biofilm bacteria from IFN-gamma-mediated macrophage killing. J Immunol 175(11): 7512–7518

Leid JG, Kerr M et al (2009) Flagellar-mediated biofilm defense mechanisms of *Pseudomonas aeruginosa* against host derived lactoferrin. Infection and Immunity 77:4559–4566

Mah TF, O'Toole GA (2001) Mechanisms of biofilm resistance to antimicrobial agents. Trends Microbiol 9(1):34–39

Mah TF, Pitts B et al (2003) A genetic basis for Pseudomonas aeruginosa biofilm antibiotic resistance. Nature 426(6964):306–310

Maw AR (1985) Age and adenoid size in relation to adenoidectomy in otitis media with effusion. Am J Otolaryngol 6(3):245–248

Nandakumar K, Obika H et al (2004) Recolonization of laser-ablated bacterial biofilm. Biotechnol Bioeng 85(2):185–189

Nivens DE, Ohman DE et al (2001) Role of alginate and its O acetylation in formation of Pseudomonas aeruginosa microcolonies and biofilms. J Bacteriol 183(3):1047–1057

O'Toole G, Kaplan HB et al (2000) Biofilm formation as microbial development. Annu Rev Microbiol 54:49–79

Ooi EH, Wormald PJ et al (2008) Innate immunity in the paranasal sinuses: a review of nasal host defenses. Am J Rhinol 22(1):13–19

Oulahal-Lagsir N, Martial-Gros A et al (2000) Ultrasonic methodology coupled to ATP bioluminescence for the non-invasive detection of fouling in food processing equipment–validation and application to a dairy factory. J Appl Microbiol 89(3):433–441

Oxley KS, Thomas JG et al (2007) Effect of ototopical medications on tympanostomy tube biofilms. Laryngoscope 117(10):1819–1824

Parkar SG, Flint SH et al (2004) Evaluation of the effect of cleaning regimes on biofilms of thermophilic bacilli on stainless steel. J Appl Microbiol 96(1):110–116

Parks QM, Young RL et al (2009) Neutrophil enhancement of Pseudomonas aeruginosa biofilm development: human F-actin and DNA as targets for therapy. J Med Microbiol 58(Pt 4): 492–502

Parsek MR, Singh PK (2003) Bacterial biofilms: an emerging link to disease pathogenesis. Annu Rev Microbiol 57:677–701

Perloff JR, Palmer JN (2004) Evidence of bacterial biofilms on frontal recess stents in patients with chronic rhinosinusitis. Am J Rhinol 18(6):377–380

Perloff JR, Palmer JN (2005) Evidence of bacterial biofilms in a rabbit model of sinusitis. Am J Rhinol 19(1):1–6

Post JC, Aul JJ et al (1996) PCR-based detection of bacterial DNA after antimicrobial treatment is indicative of persistent, viable bacteria in the chinchilla model of otitis media. Am J Otolaryngol 17(2):106–111

Post JC, Stoodley P et al (2004) The role of biofilms in otolaryngologic infections. Curr Opin Otolaryngol Head Neck Surg 12(3):185–190
Prince AA, Steiger JD et al (2008) Prevalence of biofilm-forming bacteria in chronic rhinosinusitis. Am J Rhinol 22(3):239–245
Psaltis AJ, Ha KR et al (2007) Confocal scanning laser microscopy evidence of biofilms in patients with chronic rhinosinusitis. Laryngoscope 117(7):1302–1306
Psaltis AJ, Weitzel EK et al (2008) The effect of bacterial biofilms on post-sinus surgical outcomes. Am J Rhinol 22(1):1–6
Ramadan HH, Sanclement JA et al (2005) Chronic rhinosinusitis and biofilms. Otolaryngol Head Neck Surg 132(3):414–417
Ratner AJ, Lysenko ES et al (2005) Synergistic proinflammatory responses induced by polymicrobial colonization of epithelial surfaces. Proc Natl Acad Sci USA 102(9):3429–3434
Ren D, Bedzyk LA et al (2004) Gene expression in Escherichia coli biofilms. Appl Microbiol Biotechnol 64(4):515–524
Sanclement JA, Webster P et al (2005) Bacterial biofilms in surgical specimens of patients with chronic rhinosinusitis. Laryngoscope 115(4):578–582
Sanderson AR, Leid JG et al (2006) Bacterial biofilms on the sinus mucosa of human subjects with chronic rhinosinusitis. Laryngoscope 116(7):1121–1126
Sauer K, Camper AK et al (2002) Pseudomonas aeruginosa displays multiple phenotypes during development as a biofilm. J Bacteriol 184(4):1140–1154
Senior BA, Kennedy DW et al (1998) Long-term results of functional endoscopic sinus surgery. Laryngoscope 108(2):151–157
Sheehan E, McKenna J et al (2004) Adhesion of Staphylococcus to orthopaedic metals, an in vivo study. J Orthop Res 22(1):39–43
Shirtliff ME, Mader JT et al (2002) Molecular interactions in biofilms. Chem Biol 9(8):859–871
Smith RS, Iglewski BH (2003) P. aeruginosa quorum-sensing systems and virulence. Curr Opin Microbiol 6(1):56–60
Stewart PS, Costerton JW (2001) Antibiotic resistance of bacteria in biofilms. Lancet 358(9276):135–138
Tripathi A, Conley DB et al (2004) Immunoglobulin E to staphylococcal and streptococcal toxins in patients with chronic sinusitis/nasal polyposis. Laryngoscope 114(10):1822–1826
Vallet I, Olson JW et al (2001) The chaperone/usher pathways of Pseudomonas aeruginosa: identification of fimbrial gene clusters (cup) and their involvement in biofilm formation. Proc Natl Acad Sci USA 98(12):6911–6916
Veeh RH, Shirtliff ME et al (2003) Detection of Staphylococcus aureus biofilm on tampons and menses components. J Infect Dis 188(4):519–530
Vlastarakos PV, Nikolopoulos TP et al (2007) Biofilms in ear, nose, and throat infections: how important are they? Laryngoscope 117(4):668–673
Walker TS, Tomlin KL et al (2005) Enhanced Pseudomonas aeruginosa biofilm development mediated by human neutrophils. Infect Immun 73(6):3693–3701
Wargo MJ, Hogan DA (2006) Fungal – bacterial interactions: a mixed bag of mingling microbes. Curr Opin Microbiol 9(4):359–364
Webb JS, Givskov M et al (2003) Bacterial biofilms: prokaryotic adventures in multicellularity. Curr Opin Microbiol 6(6):578–585
Whiteley M, Bangera MG et al (2001) Gene expression in Pseudomonas aeruginosa biofilms. Nature 413(6858):860–864
Woodworth BA, Antunes MB et al (2007) Murine tracheal and nasal septal epithelium for air-liquid interface cultures: a comparative study. Am J Rhinol 21(5):533–537
Wozniak DJ, Keyser R (2004) Effects of subinhibitory concentrations of macrolide antibiotics on Pseudomonas aeruginosa. Chest 125(2 Suppl):62S–69S; quiz 69S
Zuliani G, Carron M et al (2006) Identification of adenoid biofilms in chronic rhinosinusitis. Int J Pediatr Otorhinolaryngol 70(9):1613–1617

Chapter 9
Helicobacter pylori and Biofilm Formation

Lone Rasmussen and Leif Percival Andersen

9.1 Introduction

Helicobacter pylori is a Gram-negative, microaerophilic curved rod, that possesses an unipolar bundle of two to six flagella which enables it to move to a specific target with high concentration of chemoattractants, such as urea. *H. pylori* is a pathogenic bacterium, which is present in more than half of the population worldwide with a range from 20 to 30% in developed countries to 60–80% in developing countries. An infection with *H. pylori* may be asymptomatic or cause chronic gastritis, gastric, or duodenal ulcer, and it plays an important role in MALT-lymphoma and gastric cancer (Yoshiyama et al. 1999, Andersen and Wadström 2001, Amieva and El-Omar 2008, Kandulski et al. 2008). The transmission route of *H. pylori* is either oral–oral or fecal–oral and varies with geography. It is commonly believed that *H. pylori* is mainly transmitted from parents to children in developing countries whereas drinking water plays a major role in developing countries (Bruce and Maaroos 2008).

H. pylori occurs in three stages: viable spiral forms that are culturable, virulent, and can infect and induce inflammation in experimental animals; viable coccoid forms that are non-culturable, less virulent and less likely to colonize and induce inflammation in experimental animals; and a third form that consists of non-viable degenerative forms of dying *H. pylori* (Willén et al. 2000, Andersen and Rasmussen 2009).

The colonization of the gastric mucosa by *H. pylori* is dependent on at least four different basic characteristics: urease, shape, motility, and adhesins. The urease is produced by the bacteria and provides a pH-neutral microenvironment around it. This enables *H. pylori* to survive and multiply in the stomach (Marshall et al. 1990). The flagella features *H. pylori* with motility, which is an important factor when moving to and penetrating the mucin layer (Ottemann and Lowenthal 2002). Also

L.P. Andersen (✉)
Department of Infections Control 9101, Copenhagen University Hospital, Rigshospitalet, Juliane Maries Vej 18, DK-2100, Copenhagen Ø, Denmark
e-mail: lpa@rh.dk

T. Bjarnsholt et al. (eds.), *Biofilm Infections*, DOI 10.1007/978-1-4419-6084-9_9,

the spiral shape of the bacteria plays an important role when penetrating the mucin layer (Lambert et al. 1995). Once the mucin layer is penetrated, the adhesins are necessary for the bacteria to adhere to the gastric epithelial cells (Andersen 2007). There is a narrow interaction between *H. pylori* and gastric epithelial cells by cell signaling. Most well known is the Cag A protein which passes from the bacteria to the epithelial cell and among others increases filamentous material and causes alterations in the cytoskeleton of the epithelial cells (Schneider et al. 2008).

9.2 Biofilm Formation by *H. pylori*

9.2.1 Environmental

Certain environmental strains of *H. pylori* are able to form biofilm in laboratory experiments whereas others do not form biofilm (Cellini et al. 2008). They seem to convert to coccoid forms rather quickly in water supply systems. However, epidemiological studies agree that use of river or well water as drinking water is a risk factor for *H. pylori* infection compared to tap water which indicates that *H. pylori* in some long living forms are present in environmental sources (Nurgalieva et al. 2002, Karita et al. 2003, Ahmed et al. 2007).

9.2.2 Dental Plaque

H. pylori can be cultured from gastric juice from dyspeptic patients and these floating forms of *H. pylori* may be transmitted to the oral cavity by reflux or vomiting (Andersen et al. 1988, Young et al. 2000). Dental plaques in the human oral cavity have been found to harbor *H. pylori*. Electron microscopic studies have shown that both spiral and viable coccoid forms of *H. pylori* may be present in the oral cavity even though they were not cultured (Young et al. 2001). Several studies using PCR as a detection method were able to detect *H. pylori* in the mouth both in patients with a history of gastric symptoms and in patients without any history of gastric symptoms (Teoman et al. 2007, Souto and Colombo 2008). One study found *H. pylori* in subgingival biofilm in 11% of periodontally healthy patients versus 50% of patients suffering from periodontitis (Souto and Colombo 2008). Another study found *H. pylori* in the oral cavity of 60% (18/30) of patients 3 months after triple antibiotic therapy, but in the stomach in only 10% (3/30) of the same patients (Gebara et al. 2006). This may indicate that *H. pylori* survive better in the oral cavity, which may be due to some kind of protective effect by the biofilm. It is commonly believed that there may be a correlation between the prevalence of *H. pylori* found in dental plaque and the oral health status even though conclusive studies are still missing (see also Chapter 4).

9.2.3 Gastric Mucosa

Only few studies have investigated the ability of *H. pylori* to form biofilm in the gastric mucosa (Nurgalieva et al. 2002, Karita et al. 2003, Ahmed et al. 2007). All three studies agree on *H. pylori* being able to form biofilm in the gastric mucosa (see Figs. 9.1 and 9.2). Biofilms were present in *H. pylori*-positive patients and absent in *H. pylori*-negative controls. Biofilm may facilitate survival both in vivo and during transmission (Carron et al. 2006). Cottichia et al. found high biofilm density (97.3% average surface coverage) in urease-positive patients and a relative paucity of biofilm (1.6% average surface coverage) in urease negative-patients (p=0.0001). These data suggest that biofilm formation by *H. pylori* may be an important mechanism in the establishment and persistence of infection by this organism (Coticchia et al. 2006). The production of a water-insoluble biofilm by *H. pylori* may be important in enhancing resistance in host defense and antibiotics, and in microenvironmental pH homeostasis facilitating the growth and survival of *H. pylori* in vivo (Stark et al. 1999). A fourth study has reflected on the importance of biofilm in the

Fig. 9.1 Silver staining of a crypt in the gastric mucosa with numerous *Helicobacter pylori* in the lumen

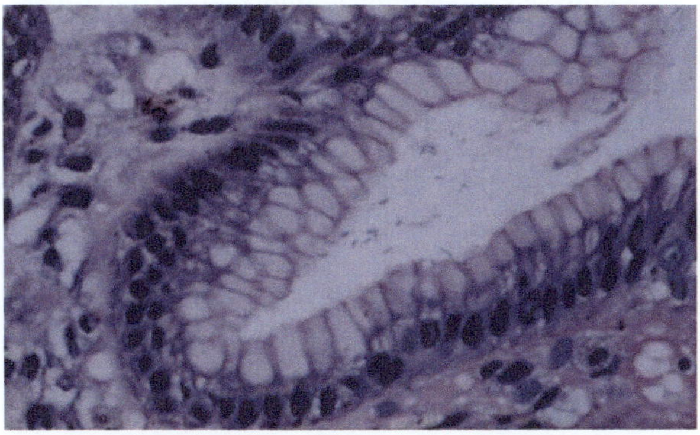

Fig. 9.2 Giemsa staining of a crypt in the gastric mucosa with numerous *Helicobacter pylori* in the lumen

treatment of *H. pylori*. Successful treatment must overcome additional barriers, such as the inoculum effect and the biofilm phenomenon. Attachment to a surface may be associated with an increase in the MIC of antibiotics (The biofilm phenomenon) (Graham 1998).

9.3 Discussion

There are several indications of *H. pylori* being able to form biofilm both in laboratory experiments, in environmental sources such as water supplies, in dental plaques, and in the human gastric mucosa. However, none of the studies seem to be really conclusive about the existence and role of biofilm formed by *H. pylori*. Some *H. pylori* strains are able to form biofilm in laboratory experiments. However, in laboratory experiments and studies on environmental sources, vital rod-shaped *H. pylori* seem to be rather short living in biofilms before they convert into coccoid forms. The ratio of rod-shaped forms of *H. pylori* compared to coccoid forms and the role of coccoid forms in biofilm have not been investigated. The same doubts about biofilm and coccoid forms of *H. pylori* occur in studies on dental plaques.

The most convincing studies for the importance of long living forms of *H. pylori* causing infection are the epidemiological studies that find river water and well water as risk factors for *H. pylori* infection as compared to tap water. However these studies do not differentiate the possible role of biofilm formation and the possible presence of coccoid forms of *H. pylori*.

Studies on the gastric mucosa agree on a difference in biofilm in patients with or without *H. pylori* infections in the gastric mucosa. However, the mucin layer of the gastric mucosa is a cop web of polysaccharides which may act as a natural "biofilm," and these studies do not investigate the interaction between *H. pylori* and the epithelial cells and their effect on the mucin layer itself as compared to a possible biofilm formation caused by *H. pylori*.

If *H. pylori* form true biofilm in the gastric mucosa and on environmental surfaces this will have serious clinical implications. First there may be a problem about the antimicrobial treatment of *H. pylori* as much higher doses of antibiotics is needed for treatment of *H. pylori* in biofilms than of free-living bacteria. This treatment may be even more difficult if *H. pylori* occur as coccoid forms in biofilms. Secondly, if *H. pylori* in biofilms/dental plaques of the oral cavity are less susceptible to treatment than those in the stomach, these will be a focus for re-infection with *H. pylori*. Thirdly, if *H. pylori* are able to form biofilm in environmental systems, such as water supplies, they may also be able to make biofilm on gastric and duodenal tubes including percutaneous enterogastric (PEG) tubes. This may cause foci where *H. pylori* are difficult or impossible to treat and to detect. Fourthly, studies have shown that *H. pylori* can survive in macrophages and it is possible that they can be brought to the blood stream this way. Other studies have indicated a relation between *H. pylori* infection and atherosclerosis. If this is true and if biofilm formation by

H. pylori is a possibility then maybe *H. pylori* can make biofilm on intravascular stents and long laying catheters and thereby increase the risk for the patient of getting atherosclerosis.

References

Ahmed KS, Khan AA, Ahmed I et al (2007) Impact of household hygiene and water source on the prevalence and transmission of *Helicobacter pylori*: a South Indian perspective. Singapore Med J ; 48(6):543–549.

Amieva MR, El-Omar EM (2008) Host-bacterial interactions in *Helicobacter pylori* infection. Gastroenterology ; 134(1):306–323.

Andersen L, Rasmussen L (2009) *Helicobacter pylori* – coccoid forms and biofilm formation. FEMS Immunol Med Microbiol ; 56(2):112–115.

Andersen LP (2007) Colonization and infection by *Helicobacter pylori* in humans. Helicobacter ; 12:12–15.

Andersen LP, Elsborg L, Justesen T (1988) *Campylobacter pylori* in peptic ulcer disease. II. Endoscopic findings and cultivation of *C. pylori*. Scand J Gastroenterol ; 23(6): 760–764.

Andersen LP, Wadström T. (2001) Basic bacteriology and culture. In: Mobley HLT, Mends GL, Hazell SL, (editors). *Helicobacter Pylori*: Physiology and genetics. ASM Press, Washington, DC; 2001., p.pp 27–38.

Bruce MG, Maaroos HI (2008) Epidemiology of *Helicobacter pylori* infection. Helicobacter ; 13 (Suppl 1):1–6.

Carron MA, Tran VR, Sugawa C et al (2006) Identification of *Helicobacter pylori* biofilms in human gastric mucosa. J Gastrointest Surg ; 10(5):712–717.

Cellini L, Grande R, Di CE et al (2008) Characterization of an *Helicobacter pylori* environmental strain. J Appl Microbiol ; 105(3):761–769.

Coticchia JM, Sugawa C, Tran VR et al (2006) Presence and density of *Helicobacter pylori* biofilms in human gastric mucosa in patients with peptic ulcer disease. J Gastrointest Surg ; 10(6):883–889.

Gebara EC, Faria CM, Pannuti C et al (2006) Persistence of *Helicobacter pylori* in the oral cavity after systemic eradication therapy. J Clin Periodontol ; 33(5):329–333.

Graham DY (1998) Antibiotic resistance in *Helicobacter pylori*: Implications for therapy. Gastroenterology ; 115(5):1272–1277.

Kandulski A, Selgrad M, Malfertheiner P (2008) *Helicobacter pylori* infection: a clinical overview. Dig Liver Dis ; 40(8):619–626.

Karita M, Teramukai S, Matsumoto S (2003) Risk of *Helicobacter pylori* transmission from drinking well water is higher than that from infected intrafamilial members in Japan. Dig Dis Sci ; 48(6):1062–1067.

Lambert JR, Lin SK, randa-Michel J (1995) *Helicobacter pylori*. Scand J Gastroenterol Suppl ; 208:33–46.

Marshall BJ, Barrett LJ, Prakash C et al (1990) Urea protects *Helicobacter-(Campylobacter)-Pylori* from the bactericidal effect of acid. Gastroenterology ; 99(3): 697–702.

Nurgalieva ZZ, Malaty HM, Graham DY et al (2002) *Helicobacter pylori* infection in Kazakhstan: effect of water source and household hygiene. Am J Trop Med Hyg ; 67(2): 201–206.

Ottemann KM, Lowenthal AC (2002) *Helicobacter pylori* Uses motility for initial colonization and to attain robust infection. Infect Immun ; 70(4):1984–1990.

Schneider S, Weydig C, Wessler S (2008) Targeting focal adhesions: *Helicobacter pylori*-host communication in cell migration. Cell Commun Signal ; 6:2.

Souto R, Colombo APV (2008) Detection of *Helicobacter pylori* by polymerase chain reaction in the subgingival biofilm and saliva of non-dyspeptic periodontal patients. J Periodontol ; 79(1):97–103.

Stark RM, Gerwig GJ, Pitman RS et al (1999) Biofilm formation by *Helicobacter pylori*. Lett Appl Microbiol ; 28(2):121–126.

Teoman I, Ozmeric N, Ozcan G et al (2007) Comparison of different methods to detect *Helicobacter pylori* in the dental plaque of dyspeptic patients. Clin Oral Investig ; 11(3): 201–205.

Willén R, Carlén B, Wang X et al (2000) Morphologic conversion of *Helicobacter pylori* from spiral to coccoid form. Scanning (SEM) and transmission electron microscopy (TEM) suggest viability. Ups J Med Sci ; 105(1):31–40.

Yoshiyama H, Nakamura H, Kimoto M et al (1999) Chemotaxis and motility of *Helicobacter pylori* in a viscous environment. J Gastroenterol ; 34 (Suppl 11):18–23.

Young KA, Akyon Y, Rampton DS et al (2000) Quantitative culture of *Helicobacter pylori* from gastric juice: the potential for transmission. J Med Microbiol ; 49(4): 343–347.

Young KA, Allaker RP, Hardie JM (2001) Morphological analysis of *Helicobacter pylori* from gastric biopsies and dental plaque by scanning electron microscopy. Oral Microbiol Immunol ; 16(3):178–181.

Chapter 10
Pseudomonas aeruginosa Biofilms in the Lungs of Cystic Fibrosis Patients

Niels Høiby, Helle Krogh Johansen, Claus Moser, Oana Ciofu, Peter Østrup Jensen, Mette Kolpen, Lotte Mandsberg, Michael Givskov, Søren Molin, and Thomas Bjarnsholt

10.1 Cystic Fibrosis

The consequence of the mutations in the CFTR gene is malfunction of the chloride channel in cystic fibrosis (CF) patients, which leads to decreased volume of the paraciliary fluid in the lower respiratory tract, and that in turn leads to impaired mucociliary clearance of inhaled microbes (Boucher 2004). This impairment of the non-inflammatory defense mechanism of the respiratory tract leads to early recruitment of the inflammatory defense mechanisms, e.g., polymorphonuclear leukocytes (PMN) and antibodies (Armstrong et al. 1995, 2005, Høiby et al. 2001). CF patients, therefore, from early childhood suffer from recurrent and chronic respiratory tract infections characterized by PMN inflammation. In spite of the inflammatory response and intensive antibiotic therapy, however, infections caused by *P. aeruginosa* persist and lead to respiratory failure and lung transplantation or death of the patients (Frederiksen et al. 1999) (Fig 10.1). Evolutionary adaptive mechanisms of *P. aeruginosa* exist which explain why this pathogen is able to survive and persist for several decades in the respiratory tract of CF patients in spite of the defense mechanisms of the host and intensive antibiotic therapy. *P. aeruginosa* is able to survive due to adaptation (1) to the inflammatory defense mechanism, (2) to the respiratory zone of the lungs, (3) to the conductive zone of the lungs, and (4) to the antibiotic therapy, and the adaptive mechanism is biofilm formation (Hoffmann et al. 2005b, Høiby 2006, Jelsbak et al. 2007, Bjarnsholt et al. 2009).

N. Høiby (✉)
University of Copenhagen, Faculty of Health Sciences, Department of International Health, Immunology and Microbiology, Blegdamsvej 3B, DK-2200 Copenhagen, Denmark; H:S Rigshospitalet, Department for Clinical Microbiology, afsnit 9301, Juliane Maries Vej 22, DK-2100 Copenhagen Ø, Denmark
e-mail: hoiby@hoibyniels.dk

T. Bjarnsholt et al. (eds.), *Biofilm Infections*, DOI 10.1007/978-1-4419-6084-9_10,

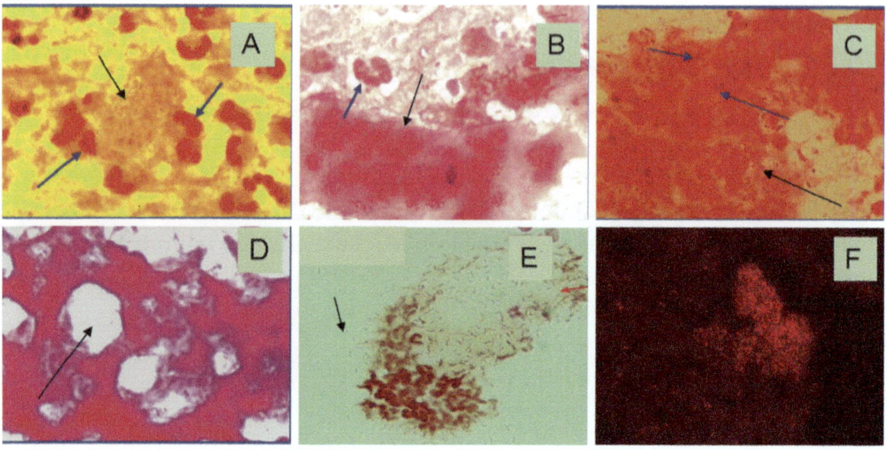

Fig. 10.1 Diversity of *P. aeruginosa* biofilms in sputum. Sputum samples from 77 chronic *P. aeruginosa* infected CF patients were examined by Gram stain and PNA FISH. The exclusive presence of *P. aeruginosa* were verified by Standard culturing and *P. aeruginosa* specific PNA FISH (**f**). A great diversity of the organization of *P. aeruginosa* was detected, both compact biofilm forming microlonies [frame **a–f** (*blue arrows point at leukocytes, black arrows point at bacterial aggregates*)] and non-adhered planktonic bacteria [**e** (*arrow*)] were observed. The organization of the PMNs was also diverse with PMNs surrounding the biofilms (**a**), distant from the biofilms (**b**) and in a very few cases appeared inside the biofilms (**c**). The biofilm microcolonies were mostly very compact; however, some samples were perforated with "holes" mimicking water-filled channels visible in some mature in vitro biofilms [**d** (*arrow point at hole*)]. Adapted from Bjarnsholt et al. (2009) and reproduced with permission from John Wiley & Sons, Inc

10.2 Survival of *P. aeruginosa* by Adaptation to the Inflammatory Defense System – The Fundamental Biofilm Strategy

Bronchoalveolar lavage studies in newly diagnosed CF infants have shown that in case of infection caused by virus or bacteria, there is a significantly increased numbers of PMNs and alveolar macrophages, whereas this is not the case in non-infected CF infants and normal children (Armstrong et al. 1995, 1997, and 2005). When PMNs try to phagocytose bacteria like *P. aeruginosa* a metabolic burst is produced which leads to liberation of highly reactive oxygen species (ROS) which will kill or induce mutations in the bacteria and damage the surrounding tissue (Miller and Britigan 1997, Egestein et al. 2008). Such oxygen radical damage has been detected in CF lungs (Hull et al. 1997). We have shown that hydrogen peroxide or activated PMNs can induce the same mutation in the *mucA* gene of *P. aeruginosa* PAO1 (Mathee et al. 1999), and we have found mutations in that gene in 93% of mucoid *P. aeruginosa* from Scandinavian CF patients (Ciofu et al. 2007) in accordance with other publications (Boucher et al. 1996, Bragonzi et al. 2006). This mutation changes the *P. aeruginosa* PAO1 strain to the characteristic mucoid phenotype (Mathee et al. 1999). Alginate is an oxygen radical scavenger and protects *P. aeruginosa* against phagocytosis and clearance from the lungs, so that the mucoid phenotype is better protected than the non-mucoid phenotype against the inflammatory defense mechanisms of the host (Song et al. 2003), and this is in accordance with our findings that a non-mucoid clinical strain had higher oxydative damage and mutation frequency than the mucoid isogenic strain (Ciofu et al. 2005). When mucoid phenotypes of *P. aeruginosa* are isolated from sputum of CF patients, biofilm growth can be detected by microscopy of Gram-stained smears of sputum from these patients (Fig. 10.1) and the identity of the rods have been assured by FISH (Bjarnsholt et al. 2009). In vitro experiments on biofilms growing in flow-cells and animal experiments suggest that resistance of non-mucoid *P. aeruginosa* biofilms to PMNs and the antibiotic tobramycin is dependent on production of quorum sensing (QS) regulated virulence factors, notably rhamnolipid, and that QS-mutants as well as use of QS inhibitors (QSI) such as garlic extract makes the biofilm susceptible to PMN attack and killing as well as eradication by tobramycin (Bjarnsholt et al. 2005a, b, Hoffmann et al. 2007, Jensen et al. 2007). Similar results have been obtained in animal experiments by treatment of the *P. aeruginosa* lung infection (mucoid strain) with azithromycin, which also inhibits quorum sensing (Hoffmann et al. 2007).

In conclusion, *P. aeruginosa* adaptates to the inflammatory defense system in the respiratory tract of CF patients by forming mucoid biofilms. Possible prophylactic measures could be use of anti-inflammatory drugs or anti-oxidant therapy to prevent ROS induced mutations in the *mucA* gene or use of macrolides such as azithromycin in doses which inhibit quorum sensing and also alginate synthesis as suggested by the in vitro results and animal experiments (Kobayashi 1995, Tateda et al. 2001, Hoffmann et al. 2007). In addition, QSI treatment of biofilm in the CF lungs may be a promising research subject in the future. Currently, however, the only available alternative is early aggressive antibiotic eradication therapy of the initial *P. aeruginosa* colonization caused by non-mucoid strains in order to prevent progression to chronic biofilm infection (Döring and Høiby 2004) which can be distinguished from intermittent colonization by the antibody response to *P. aeruginosa* and by the characteristic mucoid phenotype (Fig. 10.2) (Pressler et al. 2006, Proesmans et al. 2006).This therapy has successfully been used in the Danish CF Centre since 1989 and can prevent about 80% of chronic *P. aeruginosa* lung infection in CF patients (Høiby et al. 2005) and does not lead to resistance in contrast to treatment of established chronic biofilms of *P. aeruginosa* (Hansen et al. 2008, Johansen et al. 2008). Once the chronic biofilm infection is established, the only efficient therapy is chronic suppressive, maintenance therapy with regular anti-pseudomonal antibiotic courses systemically and/or nebulized in order to maintain the lung function for years (Döring et al. 2000).

10.3 The Conductive and the Respiratory Zones of the Lungs

The lungs consist of the smaller conductive zone and the larger respiratory zone. The respiratory zone (appr. 3000 ml, 95% of the lung volume) includes respiratory bronchioles, alveolar ducts, and alveolar sacs (Westh 2001, Høiby 2006). This part of the lungs has no cilia, no goblet cells, and no submucosal glands, and the

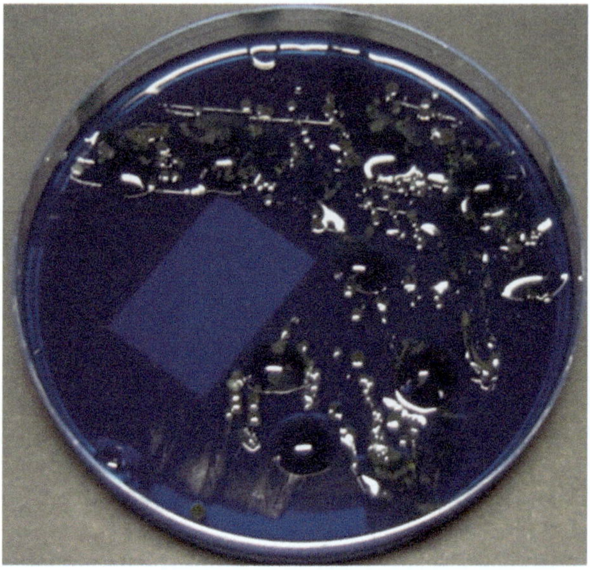

Fig. 10.2 Bacteriology of sputum from a cystic fibrosis patient with chronic *P. aeruginosa* biofilm lung infection for many years. Growth of mucoid, non-mucoid, and small colony variants of *P. aeruginosa* on a selective medium for Gram-negative rods (*Blue plate*, State Serum Institute, Denmark)

defense system consists of, e.g., alveolar macrophages and defensins. All the venous blood of the body passes through the capillaries of the alveoles which consist of a nearly continuous sheet of blood and only a very thin barrier is present between the air and the blood. The smaller conducting zone (appr. 150 ml, 5% of the lung volume) includes trachea, the bronchi, and the terminal bronchioles. This part of the lungs has cilia, goblet cells, and submucal glands, and has an ordinary arterial blood supply from aorta. The mucus is produced in the respiratory zone and the major defense system consists is the muco-ciliary escalator (Westh 2001). Nebulized tobramycin and colistin and other antibiotics are widely used to treat *P. aeruginosa* lung infection in CF patients (Johansen et al. 2008). Very high concentrations of these drugs are obtained in the conductive zone (sputum), whereas very little actually reach the respiratory zone, since the measurable concentration in serum, which reflect the amount in the respiratory zone, is very low (Levy 1984, Ramsey et al. 1999, Le Brun 2001, Geller et al. 2002, Gibson et al. 2003, Ratjen et al. 2006). On the contrary, when antibiotics are administered intravenously or orally, very low concentrations are found in sputum, but high concentrations in respiratory tissue because the whole dose of, e.g., an intravenous bolus of antibiotics is transported directly by the blood to the alveolar capillaries before being distributed to the rest of the body (Permin et al. 1983, Le Brun 2001). Since both the respiratory and the conductive zones of the lungs are infected with *P. aeruginosa*, there is both in vitro and in vivo farmacokinetic/farmacodynamic evidence for using combined systemic and nebulized antibiotics in CF patients (Jensen et al. 1987, Valerius et al. 1991, Pamp et al. 2008).

10.3.1 *Survival of* P. aeruginosa *by Adaptation to the Respiratory Zone of the Lungs*

The condition inside the airspace of the respiratory zone is aerobic (13% O_2, 5% CO_2) if the air supply and the blood supply is normal or if the air supply is normal but the blood supply is absent (20% O_2, 0.003% CO_2). If there is mucus plugging and therefore no airflow, then there is microaerophilic condition (5% O_2, 6% CO_2) (Westh 2001). In case of abscess formation, the condition is anaerobic and characterized by necrosis of the cells and tissue. *P. aeruginosa* can grow under all these conditions but the generation time is shorter in an aerobic atmosphere where O_2 is the final electron acceptor due to the much higher ATP yield compared to anaerobic conditions where NO_3^- from, e.g., alveolar macrophages or PMNs (Cedergren et al. 2003) (iNOS -> NO + ROS (O_2 -> NO_3) is the final electron acceptor (Hassett et al. 2002). PNMs can survive for several days in anaerobic conditions (Kolpen et al. 2009).

Mucoid *P. aeruginosa* biofilms are located in the respiratory zone of CF lungs according to autopsy findings (Fig. 10.3) of lungs from young CF patients who have not been intensively treated by antibiotics (Bjarnsholt et al. 2009). In explanted lungs from intensively treated older CF patients the *P. aeruginosa* biofilms are rare in the respiratory zone probably due to the effect of the antibiotic therapy (Fig. 10.4) (Bjarnsholt et al. 2009). A pronounced antibody response against alginate is produced by the patients (Pedersen et al. 1990a, b, 1992). This is in accordance with the immunological role of alveolar macrophages as antigen presenting cells in the response to offending pulmonary pathogens. The PMNs are recruited from the capillaries of the alveoles to combat the bacterial infection (Craig et al. 2009). The pronounced PMN-dominated inflammation which surrounds biofilm-containing alveoles leads to tissue damage and decrease of lung function (Goldstein and Döring 1986). Detached biofilm-containing alveoles can sometimes be seen in sputum (Høiby 2006), and degradation products from elastin and collagen can be detected in the urine (Bruce et al. 1985, Ammitzbøll et al. 1988). The healthy lungs consist of 300,000,000 alveoles (Westh 2001), and a 1–2% decline of pulmonary function per year implies, therefore, a loss of $3–6\times10^6$ alveoles/year = 8000–16000 alveoles per day. The pathogenesis of the tissue damage is immune-complex mediated inflammation promoted by the pronounced antibody response (Høiby et al. 2001) which is characteristic for a Th-2 polarized immune response (Johansen 1996). The *P. aeruginosa* biofilm infection is a focal infection, where terminal tissue damage can be detected by high resolution CT scan in chronically infected CF patients even with normal lung function (Tiddens 2002). Gradually, however, the focal infection spreads to new areas of the lungs and the respiratory function declines in spite of antibiotic therapy. An important observation is, however, that there are many PMNs in the tissue surrounding the mucoid *P. aeruginosa* biofilms whereas there are no or few PMNs inside the biofilms (Bjarnsholt et al. 2009) in accordance with the protective activity of alginate and the production of *P. aeruginosa* rhamnolipid (Jensen et al. 2007) (Figs. 10.1, 10.3, 10.4, and 10.5). No non-mucoid planktonic *P. aeruginosa* cells are seen in the respiratory

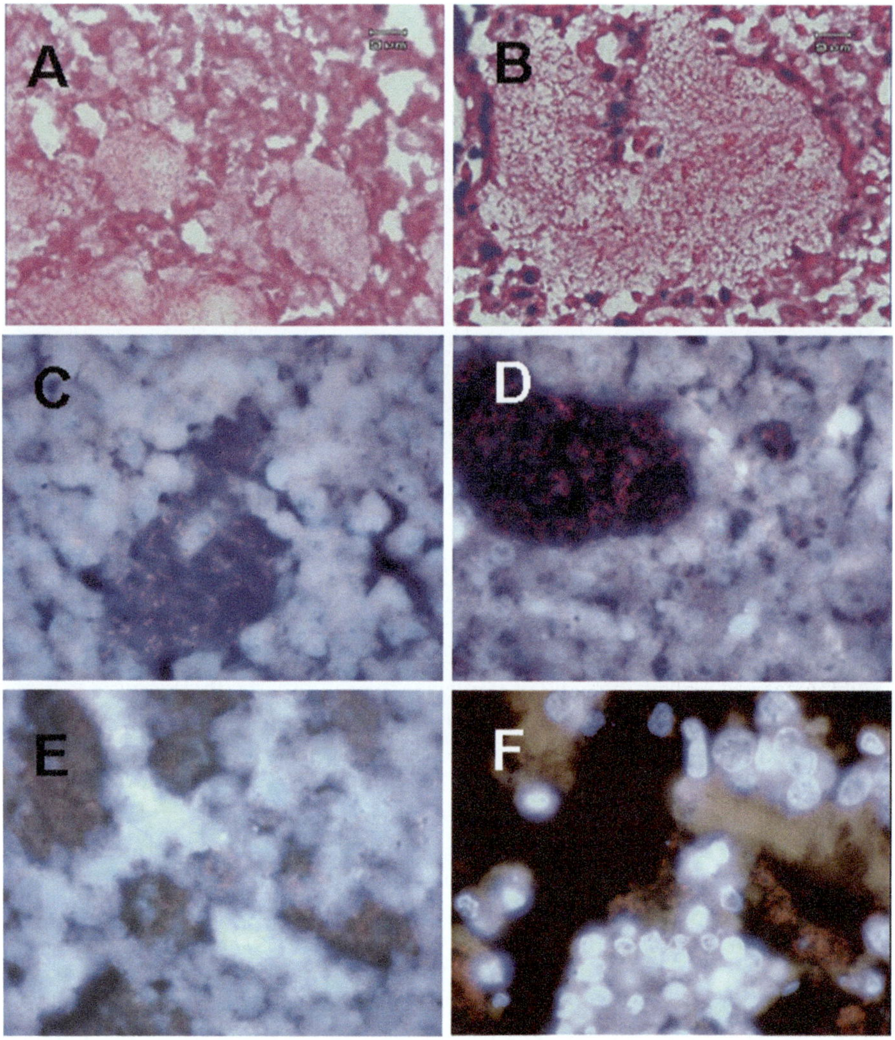

Fig. 10.3 Microscopic investigation of non-intensive treated chronically *P. aeruginosa* infected CF lungs. (**a, b**) Massive bacterial biofilms are detected by Gram stain in both the conductive (**a**) and respiratory zone (**b**). (**c, d**) Intraluminal *P. aeruginosa* biofilms surrounded by PMNs (**e, f**) *P. aeruginosa* biofilms in the respiratory zone, visualized by PNA FISH and DAPI. *P. aeruginosa* biofilms are observed in vast amounts, both in the conductive zone and in the respiratory zone. Most of the tissue in both zones is extensively destroyed. Adapted from Bjarnsholt et al. (2009) and reproduced with permission from John Wiley & Sons, Inc

zone. In conclusion, *P. aeruginosa* adaptates to the respiratory zone by forming mucoid biofilms which survives for decades in spite of the inflammatory response, whereas the lung tissue is gradually destroyed. Possible prophylactic measures are those which are suggested above to prevent adaptation to the inflammatory response.

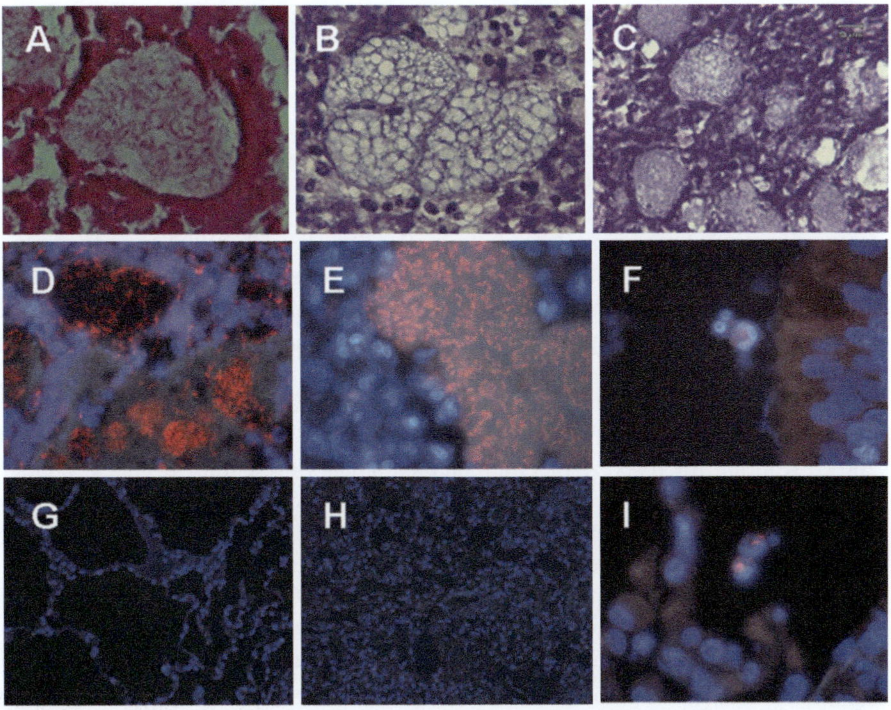

Fig. 10.4 *P. aeruginosa* biofilms are contained in the conductive zone, only very few bacteria are detected in the respiratory zone and here they are phagocytosed. (**a**) Bacteria in biofilm within a bronchia visualized with Gram stain (CF male, 41 years of age, chronic *P. aeruginosa* mucoid and non-mucoid infection for 28 years, 46 precipitating antibodies, 114 two-week anti-*P. aeruginosa* treatment courses), (**b, c**) HE stain of bacteria filled bronchiole. H, hollow zones with grid like structures – probably originating from the alginate component of the biofilms – are observed were bacteria are detected by Gram stain and PNA FISH. (**d, e**) Intraluminal *P. aeruginosa* biofilms surrounded by PMNs visualized by PNA FISH and DAPI. (**f**) Intact bronchi wall. (**g, h**) shows increasing consolidation of alveoli. (**i**) Single phagocytosed *P. aeruginosa* in the respiratory zone. Adapted from et al. (2009) and reproduced with permission from John Wiley & Sons, Inc

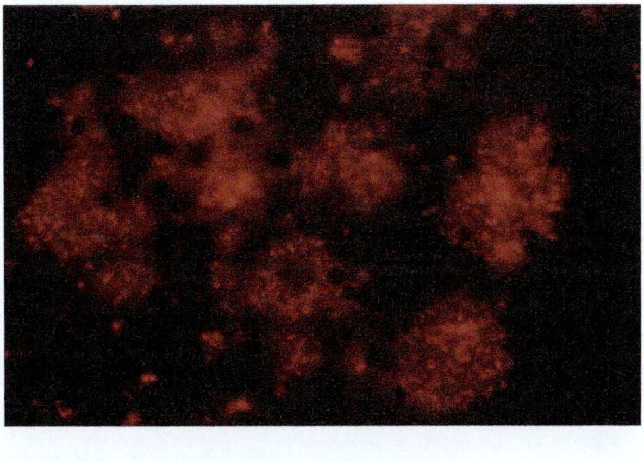

Fig. 10.5 The extrapolymeric substances within and surrounding the biofilms visualized by HE and Gram stain and PNA FISH was identified to be alginate by alginate specific immuno staining. Adapted from Bjarnsholt et al. (2009) and reproduced with permission from John Wiley & Sons, Inc

10.3.2 Survival of **P. aeruginosa** *by Adaptation to the Conductive Zone of the Lungs*

The conductive zone of the adult lungs consists of 16 generations of bronchi from trachea to the terminal bronchioles (Westh 2001, Høiby 2006). Only the 4–5 proximal generations can be directly observed by bronchoscopy in vivo whereas the distal generations can be indirectly investigated by lavage or directly in lungs from transplanted or succumbed patients. Worlitzsch et al. (2002) and Bjarnsholt et al. (2009) (Fig. 10.4) have shown that *P. aeruginosa* in the conductive zone is mainly localized inside sputum in biofilms (microcolonies), whereas very few bacteria are localized at the epithelial surface of the bronchi. Furthermore, worlitzsch showed that there are anaerobic conditions in sputum. This is mainly caused by the respiratory burst of the PMNs which are stimulated by *P. aeruginosa* in sputum and to a minor degree of the oxygen consumption by *P. aeruginosa* (Kolpen et al. 2009). Routine microscopy of Gram-stained smears of sputum from CF patients shows that besides mucoid biofilms, also planktonic single *P. aeruginosa* cells are found (Høiby 1977, Bjarnsholt et al. 2009) and that is also the case in autopsy lungs from young CF patients who have not been intensively treated by antibiotics and in explanted lungs from intensively treated older CF patients (Bjarnsholt et al. 2009). Additionally, culture of sputum regularly shows mucoid, non-mucoid, and sometimes also small-colony-variant phenotypes in sputum from each individual CF patient (Høiby 1977, Haussler et al. 1999, 2004). The reason for this characteristic phenotypic variation has been unclear for many years, but the elegant experiments with *P. fluorescens* published by Rainey and Travisona (Rainey and Travisano 1998) and with *P. aeruginosa* by Wyckoff et al. (2002) and by my group using *P. aeruginosa* isolates from a CF patient (Hoffmann et al. 2005a) have shown that such phenotypic variation occurs, when different niches are present in the growth media. The fundamental driving forces for the phenotypic variation are mutations (or recombinations) caused by, e.g., DNA-damaging free radicals and the bacterial SOS response, followed by competition and selection for different niches in a spatially heterogenous environment which, in turn, leads to adaptive divergence (Spiers et al. 2000). Such different niches occur during stationary batch culture in Erlenmeyer flasks, where oxygen is plentiful at the surface whereas oxygen is restricted at the bottom. After 2–3 days of culture this leads to split off of new specialist phenotypes of *P. aeruginosa* in addition to the wild phenotype. This does not occur in shaken batch cultures where homogenous conditions exist in the flasks. Similar results have been obtained by us (Hoffmann et al. 2005a) with a clinical mucoid *P. aeruginosa* CF strain which split off a non-mucoid phenotype and sometimes also a small-colony-variants (Fig. 10.2) which are also characteristic for CF patients (Haussler et al. 1999, 2004). The growth rate in vivo in CF sputum of *P. aeruginosa* has been shown to cover a wide range but to be much slower than the mucoid phenotype has been shown to be unstable during anaerobic conditions and the advantage in vitro (Yang et al. 2008). The non-mucoid phenotype is thought to be due to its flagella-mediated motility which is negatively regulated by the sigma factor *algT* which also positively regulates the alginate biosynthesis (Wyckoff et al. 2002). Non-mucoid

P. aeruginosa are therefore able to move to the aerobic surface, where the aerobic growth gives the selective advantage of a faster growth rate. As mentioned above, most of mucoid *P. aeruginosa* strains from CF patients have a mutation/insertion in *mucA* and this is also the case in the majority of non-mucoid (revertant) isolates from the same patients (Boucher et al. 1996, Bragonzi et al. 2006, Ciofu et al. 2007). Some but not all of the non-mucoid revertant phenotypes had mutations/insertions in *algT* which could explain the phenotype (Ciofu et al. 2007). Furthermore, many of the non-mucoid revertants are QS deficient and therefore not virulent (Lee et al. 2005). Non-mucoid phenotypes generally do not give raise to a pronounced antibody response in the few patients who only are colonized with these phenotypes (Pedersen et al. 1990a, 1992). This is probably due to their location in sputum in the conductive zone far from the key cells of the immune system. Likewise, non-mucoid phenotypes of *P. aeruginosa* have been shown not to be associated with poor prognosis or deteriorated lung function (Pedersen et al. 1992) and this is in accordance with their location in sputum in the conductive zone (Worlitzsch et al. 2002), where they probably merely induce increased sputum production and obstruction. Furthermore, whereas mucoid *P. aeruginosa* become increasingly virulent as time goes by during the infection, this is opposite to their non-mucoid revertants (Moser et al. 2009). The ability of non-mucoid revertants to produce biofilms in vitro decreases as time goes by during the infection this is not the case with mucoid phenotypes (Lee et al. 2005, 2009, submitted). According to these results, therefore, mucoid biofilms and live PMNs from the aerobic respiratory zone are co-transported to the anaerobic sputum in the conductive zone, and this is where the non-mucoid phenotypes split off. During this transportation PMNs are constantly activated and a respiratory burst is generated. The presence of non-mucoid *P. aeruginosa* cells which have been engulfed by PMNs and released DNA and elastase and myeloperoxidase and oxygen radicals from the PMNs (Brandt et al. 1995, Shah et al. 1995, Ratjen and Tummler 1999, Bjarnsholt et al. 2009) in sputum of the conductive zone show that PMNs are actually activated in the conductive zone (Kolpen et al. 2009). In conclusion, *P. aeruginosa* adaptates to the conductive zone by splitting off non-mucoid revertants which probably do not contribute much to the tissue damage and therefore do not require special therapeutic or prophylactic attention.

10.4 Survival of *P. aeruginosa* Biofilms by Adaptation to the Antibiotic Therapy

In spite of the intensive antibiotic therapy, the chronic *P. aeruginosa* lung infection is rarely if ever eradicated. The main reason is most likely the biofilm way of growth (Anwar et al. 1992a, b, Smith et al. 2003, Borriello et al. 2004, Keren et al. 2004, Moskowitz et al. 2004, Haagensen et al. 2007) (see also Chapter 13), but the frequent occurrence of multiply resistant *P. aeruginosa* in CF patients implies that conventional resistance mechanisms probably also play a role (Ciofu et al. 2001). There are many reports of beta-lactam-resistant *P. aeruginosa* strains from CF patients (Giwercman et al. 1990, 1992, Denton et al. 1996, Dibdin et al. 1996) due to stable or partly stable de-repression of chromosomal beta-lactamase (Bagge et al. 2002), multiply mutations causing resistance to ciprofloxacin (Jalal et al. 2000), frequent occurrence of resistance to tobramycin due to upregulated efflux pumps (Saiman et al. 1996, Burns et al. 1999, WestbrockWadman et al. 1999, MacLeod et al. 2000, Islam et al. 2009), and also resistance to colistin caused by tolerance or mutations which leads to production of modified LPS by addition of 4-aminoarabinose to lipid A (Denton et al. 2002, Moskowitz et al. 2004, Johansen et al. 2008, Pamp et al. 2008). We have shown that biofilm growing *P. aeruginosa* responds to beta-lactam antibiotics by increased production of chromosomal beta-lactamase (Bagge et al. 2000, 2002, 2004a, b) and that non-mucoid phenotypes are more resistant (higher MIC to antibiotics) (Ciofu et al. 2001) and have higher oxygen radical damage measured as 8-oxodG/10^6 dG than mucoid phenotypes from the same patients (Ciofu et al. 2005). An important reason for the occurrence of multiply resistant *P. aeruginosa* strains in CF patients is the high frequency of mutator strains as shown by Oliver et al. (2000) and confirmed by us (Ciofu et al. 2005) and others (Hoboth et al. 2009, Wolter et al. 2009). Mutator strains are characterized by defects of their DNA repair enzymes and the consequence of this is a much higher frequency of mutations compared to wild-type strains (Oliver et al. 2002). In *P. aeruginosa* strains from Danish CF patients the responsible mutations are found in the *mutT* or *mutY* genes of the repair enzymes (Mandsberg et al. 2009). Such mutator strains of *P. aeruginosa* from CF patients are frequently multiply resistant to antibiotics (Oliver et al. 2000, Macia et al. 2004, 2005, Ciofu et al. 2005). Biofilm growing *P. aeruginosa* have higher mutation rate than planktonically growing cells (Driffield et al. 2008). Furthermore, we have shown that such mutator strains generally have increased levels of DNA damage which, additionally, may be caused by ROS from activated PMNs during the chronic infection. We have also found that no mutator strains were detectable during the first few years of the chronic infection but after 5 years the prevalence of mutator strains increased with time (Ciofu et al. 2005). By population analysis for mutators in sputum of CF patients we found a prevalence of 54% (Ciofu et al. 2005). *P. aeruginosa*, therefore, adaptates to the intensive antibiotic therapy by producing mucoid biofilms and by becoming mutators. In conclusion, *P. aeruginosa* adaptates to the intense antibiotic therapy by forming mucoid biofilms and by conventional resistance mechanisms which act synergistically to allow *P. aeruginosa* to survive for decades in spite of the antibiotic therapy, whereas the lung tissue is gradually destroyed. Possible prophylactic measures are those which are suggested above to prevent adaptation to the inflammatory response but additionally, we have shown that induction of neutralizing antibodies against chromosomal beta-lactamase of *P. aeruginosa* improves the clinical outcome of antibiotic therapy with beta-lactam antibiotics (Ciofu et al. 1999, 2002, Ciofu 2003). This strategy may therefore be utilized clinically in the future.

10.5 Evolutionary Implications of the Adaptability of *P. aeeruginosa*

The *P. aeruginosa* PAO1 strain was sequenced by Stover et al. (2000). It has 6.2 million base pairs and 5,570 genes among which is a high proportion of regulatory, catabolic, transport, efflux, and chemotaxis genes which explains much of its adaptability (Stover et al. 2000). When investigating clinical strains from chronic lung infection of CF patients, there are some obvious differences from the PAO1 strain as described above. CF strains are alginate producers that grow in biofilms and split off non-mucoid and small colony variant phenotypes; they are often mutators, have accumulated mutations, and have become auxotrophs. We have found that clinical CF strains in most patients persist in the patients (Jelsbak et al. 2007) and grow much slower than PAO1: mean generation time is 2–3 fold slower than environmental isolates in laboratory media and in vivo in the lungs of CF patients the generation time is 100–200 min (Yang et al. 2008). When such clinical strains have survived in CF lungs for 30 years, they may have experienced 65,000 divisions during which time niche specialists have evolved which are well adaptated to survive in CF lungs but may have less capability to survive elsewhere. Compared to the evolution of our own species, the genetic and phenotypic difference between PAO1 and CF *P. aeruginosa* strains may be comparable to the distance between *Homo erectus* and modern *Homo sapiens*.

We have also found that the biofilm forming ability in vitro decreased by time when subsequent non-mucoid isolates were studied of the same clone of *P. aeruginosa* colonizing CF patients for many years (Lee et al. 2005). This change was associated with a loss of motility and with decreased production of virulence factors and quorum sensing molecules, and the isolates also became hypermutable (Lee et al. 2005). This was not the case with mucoid isolates of the same clone. These findings are in accordance with the result of Smith et al. (2006) who studied a series of subsequent non-mucoid isolates during 8 years of infection in a CF patient. They found numerous genetic adaptations during the 8-year period. Sequencing of the genome of early and late mucoid and non-mucoid isolates of *P. aeruginosa* from individual CF patients is therefore urgently needed to investigate the evolution of the *P. aeruginosa* genome in CF lungs during decades of chronic biofilm infection in spite of immune response and antibiotic therapy.

10.6 Clinical Consequences of *P. aeruginosa* Biofilms in CF Lungs

The clinical consequences of the impressive evolutionary adaptation of *P. aeruginosa* in the CF lungs are illustrated by a clinical example. We have examined explanted lungs from a 42-year-old CF male, with chronic *P. aeruginosa* infection during 28 years (Bjarnsholt et al. 2009). The patient developed a pronounced antibody response already from the beginning of the infection in 1977 and at the same time he began to be treated with regular suppressive antibiotic therapy every 3 months. He had been treated with a total of 114 2-week courses of intravenous tobramycin (total 1 kg) and anti-pseudomonal beta-lactam antibiotics (10 kg) and additionally daily nebulized colistin since 1987 (1 kg). In spite of this massive immunologic and therapeutic attack on his mucoid *P. aeruginosa* strain it continued to multiply and survive in biofilms in his lungs surrounded by activated PMNs which had destroyed his lungs (Fig. 10.4) (Bjarnsholt et al. 2009).

10.7 Prophylaxis and Treatment of *P. aeruginosa* Biofilms in CF Lungs

The currently used methods for preventing chronic *P. aeruginosa* biofilms in CF lungs are (1) prevention of cross-infection from other already chronically infected CF patients by isolation techniques and hygienic measures (Høiby and Pedersen 1989), (2) early aggressive eradication therapy of intermittent colonization by means of oral ciprofloxacin and nebulized colistin for 3 weeks or even better for 3 months or by using nebulized tobramycin as monotherapy (Döring and Høiby 2004), and (3) daily nebulized DNase (Pulmozyme) (Frederiksen B. et al. 2006). These three methods, which are combined in most CF centers, are successful and cost-efficient and has completely changed the epidemiology of chronic *P. aeruginosa* lung infection in CF patients from being very common in CF children to being predominantly a problem for adult patients, and no major problems of resistance to the antibiotics have been recorded (Frederiksen et al. 1999, Baumann et al. 2003, Hansen et al. 2008). Furthermore, although vaccines against *P. aeruginosa* have been developed and undergone clinical trials, they have not been further developed due to the success of the early, aggressive eradication therapy (Lang et al. 1995).

The recommended method for treatment of chronic *P. aeruginosa* biofilm infection is chronic suppressive antibiotic therapy (Döring et al. 2000), which is started when the chronic infection is diagnosed (continuous colonization at the monthly bacteriological examination for 6 months and/or increased level of antibodies against *P. aeruginosa* ($\geq$ 2 precipitating antibodies is the threshold in the Danish CF center in Copenhagen)) (Høiby et al. 1977). In some CF centers, which only see the patients every 3–4 months and do not have access to antibody measurement, $\geq$ 50% of the bacteriological examinations within a year is defined as chronic infection (Proesmans et al. 2006). The chronic suppressive therapy consists of daily nebulized colistin or tobramycin for the rest of the patient's life combined with either regular 2-week courses every 3 months of intravenous anti-pseudomonas antibiotics (combination therapy of two antibiotics: tobramycin or colistin + ceftazidime, or piperacillin/tazobactam, or carbapenem, or aztreonam, or ciprofloxacin) or ad hoc intravenous therapy when clinical deteriorations occur (Döring et al. 2000).

Additionally, DNase is inhaled every day to reduce the viscosity of the DNA-containing sputum (Frederiksen et al. 2006), and oral azithromycin is given to the patients continuously since it has been shown to further maintain the pulmonary function possibly due to a combination of its quorum-sensing inhibitory effect and its anti-inflammatory effect (Hansen et al. 2005) (see also Chapter 14). The chronic suppressive therapy – which is also called maintenance therapy – has successfully been able to maintain the pulmonary function or slow the decline of the pulmonary function and prolong the life of the patients for many years (Frederiksen et al. 1996). As mentioned above, the side-effects of the maintenance therapy is high level of conventional resistance mechanisms in the persisting strains and high level of allergy to the beta-lactam antibiotics (Koch et al. 1991, Ciofu et al. 1994).

References

Ammitzbøll T, Pedersen SS, Espersen F et al (1988) Excretion of urinary collagen metabolites correlates to severity of pulmonary disease in cystic fibrosis. Acta Paediatr Scand 77:842–846

Anwar H, Strap JL, Chen K et al (1992a) Dynamic interactions of biofilms of mucoid Pseudomonas aeruginosa with tobramycin and piperacillin. Antimicrob Agents Chemother 36:1208–1214

Anwar H, Strap JL, Costerton JW (1992b) Establishment of aging biofilms: possible mechanism of bacterial resistance to antimicrobial therapy. Antimicrob Agents Chemother 36:1347–1351

Armstrong DS, Grimwood K, Carlin JB et al (1997) Lower airway inflammation in infants and young children with cystic fibrosis. Am J Respir Crit Care Med 156:1197–1204

Armstrong DS, Grimwood K, Carzino R et al (1995) Lower respiratory infection and inflammation in infants with newly diagnosed cystic fibrosis. Br Med J 310:1571–1572

Armstrong DS, Hook SM, Jamsen KM et al (2005) Lower airway inflammation in infants with cystic fibrosis detected by newborn screening. Ped Pulmonol 40:500–510

Bagge N, Ciofu O, Hentzer M et al (2002) Constitutive high expression of chromosomal beta-lactamase in Pseudomonas aeruginosa caused by a new insertion sequence (IS1669) located in ampD. Antimicrob Agents Chemother 46:3406–3411

Bagge N, Ciofu O, Skovgaard LT et al (2000) Rapid development in vitro and in vivo of resistance to ceftazidime in biofilm-growing Pseudomonas aeruginosa due to chromosomal beta-lactamase. Apmis 108:589–600

Bagge N, Hentzer M, Andersen JB et al (2004a) Dynamics and spatial distribution of beta-lactamase expression in Pseudomonas aeruginosa biofilms. Antimicrob Agents Chemother 48:1168–1174

Bagge N, Schuster M, Hentzer M et al (2004b) Pseudomonas aeruginosa biofilms exposed to imipenem exhibit changes in global gene expression and beta-lactamase and alginate production. Antimicrob Agents Chemother 48:1175–1187

Baumann U, Stocklossa C, Greiner W et al (2003) Cost of care and clinical condition in paediatric cystic fibrosis patients. J Cystic Fibrosis 2:84–90

Bjarnsholt T, Jensen P-Ø, Burmølle M et al (2005a) Pseudomonas aeruginosa tolerance to tobramycin, hydrogen peroxide and polymorphonuclear leukocytes is quorum-sensing dependent. Microbiology 151:373–383

Bjarnsholt T, Jensen PØ, Fiandaca MJ et al (2009) Pseudomonas aeruginosa biofilms in the respiratory tract of cystic fibrosis patients. Pediatr Pulmonol 44:547–558

Bjarnsholt T, Jensen PØ, Rasmussen TB et al (2005b) Garlic blocks quorum sensing and promotes rapid clearing of pulmonary Pseudomonas aeruginosa infections. Microbiology 151:3873–3880

Borriello G, Werner E, Roe F et al (2004) Oxygen limitation contributes to antibiotic tolerance of Pseudomonas aeruginosa in biofilms. Antimicrob Agents Chemother 48:2659–2664

Boucher JC, Martinezsalazar J, Schurr MJ et al (1996) Two distinct loci affecting conversion to mucoidy in Pseudomonas aeruginosa in cystic fibrosis encode homologs of the serine protease HtrA. J Bacteriol 178:511–523

Boucher RC (2004) New concepts of the pathogenesis of cystic fibrosis lung disease. Eur Resp J 23:146–158

Bragonzi A, Wiehlmann L, Klockgether J et al (2006) Sequence diversity of the mucoid mucABD locus in Pseudomonas aeruginosa isolates from patients with cystic fibrosis. Microbiology 152:3261–3269

Brandt T, Breitenstein S, Vonderhardt H et al (1995) DNA concentration and length in sputum of patients with cystic fibrosis during inhalation with recombinant human DNase. Thorax 50: 880–882

Bruce MC, Poncz L, Klinger JD et al (1985) Biochemical and pathologic evidence for proteolytic destruction of lung connective tissue in cystic fibrosis. Am Rev Respir Dis 132:529–535

Burns JL, VanDalfsen JM, Shawar RM et al (1999) Effect of chronic intermittent administration of inhaled tobramycin on respiratory microbial flora in patients with cystic fibrosis. J Infec Dis 179:1190–1196

Cedergren J, Follin P, Forslund T et al (2003) Inducible nitric oxide synthase (NOS II) is constitutive in human neutrophils. APMIS 111:963–968

Ciofu O (2003) Pseudomonas aeruginosa chromosomal beta-lactamase in patients with cystic fibrosis and chronic lung infection – Mechanism of antibiotic resistance and target of the humoral immune response. Apmis 111:4–47

Ciofu O, Bagge N, Hoiby N (2002) Antibodies against beta-lactamase can improve ceftazidime treatment of lung infection with beta-lactam-resistant Pseudomonas aeruginosa in a rat model of chronic lung infection. Apmis 110:881–891

Ciofu O, Fussing V, Bagge N et al (2001) Characterization of paired mucoid/non-mucoid Pseudomonas aeruginosa isolates from Danish cystic fibrosis patients: antibiotic resistance, beta-lactamase activity and RiboPrinting. J Antimicrob Chemother 48:391–396

Ciofu O, Giwercman B, Pedersen SS et al (1994) Development of antibiotic resistance in Pseudomonas aeruginosa during two decades of antipseudomonal treatment at the Danish CF center. APMIS 102:674–680

Ciofu O, Lee B, Johannesson M et al (2008) Investigation of the algT operon sequence in mucoid and non-mucoid Pseudomonas aeruginosa isolates from 115 Scandinavian patients with cystic fibrosis and in 88 in vitro non-mucoid revertants. Scandinavian Cystic Fibrosis Study Consortium. Microbiology. 154(Pt 1):103–113

Ciofu O, Petersen TD, Jensen P et al (1999) Avidity of anti-P aeruginosa antibodies during chronic infection in patients with cystic fibrosis. Thorax 54:141–144

Ciofu O, Riis B, Pressler T et al (2005) Occurrence of hypermutable P. aeruginosa in cystic fibrosis patients is associates with the oxidative stress caused by chronic lung inflammation. Antimicrob Agents Chemother 49:2276–2282

Craig A, Mai J, Cai S et al (2009) Neutrophil recruitment to the lungs during bacterial pneumonia. Infect Immu 77:568–575

Denton M, Kerr K, Mooney L et al (2002) Transmission of Colistin-resistant Pseudomonas aeruginosa between patients attending a pediatric cystic fibrosis center. Pediatr Pulmonol 34:257–261

Denton M, Todd NJ, Littlewood JM (1996) Role of anti-pseudomonal antibiotics in the emergence of Stenotrophomonas maltophilia in cystic fibrosis patients. Eur J Clin Microbiol Infect Dis 15:402–405

Dibdin GH, Assinder SJ, Nichols WW et al (1996) Mathematical model of beta-lactam penetration into a biofilm of Pseudomonas aeruginosa while undergoing simultaneous inactivation by released beta-lactamases. J Antimicrob Chemother 38:757–769

Driffield K, Miller K, Bostock M et al (2008) Increased mutability of Pseudomonas aeruginosa in biofilms. J Antimicrob Chemother 61:1053–1056

Döring G, Conway SP, Heijerman HGM et al (2000) Antibiotic therapy against Pseudomonas aeruginosa in cystic fibrosis: a European consensus. Eur Respir J 16:749–767

Döring G, Høiby N (2004) Early intervention and prevention of lung disease in cystic fibrosis: a European consensus. J Cystic Fibrosis 3:67–91

Egestein A, Schmidt A, Herwald H (2008) Trends in Innate Immunity. Karger, Basel

Frederiksen B, Koch C, Høiby N (1999) The changing epidemiology of Pseudomonas aeruginosa infection in Danish cystic fibrosis patients, 1974–1995. Pediatr Pulmonol 28:159–166

Frederiksen B, Lanng S, Koch C et al (1996) Improved survival in the Danish cystic fibrosis centre: results of aggressive treatment. Pediatr Pulmonol 21:153–158

Frederiksen B, Pressler T, Hansen A et al (2006) Effect of aerosolised rhDNase (Pulmozyme®) on pulmonary colonization in patients with cystic fibrosis. Acta Paediatrica 95:1070–1074

Frederiksen B., Pressler T., Hansen A. et al (2006) Effect of aerosolised rhDNase (Pulmozyme®) on pulmonary colonization in patients with cystic fibrosis. Acta Paediatrica 95:1070–1074

Geller DE, Pitlick WH, Nardella PA et al (2002) Pharmacokinetics and bioavailability of aerosolized tobramycin in cystic fibrosis. Chest 122:219–226

Gibson RL, Emerson J, McNamara S et al (2003) Significant microbiological effect of inhaled tobramycin in young children with cystic fibrosis. Amer J Respir Crit Care Med 167: 841–849

Giwercman B, Lambert PA, Rosdahl VT et al (1990) Rapid emergence of resistance in Pseudomonas aeruginosa in cystic fibrosis patients due to in vivo selection of stable partially derepressed beta-lactamase producing strains. J Antimicrob Chemother 26: 247–259

Giwercman B, Meyer C, Lambert PA et al (1992) High-level beta-lactamase activity in sputum samples from cystic fibrosis patients during antipseudomonal treatment. Antimicrob Agents Chemother 36:71–76

Goldstein W, Döring G (1986) Lysosomal enzymes from polymorphonuclear leukocytes and proteinase inhibitors in patients with cystic fibrosis. Am Rev Respir Dis 134:49–56

Hansen CR, Pressler T, Høiby N et al (2008) Early aggressive eradication therapy for intermittent Pseudomonas aeruginosa airway colonization in cystic fibrosis patients: 15 years experience. J Cystic Fibrosis 7:523–530

Hansen CR, Pressler T, Koch C et al (2005) Long-term azithromycin treatment of cystic fibrosis patients with chronic P. aeruginosa infection; an observational cohort study. J Cystic Fibrosis 4:35–40

Hassett DJ, Cuppoletti J, Trapnell B et al (2002) Anaerobic metabolism and quorum sensing by Pseudomonas aeruginosa biofilms in chronically infected cystic fibrosis airways: rethinking antibiotic treatment strategies and drug targets. Advan Drug Delivery Rev 54:1425–1443

Haussler S (2004) Biofilm formation by the small colony variant phenotype of Pseudomonas aeruginosa. Environ Microbiol 6:546–551

Haussler S, Tummler B, Weissbrodt H et al (1999) Small-colony variants of Pseudomonas aeruginosa in cystic fibrosis. Clin Infect Dis 29:621–625

Hoboth C, Hoffmann R, Eichner A et al (2009) Dynamics of adaptive microevolution of hypermutable pseudomonas aeruginosa during chronic pulmonary infection in patients with cystic fibrosis. J Infect Dis 200:118–130

Hoffmann N, Lee B, Hentzer M et al (2007) Azithromycin blocks quorum sensing and alginat polymer formation and increases the sensitivity to serum and stationary gowth phase killing of P. aeruginosa and attenuates chronic P. aeruginosa lung infection in Cftr-/- mice. Antimicrob Agents Chemother

Hoffmann N, Rasmussen TB, Jensen PO et al (2005a) Novel mouse model of chronic Pseudomonas aeruginosa lung infection mimicking cystic fibrosis (Vol 73, pg 2504, 2005). Infec Immunity 73:5290

Hoffmann N, Rasmussen TB, Jensen PØ et al (2005b) Novel mouse model of chronic Pseudomonas aeruginosa lung infection mimicking cystic fibrosis. Infect Immun 73:2504–2514

Hull J, Vervaart P, Grimwood K et al (1997) Pulmonary oxidative stress response in young children with cystic fibrosis. Thorax 52:557–560

Høiby N (1977) Pseudomonas aeruginosa infection in cystic fibrosis. Diagnostic and prognostic significance of Pseudomonas aeruginosa precipitins determined by means of crossed immunoelectrophoresis. A survey. Acta Pathol Microbiol Scand Suppl 262 (C):3–96

Høiby N (2006) *P. aeruginosa* in cystic fibrosis patients resists host defenses, antibiotics. Microbe (ASM) 1:571–577

Høiby N, Flensborg EW, Beck B et al (1977) Pseudomonas aeruginosa infection in cystic fibrosis. Diagnostic and prognostic significance of Pseudomonas aeruginosa precipitins determined by means of crossed immunoelectrophoresis. Scand J Resp Dis 58:65–79

Høiby N, Frederiksen B, Pressler T (2005) Eradication of early Pseudomonas aeruginosa infection. J Cystic fibrosis 4:49–54

Høiby N, Johansen HK, Moser C et al (2001) Pseudomonas aeruginosa and the biofilm mode of growth. Microb Infect 3:1–13

Høiby N, Pedersen SS (1989) Estimated risk of cross-infection with Pseudomonas aeruginosa in Danish Cystic Fibrosis patients. Acta Paediat Scand 78:395–404

Haagensen J, Klausen M, Ernst RK et al (2007) Differentiation and distribution of colistin- and sodium dodecyl sulfate-tolerant cells in Pseudomonas aeruginosa biofilms. J Bacteriol 189: 28–37

Islam S, Oh H, Jalal S et al (2009) Chromosomal resistance mechanisms for aminoglycosides in Pseudomonas aeruginosa cystic fibrosis isolates. Clin Microbiol Infect 15:60–66

Jalal S, Ciofu O, Høiby N et al (2000) Molecular mechanisms of fluoroquinolone resistance in Pseudomonas aeruginosa isolates from cystic fibrosis patients (Vol 44, pg 710, 2000). Antimicrob Agents Chemother 44:1410

Jelsbak L, Johansen HK, Frost A-L et al (2007) Molecular epidemiology and dynamics of Pseudomonas aeruginosa populations in lungs of cystic fibrosis patients. Infect Immun 75:2214–2224

Jensen PØ, Bjarnsholt T, Phipps R et al (2007) Rapid necrotic killing of polymorphonuclear leukocytes is caused by quorum-sensing-controlled production of rhamnolipid by Pseudomonas aeruginosa. Microbiology 153:1329–1338

Jensen T, Pedersen SS, Garne S et al (1987) Colistin inhalation therapy in cystic fibrosis patients with chronic Pseudomonas aeruginosa lung infection. J Antimicrob Chemother 19:831–838

Johansen HK (1996) Potential of preventing Pseudomonas aeruginosa lung infections in cystic fibrosis patients: experimental studies in animals. APMIS 104:5–42

Johansen HK, Moskowitz SM, Ciofu O et al (2008) Spread of colistin-resistant non-mucoid Pseudomonas aeruginosa among chronically infected Danish cystic fibrosis patients. J Cystic Fibrosis 7:391–397

Keren I, Kaldalu N, Spoering A et al (2004) Persister cells and tolerance to antimicrobials. Fems Microbiol Lett 230:13–18

Kobayashi H (1995) Biofilm disease: its clinical manifestation and therapeutic possibilities of macrolides. Am J Med 99:S26–S30

Koch C, Hjelt K, Pedersen SS et al (1991) Retrospective clinical study of hypersensitivity reactions to aztreonam and six other beta-lactam antibiotics in cystic fibrosis patients receiving multiple treatment courses. Rev Infect Dis 13:S608–S611

Kolpen M, Hansen CR, Bjarnsholt T et al (2010) Polymorphonuclear leukocytes consume oxygen in sputum from chronic *Pseudomonas aeruginosa* pneumonia in cystic fibrosis. Thorax 65(1):57–62

Lang AB, Schaad UB, Rudeberg A et al (1995) Effect of high-affinity anti-Pseudomonas aeruginosa lipopolysaccharide antibodies induced by immunization on the rate of Pseudomonas aeruginosa infection in patients with cystic fibrosis. J Pediatr 127:711–717

Le Brun PPH (2001) Optimization of antibiotic inhalation therapy in cystic fibrosis. Studies on nebulized tobramycin. Development of a colistin dry powder inhaler system. University of Groningen, Holland

Lee B, Haagensen JAJ, Ciofu O et al (2005) Heterogeneity of biofilms formed by non-mucoid Pseudomonas aeruginosa isolates from patients with cystic fibrosis. J Clin Microbiol 43: 5247–5255

Levy J, Smith AL, Koup JR, Williams-Warren, J, Ramsey, B (1984) Disposition of tobramycin in patients with cystic fibrosis: a prospective controlled study. J Pediat 105:117–124

Macia MD, Blanquer D, Togores B et al (2005) Hypermutation is a key factor in development of multiple-antimicrobial resistance in Pseudomonas aeruginosa strains causing chronic lung infections. Antimicrob Agents Chemother 49:3382–3386

Macia MD, Borrell N, Perez JL et al (2004) Detection and susceptibility testing of hypermutable Pseudomonas aeruginosa strains with the Etest and disk diffusion. Antimicrob Agents Chemother 48:2665–2672

MacLeod DL, Nelson LE, Shawar RM et al (2000) Aminoglycoside-resistance mechanisms for cystic fibrosis Pseudomonas aeruginosa isolates are unchanged by long-term, intermittent, inhaled tobramycin treatment. J Infec Dis 181:1180–1184

Mandsberg LF, Ciofu O, Kirkby N et al (2009) Antibiotic resistance in P. aeruginosa strains with increased mutation frequency due to inactivation of the DNA oxidative repair system. Antimicrob Agents Chemother 53:2483–2491

Mathee K, Ciofu O, Sternberg C et al (1999) Mucoid conversion of Pseudomonas aeruginosa by hydrogen peroxide: a mechanism for virulence activation in the cystic fibrosis lung. Microbiology 145:1349–1357

Miller RA, Britigan BE (1997) Role of oxidants in microbial pathophysiology. Clin Microbiol Rev 10:1–18

Moser C, van Gennip M, Bjarnsholt T et al (2009) Novel experimental Pseudomonas aeruginosa lung infection model mimicking long-term host-pathogen interactions in cystic fibrosis. APMIS 117:95–107

Moskowitz SM, Foster JM, Emerson J et al (2004) Clinically feasible biofilm susceptibility assay for isolates of Pseudomonas aeruginosa from patients with cystic fibrosis. J Clin Microbiol 42:1915–1922

Oliver A, Baquero F, Blazquez J (2002) The mismatch repair system (MutS, mutL and uvrD genes) in Pseudomonas aeruginosa: molecular characterization of naturally occurring mutants. Mol Microbiol 43:1641–1650

Oliver A, Canton R, Campo P et al (2000) High frequency of hypermutable Pseudomonas aeruginosa in cystic fibrosis lung infection. Science 288:1251–1253

Pamp SJ, Gjermansen M, Johansen HK et al (2008) Tolerance to the antimicrobial peptide colistin in Pseudomonas aeruginosa biofilms is linked to metabolically active cells, and depends on the prm and mexAB-oprM genes. Mol Microbiol 68:223–240

Pedersen SS, Espersen F, Høiby N et al (1990a) Immunoglobulin-A and immunoglobulin-G antibody responses to alginates from *Pseudomonas aeruginosa* in patients with cystic fibrosis. J Clin Microbiol 28:747–755

Pedersen SS, Høiby N, Espersen F et al (1992) Role of alginate in infection with mucoid Pseudomonas aeruginosa in cystic fibrosis. Thorax 47:6–13

Pedersen SS, Kharazmi A, Espersen F et al (1990b) Pseudomonas Aeruginosa alginate in cystic fibrosis sputum and the inflammatory response. Infect Immun 58:3363–3368

Permin H, Koch C, Høiby N et al (1983) Ceftazidime treatment of chronic Pseudomonas aeruginosa respiratory tract infection in cystic fibrosis. J Antimicrob Chemother 12(supplementum A):313–323

Pressler T, Frederiksen B, Skov M et al (2006) Early rise of anti-Pseudomonas antibodies and a mucoid phenotype of Pseudomonas aeruginosa are risk factors for development of chronic lung infection – a case control study. J Cystic Fibrosis 5:9–15

Proesmans M, Balinska-Miskiewicz W, Dupont L et al (2006) Evaluating the "Leeds criteria" for Pseudomonas aeruginosa infection in a cystic fibrosis center. Eur Respir J 27:937–943

Rainey PB, Travisano M (1998) Adaptive radiation in a heterogenous environment. Nature 394:6 9–72

Ramsey BW, Pepe MS, Quan JM et al (1999) Intermittent administration of inhaled tobramycin in patients with cystic fibrosis. N Engl J Med 340:23–30

Ratjen F, Rietschel E, Kasel D et al (2006) Pharmacokinetics of inhaled colistin in patients with cystic fibrosis. J Antimicrob Chemother 57:306–311

Ratjen F, Tummler B (1999) Comparison of the in vitro and in vivo response to inhaled DNase in patients with cystic fibrosis. Thorax 54:91
Saiman L, Mehar F, Niu WW et al (1996) Antibiotic susceptibility of multiply resistant Pseudomonas aeruginosa isolated from patients with cystic fibrosis, including candidates for transplantation. Clin Infect Dis 23:532–537
Shah PL, Scott SF, Fuchs HJ et al (1995) Medium term treatment of stable stage cystic fibrosis with recombinant human DNase I. Thorax 50:333–338
Smith AL, Fiel SB, MayerHamblett N et al (2003) Susceptibility testing of Pseudomonas aeruginosa isolates and clinical response to parenteral antibiotic administration – Lack of association in cystic fibrosis. Chest 123:1495–1502
Smith EE, Buckley DG, Wu Z et al (2006) Genetic adaptation by Pseudomonas aeruginosa to the airways of cystic fibrosis patients. Proc Natl Acad Sci USA 103(22):8487–8492
Song ZJ, Wu H, Ciofu O et al (2003) Pseudomonas aeruginosa alginate is refractory to Th1 immune response and impedes host immune clearance in a mouse model of acute lung infection. J Med Microbiol 52:731–740
Spiers AJ, Buckling A, Rainey PB (2000) The causes of Pseudomonas diversity. Microbiology 146:2345–2350
Stover CK, Pham XQ, Erwin AL et al (2000) Complete genome sequence of Pseudomonas aeruginosa PAO1, an opportunistic pathogen. Nature 406:959–964
Tateda K, Comte R, Pechere JC et al (2001) Azithromycin inhibits quorum sensing in Pseudomonas aeruginosa. Antimicrob Agents Chemother 45:1930–1933
Tiddens HAWM (2002) Detecting early structural lung damage in cystic fibrosis. Pediatr Pulmonol 34:228–231
Valerius NH, Koch C, Hoiby N (1991) Prevention of chronic Pseudomonas aeruginosa colonisation in cystic fibrosis by early treatment. Lancet 338:725–726
WestbrockWadman S, Sherman DR, Hickey MJ et al (1999) Characterization of a Pseudomonas aeruginosa efflux pump contributing to aminoglycoside impermeability. Antimicrob Agents Chemother 43:2975–2983
Westh JB (2001) Pulmonary physiology and pathophysiology. Lippincott Williams & Wilkins, Philadelphia, PA
Wolter DJ, Black JA, Lister PD et al (2009) Multiple genotypic changes in hypersusceptible strains of Pseudomonas aeruginosa isolated from cystic fibrosis patients do not always correlate with the phenotpe. J Antimicrob Chemother 64:294–300
Worlitzsch D, Tarran R, Ulrich M et al (2002) Effects of reduced mucus oxygen concentration in airway Pseudomonas infections of cystic fibrosis patients. J Clin Invest 109:317–325
Wyckoff TJO, Thomas B, Hassett DJ et al (2002) Static growth of mucoid Pseudomonas aeruginosa selects for non-mucoid variants that have acquired flagellum-dependent motility. Microbiology Sgm 148:3423–3430
Yang L, Haagensen JAJ, Jelsbak L et al (2008) In situ growth rates and biofilm development of Pseudomonas aeruginosa populations in chronic lung infection. J Bacteriol 190:2767–2776

Chapter 11
Innate Immune Response to Infectious Biofilms

Peter Østrup Jensen and Claus Moser

11.1 Introduction

The term "innate immunity" was first promoted by Ilya Metchnikoff to describe the ability of phagocytes to protect the host from infection. For his achievements in innate immunity, Metchnikoff together with Paul Ehrlich, the father of adaptive immunity, were awarded the Nobel prize in 1908. This event made it official: immunology had become an academic discipline in its own right. However, it was soon to be realized that the new discipline was born with dichotomous thinking in the attempt to decide whether the nature of a particular immune response is innate or adaptive. Over the years mutually exclusive perceptions of innate and adaptive immunity have divided immunologists, but the present prevailing view appreciates that the principles of Metchnikoff and Ehrlich are complementary (Kaufmann 2008, Nathan 2008). Ironically, based on their contemporary observations, already the two befriended laureates were actually valuing their findings as being complementary mechanisms of immunity (Nobel Lectures). The brilliant approach of Metchnikoff to study living, transparent animals undisputedly revealed the spontaneous afflux of mobile host cells toward points of infectious microbes. Though these mobile cells play no part in the digestion of food, the capability of the mobile cells to store foreign bodies (Nobel Lectures) has provided a rich food for thoughts to inspire research, which has resulted in identifying several underlying mechanisms of phagocytosis and additional mechanisms with a natural reaction to intruding microorganisms. Considering the immune response towards biofilm this book aims to display both the innate and adaptive response as well as their complementary interphase. This chapter is particularly focused on the innate immune response to infectious biofilms. In this respect, it should be noted that while Metchnikoff apparently conducted his experiments with planktonic microorganism; much less is known about the innate response to a pure biofilm infection. This is probably caused by the laborious process of demonstrating biofilms in infected patients, by

P.Ø. Jensen (✉)
H:S Rigshospitalet, Department for Clinical Microbiology, afsnit 7602, Juliane Maries Vej 22, DK-2100 Copenhagen Ø, Denmark
e-mail: peter.oestrup.jensen@rh.regionh.dk

T. Bjarnsholt et al. (eds.), *Biofilm Infections*, DOI 10.1007/978-1-4419-6084-9_11,

difficulties establishing biofilm infections without the nearly inevitable shedding of planktonic cells, and by the relatively short period of time in which the immune response to biofilm has been studied.

11.2 Definition

Innate immunity consists of germline encoded, nonclonal mechanisms that provide nonspecific protection against pathogens. All animals possess innate immunity systems with the capacity to recognize and eradicate intruding microorganisms by mechanisms that are not influenced inherently by repeated encounters with infectious intruders (Kimbrell 2001). Several innate antimicrobial systems have been recognized and may be divided into resident and mobile barriers. Biofilm infections have mainly been reported from patients, whose resident barriers are defective or have been bypassed by the introduction of artificial surfaces, such as implants, or by the appearance of endogenous inert surfaces like dead bone. While the mobile barriers may be considered to effectuate the major contribution of the innate immune response to the pathogenesis during biofilm infections. Excellent reviews of the innate response to infections are already broadly available, but their main focus has not been the particular innate response to biofilm infections. Therefore, the studies included in this chapter are mainly selected and discussed according to their ability to provide evidence for innate responses to biofilm infections as opposed to the response to infections with planktonic bacteria.

11.3 Recognition of Biofilms

The innate immune system is the first line of the defensive mechanisms for protecting the host from invading microbial pathogens (Akira et al. 2001). The ability to detect intruding microorganisms is aided by pattern recognition receptors (PRRs) that recognize conserved microbial pathogen-associated molecular patterns (PAMPs) and signal their presence resulting in activation of host response. Even though several classes of PRRs and their ligands are known, PRRs specific for biofilm-growing microorganisms have not yet been identified. This makes highlighting of PRRs with particular importance for microbial biofilms as opposed to planktonic populations, a dubious task, and rather suggests the relevance of briefly reviewing the major PRR types, which due to the resilience of biofilms may be subject to continuous stimulation resulting in activation and modulation of both the innate and the adaptive immune response.

11.3.1 Secreted PRRs

Several PRRs are secreted from the cells that produce them including complement receptors, collectins, serum amyloid, C-reactive protein, and peptidoglycan recognition proteins. Of these receptors the interaction with biofilms has been most intensely studied for the complement receptors.

11.3.2 Complement Receptors

Activation of the complement cascade may be triggered by interactions between the PRRs of the complement system and their corresponding PAMPs. At least five soluble complement receptors [C1, C2, C3, C4, and Mannan-binding lectin (MBL)] scout for PAMPs. Binding of PRRs of the complement system to PAMPs has the potential to initiate the classical, the lectin, and the alternative pathways. All these cascades depend on subsequent enzymatic activation of C3 and C5 to exert their terminal actions, including mediation of inflammation, opsonization of pathogens, and lysis of pathogens by formation of the membrane-attack complex (Rambach et al. 2008).

A role of the complement system during biofilm infections is far from firmly established. Apparently, no cases of biofilm infections have been reported for patients with complement deficiency so far. This may rely on difficulties in establishing the presence of biofilms in clinical samples, but may also, in part, be due to the ability of biofilm growing microorganisms to obtain resistance against the complement system and thus to establish biofilm infections in spite of activation of the complement systems. In cystic fibrosis (CF) (see Chapter 10), activated complement (C3c) was more frequent in the sputum from patients with chronical *P. aeruginosa* lung infection (Schiøtz et al. 1979). An in vitro study has suggested lipopolysaccharide as a main mediator of complement activation by *P. aeruginosa*. However, it could not be demonstrated that complement activation was only due to biofilm formation since planktonic bacteria activated the complement system more than biofilm bacteria (Jensen et al. 1993) and both planktonic and biofilm growing *P. aeruginosa* are frequently found in the same samples from CF patient (see Chapter 10). In fact, purified alginate, the main component of the matrix of the virulent mucoid biofilms, from *P. aeruginosa* was unable to induce substantial complement activation (Pedersen et al. 1990a). Nevertheless, protection against complement opsonization may be obtained in mucoid *P. aeruginosa* biofilms due to the high content of alginate with O-acetylation (Pier et al. 2001). Furthermore, evasion from binding to the soluble complement receptor, MBL, and from the subsequent complement activation has been demonstrated in *P. aeruginosa* isolated from CF sputum samples (Davies et al. 2000). This evasion from the lectin-pathway by *P. aeruginosa* may explain why colonization with *P. aeruginosa* was a more frequent finding in MBL-sufficient CF patients than in MBL-deficient CF patients (Carlsson et al. 2005). In addition to the protection against the complement system obtained by biofilm formation of *P. aeruginosa*, *Mycoplasma pulmonis* may also acquire protection from lytic effects of the complement system by biofilm generation in vitro (Simmons et al. 2007) and the masking of underlying cell wall lipids with glycopeptidolipid, due to biofilm formation, may facilitate colonization with *Mycobacterium abscessus* by preventing binding of MBL and the consequent stimulation of alveolar macrophages through interactions with TLR2 (Rhoades et al. 2009). On the contrary, sputum isolates of *Burkholderia cepacia*, which may also grow as biofilms in CF airways, were able to bind MBL resulting in complement activation (Davies et al. 2000).

11.3.3 Membrane Bound Receptors

The list of membrane bound PRRs is constantly growing and may be divided into Toll-like receptors (TLRs), scavenger receptors, C-type lectin receptors, mannose receptors, dendritic cell-specific intercellular adhesion molecule-grabbing nonintegrin, and complement receptors. No member of any of these classes of PRRs has so far been demonstrated to mount a response specific for biofilm infections. But, since recent studies have demonstrated TLR-mediated responses to matrix components of biofilms and to bacterial products of both planktonic and biofilm infections, TLRs are particularly emphasized in this chapter.

11.3.4 Toll-Like Receptors (TLRs)

The TLRs are integral membrane glycoproteins with cytoplasmic tails that transmit intracellular signalling. The extracellular region of TLRs contains leucine-rich repeat motifs, which are thought to be involved directly in the recognition of various pathogens (Akira et al. 2004). In human, ten different TLRs have been identified and genetic studies have revealed their respective ligands (Takeda 2005). PAMP recognition by TLRs provokes rapid activation of two signalling pathways: a MyD88-dependent pathway with fast activation of nuclear factor-κB NFκB and mitogen-activated protein kinase (MAPK) that promotes generation of pro-inflammatory cytokines, and a MyD88-independent pathway with slow activation of NFκB and MAPK associated with the induction of IFN-beta and IFN-inducible genes and maturation of dendritic cells (Akira et al. 2004, Parker et al. 2007).

The importance of the involvement of TLRs in the host defence against biofilms is less clear than the significance of TLRs during acute (planktonic) infections. In mice lacking MyD88, that transduces a core set of TLR-induced signals; impaired early clearance of planktonic *P. aeruginosa* has been demonstrated (Power et al. 2004, Skerrett et al. 2005) while the later host response was independent of MyD88 (Power et al. 2006). Evidence for the involvement of TLRs in clinical biofilm infections has mainly accumulated from chronically infected CF patients (see Chapter 10) and from dental plaques (see Chapter 4). Because the pathogen recognition capacity of neutrophils mostly rely on TLRs (Parker et al. 2005) investigations of the expression of TLRs on the neutrophils from the airways of chronically infected CF patients have recently emerged. A higher number of TLR4-positive cells, mainly neutrophils, were found in the submucosa of patients with CF (Hauber et al. 2005). This may be ascribed to increased endobronchial accumulation of neutrophils that have increased TLR4 expression (Petit-Bertron et al. 2008) rather than increased TLR4 expression on the neutrophils since TLR5 was the only MyD88-dependent TLR that was increased on neutrophils from the lungs of chronically infected CF patients (Koller et al. 2008). This elevated TLR5 expression on the neutrophils was probably mediated by IL-8, TNF-α, G-CSF, and by the interaction of TLR1 and TLR2 resulting from the binding to the bacterial lipoprotein (Koller et al. 2008). Whether the expression of the flagellin receptor, TLR5 (Hayashi et al. 2001), results in an increased innate response against the biofilms in the CF lungs is difficult to decide because flagella were absent in biofilm growing mucoid *P. aeruginosa* isolates from CF airways (Garrett et al. 1999). But the TLR5-mediated enhanced phagocytosis may contribute to the host defence against the planktonic, flagellin-intact *P. aeruginosa* subpopulations in the CF airways. In fact, apparently only planktonic bacteria were engulfed by PMNs in explanted lungs and sputum from chronically infected CF patients (Bjarnsholt et al. 2009, Kolpen et al. 2010), and acute lung infections with flagellin-defective planktonic *P. aeruginosa* were cleared later in mice (Balloy et al. 2007). In the quest for identifying a biofilm specific innate response the ability of bacterial DNA, being a matrix component of biofilms (Whitchurch et al. 2002), to activate neutrophils has been examined. Recent studies have established activation of neutrophils by bacterial DNA through TLR9-independent mechanisms resulting in upregulation of intracellular signalling pathways and IL-8 production (Alvarez et al. 2006, Fuxman Bass et al. 2008). As opposed to the intracellular distribution of the well-established DNA recognizing TLR9 (Hemmi et al. 2000), the triggering of the neutrophils by bacterial DNA involved a surface molecule (Trevani et al. 2003) that has not yet been identified. Some DNA may, however, also be released by planktonic bacteria (Allesen-Holm et al. 2006) and, therefore, the establishment of the neutrophil response to the biofilm matrix component, DNA, as a specific innate response to biofilms rely on the outcome of studies involving neutrophil activation, identification, and distribution of the involved surface molecule during interactions with biofilm growing bacteria and planktonic bacteria. Increased alginate content is another hallmark of the matrix in the mucoide *P. aeruginosa* biofilm and is considered the strongest virulence factor in chronically infected CF patients (See Chapter 10). Though neutrophils are able to respond to alginate by an increased respiratory burst (Pedersen et al. 1990a) and alginate may stimulate monocytes to produce cytokines (Otterlei et al. 1993) in vitro, the involved receptors have not been clarified. It has, however, been demonstrated that both TLR2 and TLR4 are participating in the activation of monocytes by the mannuronic acid polymeric components of alginate produced by *P. aeruginosa* (Flo et al. 2002).

TLR-pathways also provide macrophages with the ability for PAMP recognition and response, and because macrophages are distributed throughout the body they constitute a major alarm system of the innate immune response to infection (Takeda et al. 2003). But it is not clear yet whether the TLR-pathways of macrophages lead to a particular response to biofilm infections.

Dendritic cells are like macrophages derived from monocytes and equipped with functional TLR-pathways. By presenting phagocytosed and processed microbial intruders to naïve T-cells in the lymph nodes, the dendritic cells are well known to direct the acquired immune response (Kapsenberg 2003, Colonna et al. 2006). Activation of TLR induces the maturation of dendritic cells into professional antigen presenting cells with increased surface antigen presentation and expression of CD80/CD86 that promotes the interaction with CD28 on T cells, which is required for the activation and differentiation of naïve T cells (Akira et al. 2001). Recently, the involvement of TLR-pathways of dendritic cells in directing the acquired immune response towards a Th1 or Th2 response (See Chapter 12) has been demonstrated. In these studies, induction of a strong Th1 cell development by TLR-pathways of dendritic cells has been achieved when stimulating TLR4 with LPS (Kaisho et al. 2002), TLR7 with imidazoquinolines (Ito et al. 2002), and TLR9 with CpG DNA (Boonstra et al. 2003). This ability of TLR activation on dendritic cells to direct a Th1 response may also apply during microbial infection in vivo (Jankovic et al. 2002). Interestingly, the Th-response during chronic lung infection with *P. aeruginosa* biofilms in cystic fibrosis is dominated by the Th2 type (See Chapter 12). Therefore, the contribution of TLR-pathways of dendritic cells to the development of the predominant Th2 response may depend on binding of LPS to TLR4 since TLR4 was necessary for optimal development of Th2-type immune responses (Dabbagh et al. 2002). Alternatively, the maturation of the dendritic cells during chronic lung infection with *P. aeruginosa* biofilms in CF is likely to be influenced by the pro-inflammatory environment of the infected lungs where proteases, in particular neutrophil elastase, may cleave part of the extracellular receptors on the dendritic cells (Roghanian et al. 2006) and pro-inflammatory cytokines, such as G-CSF, favour the maturation of dendritic cells with a Th2 stimulating capacity (See Chapter 12).

The first resident part of the innate immune systems to encounter intruding microbes is the epithelial cell layer. Epithelial cells line the body and harbour a vast array of antimicrobial functions including release of ions, defensins, ingestion of bacteria, and recruitment of host cells (Shimono et al. 2003, Walsh et al. 2003, Cutler et al. 2006, Bartlett et al. 2009). All the required components of the TLR signalling pathway to mount a response and participate in the protection of the respiratory tract from inhaled pathogens are present in airway epithelial cells (Muir et al. 2004) and in the epithelial cells of the oral mucosa (See Chapter 4). Sensing of *P. aeruginosa* and other CF pathogens results in selective mobilization of TLR2 to the apical surface of the airway cells (Muir et al. 2004). The response to *P. aeruginosa* in human airway epithelia is also mediated by interactions between TLR5 and flagellin (Zhang et al. 2005), which, as already mentioned, is absent in the biofilm growing mucoid isolates from chronically infected CF patients (Garrett et al. 1999). Soluble components, such as lipopeptides, LPS, and bacterial DNA, that may be derived from infectious biofilms, generate a TLR-mediated response in cultures of human airway epithelial cells (Green et al. 2005), which further suggests that TLRs are involved in the epithelial innate response to biofilms.

As opposed to the awaited demonstration of specific innate receptors for biofilms, soluble receptors of the acquired immune response against alginate, which is upregulated in the virulent mucoid *P. aeruginosa* biofilms, has long been recognized in the form of alginate-specific antibodies in the sera of CF patients with chronic *P. aeruginosa* lung infection (Pedersen et al. 1990b) (See Chapter 12). Moreover, generation of specific antibodies directed against biofilm-associated upregulated cell wall components have been demonstrated in experimental osteomyelitis with *Staphylococcus aureus* (Brady et al. 2006). Due to the presence of Fc-receptors on the neutrophils these specific antibodies may mediate the recognition of biofilms by the neutrophils.

11.4 The Innate Response to Biofilms

The response to PAMPs by PRR-pathways may lead to the activation and translation of a multitude of pro-inflammatory cytokine genes and increased antimicrobial activity of the cellular components of the innate immune system (Akira et al. 2004, Parker et al. 2007). Solid arguments for a response to biofilm infections by cellular components of the mobile innate barriers are provided from cystic fibrosis and chronic wounds. Intense accumulation of polymorphonuclear leukocytes (PMNs), in particular neutrophils, at the site of the biofilms has been demonstrated in explanted lungs from chronically infected cystic fibrosis patient (Bjarnsholt et al. 2009) and in biopsies from chronic wounds (Bjarnsholt et al. 2008, Kirketerp et al. 2008). Apparently, the surrounding of the biofilms by PMNs was not dependent on the presence of detectable planktonic bacteria. In addition to attracting the PMNs, the biofilms in the endobronchial secretions in the respiratory tract of chronically infected CF patients may stimulate the respiratory burst of the summoned PMNs resulting in generation of reactive oxygen species and oxygen depletion (Kolpen et al. 2010), which is a hallmark of the infected endobronchial mucus (Wörlitzsch et al. 2002). Though the immune response in these chronic infections is likely to be affected by the acquired response (See Chapter 12), intense activation of the PMNs by biofilms that resembles the situation in infected CF airways and chronically infected wounds has been obtained in several experimental studies with no influence of the acquired immune response. When adding human PMNs to biofilm cultures, which in principle have been depleted from planktonic bacteria, the innate response of PMNs to *P. aeruginosa* biofilms resembles the situation in the patients by attachment to the biofilm and phenotypic-dependent penetration, phagocytosis, respiratory burst, and eradication (Jesaitis et al. 2003, Leid et al. 2005, Bjarnsholt et al. 2005, Jensen et al. 2007, van Gennip et al. 2009, Alhede et al. 2009). Early sampling before the establishment of the acquired immune response during experimental biofilm lung infections has also demonstrated that accumulation of activated PMNs in the airways is a part of the innate immune response to lung infections with *P. aeruginosa* biofilms (Jensen et al. 2004, 2007, Bjarnsholt et al. 2005, van Gennip et al. 2009, Alhede et al. 2009).

The chronic lung infection with *P. aeruginosa* in CF patients has inspired some of the most intensive studies of the innate response to biofilm infections. In the case of CF, the poor self-cleaning capacity of the airway epithelia, caused by thick layers of dehydrated, viscous mucus is believed to be the main reason for the frequent occurrence of chronic biofilm infections (see Chapter 10). In addition, recent in vitro data suggest that abnormal accumulation and loss of iron by CF airway epithelial cells promotes the formation of antibiotic-resistant *P. aeruginosa* biofilms (Moreau-Marquis et al. 2008). On the contrary, the main anti-bacterial mobile cellular component of the innate immune response, the neutrophils, is apparently competent in CF (Haggie et al. 2009), which is supported by the clinical observations that CF patients do not develop the acute infections normally associated with immune-deficiencies.

In this condition, the response by the PMNs has gained particular attention due the suspicion of detrimental effects of the numerous PMNs and their failure to eradicate the biofilm in the airways. The respiratory burst of the abundant PMNs in the lungs of CF patients with chronic *P. aeruginosa* lung infection is in general suspected to be a major cause of the ROS lesions of the lung tissue and may promote bacterial virulence. It is possible that the respiratory burst of the PMNs in infected airways is activated by direct contact with the bacteria, by LPS-immune complexes (Kronborg et al. 1993), or by alginate (Pedersen 1992), and the PMNs may be primed by bacterial endotoxins (Kharazmi et al. 1987) and by soluble components of the innate immune response such as, TNF-α, platelet-activating factor, leukotriene B4, and IL8 (Downey et al. 2009, Kharazmi et al. 1988, You et al. 1991, Jones et al. 2003, Jensen et al. 2006). In addition, the respiratory burst of the PMNs may be activated already during the migration in the inflamed tissue by the binary signalling from the engagement of the integrins and binding to inflammatory cytokines (Nathan et al. 1989). With 8000 L of oxygen-rich air being inhaled per day (Cantin et al. 2007) the active respiratory burst of the PMNs in the mucus is supplied with sufficient molecular oxygen to allow the activated phagocyte NADPH-oxidase to produce superoxide (Babior et al. 1973) from which several reactive oxygen species (ROS) is rapidly formed. In a recent study, we demonstrated the respiratory burst of the PMNs to be the main oxygen consuming activity in endobronchial secretions from chronically infected CF patients (Kolpen et al. 2010). Therefore, the respiratory burst of the PMNs is likely to be the major reason for the ROS lesions found in the infected CF lungs (Jones et al. 2000, van der Vliet et al. 2000, Balint et al. 2001, Kettle et al. 2004, van der Vliet 2008). Considering the short half-lives of the ROS generated by the respiratory burst of the PMNs, the tissue adjacent to the infected mucus is most likely to be exposed to PMN-derived ROS that may contribute to the damage of the bronchi seen in lungs from CF patients with chronic *P. aeruginosa* lung infection (Baltimore et al. 1989, Wiebe et al. 2006).

The ROS generated by the respiratory burst of the PMNs in the mucus may be released very close to and actually reach the bacteria, due to the intimate surrounding of the endobronchial biofilms by the PMNs (Fig 11.1a) (Bjarnsholt et al. 2009) and the uptake of planktonic bacteria (Fig 11.1b). Thus, the surviving bacteria may accumulate ROS lesions. In particular, the interaction between

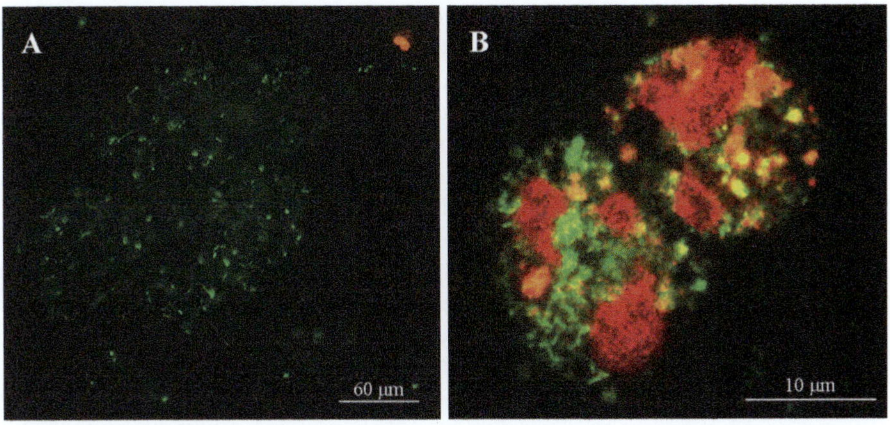

Fig. 11.1 Demonstration of biofilm-mediated protection against active PMNs in expectorated sputum from a CF patient with chronic *P. aeruginosa* lung infection. The sputum sample was stained with hydroethidine to visualize the respiratory burst in PMNs with red fluorescence and DNA was stained by the green fluorescent SYTO9. (**a**) Biofilm next to a PMN (*arrow*) with active respiratory burst failing to penetrate and to phagocytose the biofilm. (**b**) Active respiratory burst in PMNs with engulfed planktonic bacteria (*arrow*)

P. aeruginosa and phagocytosing PMNs may lead to increased content of 8-hydroxyguanine (8-oxodG) in the bacterial DNA caused by ROS from the respiratory burst (Ciofu et al. 2005). 8-oxodG lesions in DNA are highly mutagenic due to their ambiguous base-pairing properties, pairing with either A or C during DNA synthesis (Shibutani et al. 1991), and are associated with development of resistance to antibiotics (Oliver et al. 2000). Consequently, the ROS generation by PMNs may lead to the accumulation of hypermutable isolates from CF patients with chronic *P. aeruginosa* lung infection (Oliver et al. 2000) due to 8-oxodG lesions in the bacterial DNA (Ciofu et al. 2005). In addition, mutations in the *mucA* gene, caused by PMN derived ROS, may alter the virulence of *P. aeruginosa* leading to formation of mucoid phenotypes (Mathee et al. 1999), which are correlated to poor prognosis, deterioration of the lung function, and increased tissue damage (Pedersen 1992). In fact, more than 90% of the mucoid isolates from 91 Scandinavian patients with chronic *P. aeruginosa* infection contained mutations in the *mucA* gene (Ciofu et al. 2008). In addition to the generation of ROS by activated PMNs, the proteolytic activity underlying the tissue deterioration in chronically infected CF lungs is also regarded to be derived from activated PMNs due to extensive release of proteases (Suter et al. 1984, Mayer-Hamblett et al. 2007), rather than from bacterial proteases, which are subject to neutralisation by antibodies (See Chapters 10 and 12).

The accelerated oxygen depletion in the mucus by the PMNs may affect the relative distribution of anaerobic, microaerophilic, and normoxic zones in the mucus. Though the precise oxygen profile of the endobronchial mucus throughout the entire lungs is not determined, the activity of the PMNs is sensitive to the local oxygen tension. In particular, the postponing of apoptosis due to hypoxia (Hannah et al. 1995) may contribute to the accumulation of PMNs in the lung secretions thereby increasing the inflammation. Apparently, PMNs are able to produce a substantial part of the energy needed for phagocytosis by anaerobic glycolysis resulting in L-lactate formation (Borregaard et al. 1982). Accordingly, the increased glucose uptake in PMNs in CF lungs (Chen et al. 2006) and the high concentration of L-lactate in sputum from CF patients with chronic *P. aeruginosa* lung infection (Wölitzsch et al. 2007) support the presence of active PMNs in the anaerobic mucus in infected CF lungs. The bacterial response to oxygen depletion includes slow growth and cyanide secretion at microaerophilic condition (Alvarez-Ortega et al. 2007, Pessi et al. 2000) and alginate production induced by anaerobic conditions (Wölitzsch et al. 2002). Indeed, isolates with a long doubling time and production of cyanide have recently been observed in sputum from chronically infected CF patients (Yang et al. 2008, Ryall et al. 2008, Sanderson et al. 2008), while the presence of alginate producing *P. aeruginosa* has been recognized for years (Pedersen et al. 1989).

During the chronic *P. aeruginosa* lung infection, the PMNs accumulate in numbers that are balanced to the number of bacteria in the endobronchial secretions (Muhlebach et al. 1999, Kolpen et al. 2010). The mechanisms by which the host response manages to account for numbers of bacteria are not precisely known, but may involve the innate immune response. In this respect, granulocyte-colony stimulating factor (G-CSF) may control the number of circulating PMNs by stimulating proliferation and mobilization of PMNs in the bone marrow. Indeed, decreased serum levels of G-CSF were observed during depression of the lung infection with intensive antibiotic treatment and were associated with lover number of circulating PMNs and improved lung functions (Jensen et al. 2006).

Therefore, the formation of infectious biofilm in the lungs of CF patients may not only serve as an example of increased biofilm susceptibility due to defective resident barriers. But also demonstrates that the innate immune response may aid to the selection of biofilms during chronic infections.

11.5 Conclusion – Significance of the Innate Immune Response for Biofilm Formation

The mobile innate immune response may prevent formation of infectious biofilms as exemplified by the onset of infectious periodontal biofilms during neutropenia where the healthy balance between the periodontal innate immune response to the dental plaques cannot be maintained due to shortage of the bactericidal PMNs (see Chapter 4). Whether other infectious biofilms result from neutropenia or other disorders of the mobile innate immune response remains to be documented. The ability to prevent infectious biofilms by the resident innate immune barriers is evidenced in cases of defective epithelia such as chronic wounds (See Chapter 2), thermal wounds (see Chapter 16), and cystic fibrosis (see Chapter 10). Furthermore, the infectious biofilms frequently found on indwelling medical devices (see Chapters 5 and 6) may be a result of bypassing the antimicrobial capacity of epithelia by introducing inert, solid surfaces allowing the biofilm to attach and form a device-related reduction of the microbicidal activity of the summoned PMNs (see Chapter 5).

In addition to the preventive effects of the innate immune response, evidence has been presented in this chapter for promotion of infectious biofilms due to the innate immune response. As stated for CF, the response of competent PMNs may favour the presence of biofilm-growing bacteria by removing planktonic bacteria by phagocytosis and by inducing resistant biofilms due to the consequences of the respiratory burst. Whether these mechanisms apply to other infectious biofilms remains to be verified.

References

Akira S, Takeda K, Kaisho T (2001) Toll-like receptors: critical proteins linking innate and acquired immunity. Nat Immunol 2(8):675–680

Akira S, Takeda K (2004) Toll-like receptor signalling. Nat Rev Immunol 4(7):499–511

Alhede M, Bjarnsholt T, Jensen PØ et al (2009) Pseudomonas aeruginosa recognizes and responds aggressively to the presence of polymorphonuclear leukocytes. Microbiology 55(11): 3500–3508

Allesen-Holm M, Barken KB, Yang L et al (2006) A characterization of DNA release in Pseudomonas aeruginosa cultures and biofilms. Mol Microbiol 59(4):1114–1128

Alvarez ME, Fuxman Bass JI, Geffner JR et al (2006) Neutrophil signaling pathways activated by bacterial DNA stimulation. J Immunol 177(6):4037–4046

Alvarez-Ortega C, Harwood CS (2007) Responses of *Pseudomonas aeruginosa* to low oxygen indicate that growth in the cystic fibrosis lung is by aerobic respiration. Mol Microbiol 65(1):153–165

Babior BM, Kipnes RS, Curnuttte JT (1973) Biological defense mechanisms. The production by leukocytes of superoxide, a potential bactericidal agent. J Clin Invest 52(3):741–744

Balint B, Kharitonov SA, Hanazawa T et al (2001) Increased nitrotyrosine in exhaled breath condensate in cystic fibrosis. Eur Respir J 17(6):1201–1207

Balloy V, Verma A, Kuravi S et al (2007) The role of flagellin versus motility in acute lung disease caused by Pseudomonas aeruginosa. J Infect Dis 2007 196(2):89–296

Baltimore RS, Christie CD, Smith GJ (1989) Immunohistopathologic localization of *Pseudomonas aeruginosa* in lungs from patients with cystic fibrosis. Implications for the pathogenesis of progressive lung deterioration. Am Rev Respir Dis 140(6):1650–1661

Bartlett JA, Fischer AJ, McCray PB Jr (2008) Innate immune functions of the airway epithelium. In: Egesten A, Schmidt A, Herwald H (eds) Trends in Innate Immunity. Contrib Microbiol. Basel, Karger, 15:78–100

Bjarnsholt T, Jensen PØ, Burmølle M et al (2005) Pseudomonas aeruginosa tolerance to tobramycin, hydrogen peroxide and polymorphonuclear leukocytes is quorum-sensing dependent. Microbiology 151(2):373–383

Bjarnsholt T, Kirketerp-Møller K, Jensen PØ et al (2008) Why chronic wounds will not heal: a novel hypothesis. Wound Repair Regen 16(1):2–10

Bjarnsholt T, Jensen PØ, Fiandaca M J et al (2009) Pseudomonas aeruginosa biofilms in the respiratory tract of cystic fibrosis patients. Pediatr Pulmonol 44(6):547–558

Boonstra A, Asselin-Paturel C, Gilliet M et al (2003) Flexibility of mouse classical and plasmacytoid-derived dendritic cells in directing T helper type 1 and 2 cell development: dependency on antigen dose and differential toll-like receptor ligation. J Exp Med 197(1): 101–109

Borregaard N, Herlin T (1982) Energy metabolism of human neutrophils during phagocytosis. J Clin Invest 70(3):550–557

Brady RA, Leid JG, Camper AK et al (2006) Identification of Staphylococcus aureus proteins recognized by the antibody-mediated immune response to a biofilm infection. Infect Immun 74(6):3415–3426

Cantin AM, White TB, Cross CE et al (2007) Antioxidants in cystic fibrosis. Conclusions from the CF antioxidant workshop, Bethesda, Maryland, November 11–12, 2003. Free Radic Biol Med 42(1):15–31

Carlsson M, Sjöholm AG, Eriksson L et al (2005) Deficiency of the mannan-binding lectin pathway of complement and poor outcome in cystic fibrosis: bacterial colonization may be decisive for a relationship. Clin Exp Immunol 139(2):306–313

Chen DL, Ferkol TW, Mintun MA et al (2006) Quantifying pulmonary inflammation in cystic fibrosis with positron emission tomography. Am J Respir Crit Care Med 173(12): 1363–1369

Ciofu O, Riis B, Pressler T et al (2005) Occurrence of hypermutable *Pseudomonas aeruginosa* in cystic fibrosis patients is associated with the oxidative stress caused by chronic lung inflammation. Antimicrob Agents Chemother 49(6):2276–2282

Ciofu O, Lee B, Johannesson M et al (2008) Investigation of the algT operon sequence in mucoid and non-mucoid *Pseudomonas aeruginosa* isolates from 115 Scandinavian patients with cystic fibrosis and in 88 in vitro non-mucoid revertants. Microbiology 154(1):103–113

Colonna M, Pulendran B, Iwasaki A (2006) Dendritic cells at the hostpathogen interface. Nat Immunol 7(2):117–120

Cutler CW, Jotwani R (2006) Dendritic cells at the oral mucosal interface. J Dent Res 85(8): 678–689

Dabbagh K, Dahl ME, Stepick-Biek P et al (2002) Toll-like receptor 4 is required for optimal development of Th2 immune responses: role of dendritic cells. J Immunol 168(9):4524–4530

Davies J, Neth O, Alton E et al (2000) Differential binding of mannose-binding lectin to respiratory pathogens in cystic fibrosis. Lancet 355(9218):1885–1886

Downey DG, Bell SC, Elborn JS (2009) Neutrophils in cystic fibrosis. Thorax 64(1):81–88

Flo TH, Ryan L, Latz E et al (2002) Involvement of toll-like receptor (TLR) 2 and TLR4 in cell activation by mannuronic acid polymers. J Biol Chem 277(38):35489–35495

Garrett ES, Perlegas D, Wozniak DJ (1999) Negative Control of flagellum synthesis in *Pseudomonas aeruginosa* is modulated by the Alternative Sigma Factor AlgT (AlgU). J Bacteriol 181(23):7401–7404

Greene CM, Carroll TP, Smith SG et al (2005) TLR-induced inflammation in cystic fibrosis and non-cystic fibrosis airway epithelial cells. J Immunol 174(3):1638–1646

Fuxman Bass JI, Gabelloni ML, Alvarez ME et al (2008) Characterization of bacterial DNA binding to human neutrophil surface. Lab Invest 88(9):926–937

Haggie PM, Verkman AS (2009) Defective organellar acidification as a cause of cystic fibrosis lung disease: reexamination of a recurring hypothesis. Am J Physiol Lung Cell Mol Physiol 296(6):L859–L867

Hannah SK, Mecklenburgh I, Rahman GJ et al (1995) Hypoxia prolongs neutrophil survival in vitro. FEBS Lett. 372(2–3):233–237

Hauber HP, Tulic MK, Tsicopoulos A et al (2005) Toll-like receptors 4 and 2 expression in the bronchial mucosa of patients with cystic fibrosis. Can Respir J 12(1):13–18

Hayashi F, Smith KD, Ozinsky A et al (2001) The innate immune response to bacterial flagellin is mediated by Toll-like receptor 5. Nature 410(6832):1099–1103

Hemmi H, Takeuchi O, Kawai T et al (2000) A Toll-like receptor recognizes bacterial DNA. Nature 408(6813):740–745

Ito T, Amakawa R, Kaisho T et al (2002) Interferon-a and interleukin-12 are induced differentially by Toll-like receptor 7 ligands in human blood dendritic cell subsets. J Exp Med 195(11): 1507–1512

Jankovic D, Kullberg MC, Hieny S et al (2002) In the absence of IL-12, CD4+ T cell responses to intracellular pathogens fail to default to a Th2 pattern and are host protective in an IL-10-/- setting. Immunity 16(3):429–439

Jensen ET, Kharazmi A, Garred P et al (1993) Complement activation by *Pseudomonas aeruginosa* biofilms. Microb Pathog 15(5):377–388

Jensen PØ, Moser C, Kobayashi O et al (2004) Faster activation of polymorphonuclear neutrophils in resistant mice during early innate response to Pseudomonas aeruginosa lung infection. Clin Exp Immunol 137(3):478–485

Jensen PØ, Moser C, Kharazmi A et al (2006) Increased serum concentration of G-CSF in cystic fibrosis patients with chronic Pseudomonas aeruginosa pneumonia. J Cyst Fibros 5(3): 145–151

Jensen PØ, Bjarnsholt T, Phipps R et al (2007) Rapid necrotic killing of polymorphonuclear leukocytes is caused by quorum-sensing-controlled production of rhamnolipid by Pseudomonas aeruginosa. Microbiology 153(5):329–1338

Jesaitis AJ, Franklin MJ, Berglund D et al (2003) Compromised host defense on Pseudomonas aeruginosa biofilms: characterization of neutrophil and biofilm interactions J Immunol 171(8):4329–4339

Jones AM, Martin L, Bright-Thomas RJ et al (2003) Inflammatory markers in cystic fibrosis patients with transmissible Pseudomonas aeruginosa. Eur Respir J 22(3):503–506

Jones KL, Hegab AH, Hillman BC et al (2000). Elevation of nitrotyrosine and nitrate concentrations in cystic fibrosis sputum. Pediatr Pulmonol 30(2):79–85

Kaisho T, Hoshino K, Iwabe T et al (2002) Endotoxin can induce MyD88-deficient dendritic cells to support T(h)2 cell differentiation. Int Immunol 14(7):695–700

Kapsenberg ML (2003) Dendritic-cell control of pathogen-driven T-cell polarization. Nat Rev Immunol 3(12):984–993

Kaufmann SH (2008) Immunology's foundation: the 100-year anniversary of the Nobel Prize to Paul Ehrlich and Elie Metchnikoff. Nat Immunol 9(7):705–712

Kettle AJ, Chan T, Osberg I et al (2004) Myeloperoxidase and protein oxidation in the airways of young children with cystic fibrosis. Am J Respir Crit Care Med 170(12):1317–1323

Kharazmi A, Rechnitzer C, Schiøtz PO et al (1987) Priming of neutrophils for enhanced oxidative burst by sputum from cystic fibrosis patients with *Pseudomonas aeruginosa* infection. Eur J Clin Invest 17(3):256–261

Kharazmi A, Nielsen H, Bendtzen K (1988) Modulation of human neutrophil and monocyte chemotaxis and superoxide responses by recombinant TNF-alpha and GM-CSF. Immunobiology 177(4–5):363–370

Kimbrell DA, Beutler B (2001) The evolution and genetics of innate immunity. Nat Rev Genet 2(4):256–267

Kirketerp-Møller K, Jensen PØ, Fazli M et al (2008) Distribution, organization, and ecology of bacteria in chronic wounds J Clin Microbiol 46(8): 2717–2722

Kolpen M, Hansen CR, Bjarnsholt T et al (2010) Polymorphonuclear leukocytes consume oxygen in sputum from chronic Pseudomonas aeruginosa pneumonia in cystic fibrosis. Thorax 65(1):57–62

Koller B, Kappler M, Latzin P et al (2008) LR expression on neutrophils at the pulmonary site of infection: TLR1/TLR2-mediated up-regulation of TLR5 expression in cystic fibrosis lung disease. J Immunol 181(4):2753–2763

Kronborg G, Fomsgaard A, Jensen ET et al (1993) Induction of oxidative burst response in human neutrophils by immune complexes made in vitro of lipopolysaccharide and hyperimmune serum from chronically infected patients. APMIS 101(11):887–894

Leid JG, Willson CJ, Shirtliff ME et al (2005) The exopolysaccharide alginate protects Pseudomonas aeruginosa biofilm bacteria from IFN-gamma-mediated macrophage killing. J Immunol 175(11):7512–7818

Madianos PN, Bobetsis YA, Kinane DF (2005) Generation of inflammatory stimuli: how bacteria set up inflammatory responses in the gingiva. J Clin Periodontol 32(S6):57–71

Mathee K, Ciofu O, Sternberg C et al (1999) Mucoid conversion of *Pseudomonas aeruginosa* by hydrogen peroxide: a mechanism for virulence activation in the cystic fibrosis lung. Microbiology 145(6):1349–1357

Moreau-Marquis S, Bomberger JM, Anderson GG et al (2008) The DeltaF508-CFTR mutation results in increased biofilm formation by Pseudomonas aeruginosa by increasing iron availability. Am J Physiol Lung Cell Mol Physiol 295(1):L25–37

Mayer-Hamblett N, Aitken ML, Accurso FJ et al (2007) Association between pulmonary function and sputum biomarkers in cystic fibrosis. Am J Respir Crit Care Med 175(8): 822–828

Muhlebach MS, Stewart PW, Leigh MW et al (1999) Quantitation of inflammatory responses to bacteria in young cystic fibrosis and control patient. Am J Respir Crit Care Med 160(1): 186–191

Muir A, Soong G, Sokol S et al (2004) Toll-like receptors in normal and cystic fibrosis airway epithelial cells. Am J Respir Cell Mol Biol 30(6):777–783

Nathan C, Srimal S, Farber C et al (1989) Cytokine-induced respiratory burst of human neutrophils: dependence on extracellular matrix proteins and CDU/CD18 integrins. J Cell Biol 109(3):1341–49

Nathan C (2008) Metchnikoff's Legacy in 2008. Nat Immunol 9(7):695–698

Nobel Lectures (1967) Physiology or medicine 1901–1921. Elsevier, Amsterdam

Oliver A, Canton R, Campo P et al (2000) High frequency of hypermutable *Pseudomonas aeruginosa* in cystic fibrosis lung infection. Science 288(5469):1251–1254

Otterlei M, Sundan A, Skjak-Braek G et al (1993) Similar mechanisms of action of defined polysaccharides and lipopolysaccharides: characterization of binding and tumor necrosis factor alpha induction. Infect Immun 61(5):1917–1925

Parker LC, Whyte MK, Dower SK et al (2005) The expression and roles of Toll-like receptors in the biology of the human neutrophil. J Leukocyte Biol 77(6):886–892

Parker LC, Prince LR, Sabroe I (2007) Translational mini-review series on Toll-like receptors: networks regulated by Toll-like receptors mediate innate and adaptive immunity. Clin Exp Immunol 147(2):199–207

Pedersen SS, Kharazmi A, Espersen F et al (1990a) Pseudomonas aeruginosa alginate in cystic fibrosis sputum and the inflammatory response. Infect Immun 58(10):3363–3368

Pedersen SS, Espersen F, Høiby N et al (1990b) Immunoglobulin A and immunoglobulin G antibody responses to alginates from Pseudomonas aeruginosa in patients with cystic fibrosis. J Clin Microbiol 28(4):747–755

Pedersen SS (1992) Lung infection with alginate-producing, mucoid *Pseudomonas aeruginosa* in cystic fibrosis. APMIS S28:1–79

Pessi G, Haas D (2000) Transcriptional control of the hydrogen cyanide biosynthetic genes hcnABC by the anaerobic regulator ANR and the quorum-sensing regulators LasR and RhlR in *Pseudomonas aeruginosa*. J Bacteriol 182(24):6940–6949

Petersen SV, Thiel S, Jensenius JC (2001) The mannan-binding lectin pathway of complement activation: biology and disease association. Mol Immunol 38(2–3):133–149

Petit-Bertron AF, Tabary O, Corvol H et al (2008) Circulating and airway neutrophils in cystic fibrosis display different TLR expression and responsiveness to interleukin-10. Cytokine 41(1):54–60

Pier GB, Coleman F, Grout M et al (2001) Role of alginate O acetylation in resistance of mucoid Pseudomonas aeruginosa to opsonic phagocytosis. Infect Immun 69(3): 1895–1901

Power MR, Peng Y, Maydanski E et al (2004) The development of early host response to *Pseudomonas aeruginosa* lung infection is critically dependent on myeloid differentiation factor 88 in mice. J Biol Chem 279(47):49315–49322

Power MR, Marshall JS, Yamamoto M et al (2006) The myeloid differentiation factor 88 is dispensable for the development of a delayed host response to Pseudomonas aeruginosa lung infection in mice. Clin Exp Immunol 146(2):323–329

Rambach G, Würzner R, Speth C (2008) Complement: an efficient sword of innate immunity. In: Egesten A, Schmidt A, Herwald H (eds) Trends in innate immunity. Contrib Microbiol, vol 15. Karger, Basel, pp 78–100

Rhoades ER, Archambault AS, Greendyke R et al (2009) Mycobacterium abscessus Glycopeptidolipids mask underlying cell wall phosphatidyl-myo-inositol mannosides blocking induction of human macrophage TNF-alpha by preventing interaction with TLR2. J Immunol 183(3):1997–2007

Roghanian A, Drost EM, MacNee W et al (2006) Inflammatory lung secretions inhibit dendritic cell maturation and function via neutrophil elastase. Am J Respir Crit Care Med 174(11): 1189–1198

Ryall B, Davies JC, Wilson R et al (2008) *Pseudomonas aeruginosa*, cyanide accumulation and lung function in CF and non-CF bronchiectasis patients. Eur Respir J 32(3):740–747

Sanderson K, Wescombe L, Kirov SM et al (2008) Bacterial cyanogenesis occurs in the cystic fibrosis lung. Eur Respir J 32(2):329–333

Schiøtz PO, Nielsen H, Høiby N et al (1978) Immune complexes in the sputum of patients with cystic fibrosis suffering from chronic Pseudomonas aeruginosa lung infection. Acta Pathol Microbiol Scand C 86(1):37–40

Shibutani S, Takeshita M, Grollman AP (1991) Insertion of specific bases during DNA synthesis past the oxidation damaged base 8-oxodG. Nature 349(6308):431–434

Shimono M, Ishikawa T, Enokiya Y et al (2003) Biological characteristics of the junctional epithelium. J Electron Microsc (Tokyo) 52(6):627–639

Simmons WL, Dybvig K (2007) Biofilms Protect *Mycoplasma pulmonis* Cells from Lytic Effects of Complement and Gramicidin. Infect Immun 75(8):3696–3699

Skerrett SJ, Liggitt HD, Hajjar AM et al (2004) Cutting edge: myeloid differentiation factor 88 is essential for pulmonary host defence against *Pseudomonas aeruginosa* but not *Staphylococcus aureus*. J Immunol 172(6):3377–3381

Suter S, Schaad UB, Roux L et al (1984) Granulocyte neutral proteases and Pseudomonas elastase as possible causes of airway damage in patients with cystic fibrosis. J Inf Dis 149(4):523–531

Stahl PD, Ezekowizt B (1998) The mannose receptor is a pattern recognition receptor involved in host defense. Curr Opin Immunol 10(1):50–55

Strober W, Murray PJ, Kitani A et al (2006) Signalling pathways and molecular interactions of NOD1 and NOD2. Nat Rev Immunol 6(1):9–20

Takeda K, Akira S (2005) Toll-like receptors in innate immunity. Int Immunol 17(1):1–14

Takeda K, Akira S (2003) Toll receptors and pathogen resistance. Cell Microbiol 5(3):1431–1453

Trevani AS, Chorny A, Salamone G et al (2003) Bacterial DNA activates human neutrophils by a CpG-independent pathway. Eur J Immunol 33(11):3164–3174

Turner MW, Hamvas RM (2000) Mannose-binding lectin: structure, function, genetics and disease associations. Rev Immunogenet 2(3):305–322

van der Vliet A, Nguyen MN, Shigenaga MK et al (2000) Myeloperoxidase and protein oxidation in cystic fibrosis. Am J Physiol 279(3):L537–L546

van der Vliet A (2008) NADPH oxidases in lung biology and pathology: host defense enzymes, and more. Free Radic Biol Med 44(6):938–955

van Gennip M, Christensen LD, Alhede M et al (2009) Inactivation of the rhlA gene in Pseudomonas aeruginosa prevents rhamnolipid production, disabling the protection against polymorphonuclear leukocytes. APMIS 117(7):537–546

Walsh LJ (2003) Mast cells and oral inflammation. Crit Rev Oral Biol Med 14(3):188–198

Whitchurch CB, Tolker-Nielsen T, Ragas PC et al (2002) Extracellular DNA required for bacterial biofilm formation. Science 295(5559):1487

Wiebe BM, Burton CM, Milman N et al (2006) Morphometric examination of native lungs in human lung allograft recipients. APMIS 114(11):795–804

Worthley DL, Bardy PG, Mullighan CG (2005) Mannose-binding lectin: biology and clinical implications. Intern Med J 35(9):548–555

Wörlitzsch D, Bensel T, Borneff-Lipp M et al (2007) *Pseudomonas aeruginosa* and lactate in vitro and in CF sputum. Pediatr Pulmonol S30:317–317

Wörlitzsch D, Tarran R, Ulrich M et al (2002) Effects of reduced mucus oxygen concentration in airway Pseudomonas infections of cystic fibrosis patients. J Clin Invest 109(3):317–325

Yang L, Haagensen JAJ, Jelsbak L et al (2008) In situ growth rates and biofilm development of *Pseudomonas aeruginosa* populations in chronic lung infections. J Bacteriol 190(8):2767–2776
Yuo A, Kitagawa S, Kasahara T et al (1991) Stimulation and priming of human neutrophils by interleukin-8: cooperation with tumor necrosis factor and colony-stimulating factors. Blood 78(10):2708–2714
Zhang Z, Louboutin JP, Weiner DJ et al (2005) Human airway epithelial cells sense Pseudomonas aeruginosa infection via recognition of flagellin by Toll-like receptor 5. Infect Immun 73(11):7151–7160

Chapter 12
Adaptive Immune Responses and Biofilm Infections

Claus Moser and Peter Østrup Jensen

12.1 Introduction

The adaptive immune response has been developed to distinguish between self and non-self just as the innate immune response. However, in comparison with the innate immune response the adaptive immune response is characterized by a higher degree of specificity and so-called memory. Where the innate immune response is designed to recognize a broad spectrum of foreign antigens (pathogen associated molecular patterns, PAMPS) present on numerous microorganisms, e.g. peptidoglycan or flagellin even with species dependent differences (e.g. phase variation) by the pattern recognition receptors (PRR), the adaptive immune response recognizes species or even strain specific antigens. The memory is characterized by a clonal expansion of specialized subtypes of lymphocytes (effector and central memory cells) during the first exposure resulting in a significantly faster, stronger and higher affinity response as compared to the first response. In contrast, the innate response by itself cannot distinguish between a primary or subsequent exposure (Janeway and Travers 1997, Roitt et al. 2006).

Activation of the adaptive immune response is initiated simultaneously or shortly after activation of the innate immune response, although with some inertia. In accordance to what is published activation of the adaptive immune response during biofilm infections follows the same mechanisms as during infection with the same microorganism during a non-biofilm forming infection. However, since the biofilm infection persists, an odd situation appears with activation of both parts of the host response. In biofilm infections the effector mechanisms of the immune system are insufficient in eliminating the pathogen. Therefore, the difference between the adaptive immune response to a biofilm and a non-biofilm infection lies in the impaired clearance of the microorganism and the contribution of the adaptive immune responses to the pathology (Høiby et al. 2001, Brady et al. 2008, Schaudinn et al. 2009). Activation of an adaptive immune response is generally

C. Moser (✉)
H:S Rigshospitalet, Department for Clinical Microbiology, afsnit 9301, Juliane Maries Vej 22, DK-2100 Copenhagen Ø, Denmark
e-mail: moser@dadlnet.dk

T. Bjarnsholt et al. (eds.), *Biofilm Infections*, DOI 10.1007/978-1-4419-6084-9_12,

considered important for resolution of any infection especially in the pre-antibiotic era and mandatory for development of immunity against pathogens, mainly utilized in development of vaccines. In the effector mechanisms the adaptive immune response often act in synergy with the innate immune response as well as the type of the innate immune response influences on the type of the adaptive immune response generated. In biofilm infections however, the persistent infection can resist the released antibodies, chemoattracted, activated and opsonized phagocytes, as well as other components of the host response. Instead the surrounding tissue is subject to deleterious oxidative radicals and enzymes released from the host itself. In addition to various pathogen-specific virulence factors the release of proteases and other exoenzymes from the host cells can result in degradation of important surface molecules on the immune cells and thereby contribute to the impaired anti-biofilm effect of the host (Kharazmi et al. 1984, Horvat and Parrmely 1988, Theander et al. 1988, Kharazmi and Nielsen 1991, McCormick et al. 1997). Probably the host response itself is the major factor for the damage of the tissue, since neutralizing antibodies directed against a number of bacterial virulence factors during biofilm infections have been reported (Döring and Høiby 1983, Döring et al. 1985, Petersen et al. 1996). In the inherited disease cystic fibrosis (CF) the patients have been reported to develop specific antibodies against elastase, lipopolysaccharide, flagella, etc., indicating that those virulence factors are being neutralized during the chronic infection. This supports that these virulence factors may be important for the early colonization and infection, but are not directly doing any harm to the tissue. However, instead the antibodies have been shown to result in immune complexes precipitating in the parenchyma leading to activation of complement and opsonization of PMNs, and thereby indirectly inducing tissue damage (Koch and Høiby 1993, Høiby et al. 2001).

Another important point is that the increasing insight into the mechanisms and components of the adaptive immune response opens up for new treatment targets in the immune response; a pivotal fact due to the antibiotic tolerance and resistance of biofilm infections, and the resistance of biofilm infections to the native developed host responses.

The present chapter is not a new immunology chapter per se; those issues are presented thoroughly and much better in a number of basic immunology books to which we refer (Janeway and Travers 1997, Roitt et al. 2006). The present chapter will focus on the biofilm related observations of adaptive immune responses and biofilm infections. By far the majority of studies on biofilm infections and adaptive immune responses have been performed in infections with *P. aeruginosa* in patients with cystic fibrosis and therefore this subject will function as a core in the rest of the chapter with references to biofilm infections with other pathogens when appropriate.

12.2 Components of the Adaptive Immune Response

The adaptive immune system is composed of cellular and humoral parts. The cells involved are the T-cells (Thymus dependent) and the B-cells (bone marrow or Bursa Fabricii dependent). The T-cells can be divided into T-helper (Th-) cells which are CD4 (cluster of differentiation) positive cells due to the surface expression of this molecule and the cytotoxic T-cells (Tc), which are CD8 positive. Th-cells are important in accelerating activation of the adaptive immune response through accessory activation signals and cytokine production, whereas Tc-cells are functioning as killer cells of infected host cells by a system injecting apoptosis promoting factors (granzymes) into the infected host cells.

B-cells express immunoglobulins, especially IgM, on their surface and are precursors of plasma cells, which produce antibodies at the clonal level. Antibodies are immunoglobulins which in one end can bind to different antigens, and in the other end harbour the abilities for binding to receptors on the phagocytes, opsonisation and complement activation. The antibodies are divided into five classes (IgA_{1-2}, IgD, IgE, IgG_{1-4} and IgM) and differ in their function (complement activation and opsonization), affinities, site of production, which phase of the infection they are dominating (early or late/re-infection) and half-lives.

As mentioned in the introduction so-called memory cells are being generated during the primary response to an antigen. Memory B-cells with surface expression of IgG, IgA or IgE with an increase in specific cells of 10–100 times and with a much higher degree of affinity are generated. In addition both effector and central memory T-cells are being produced during a primary response with an increase in number of 100–1000 times and more antigen specific T-cells are being produced (Janeway and Travers 1997, Roitt et al. 2006).

12.3 Activation of the Adaptive Immune Response the Dendritic Cells

Two major cell types are often considered as links between the innate and the adaptive immune response. Macrophages (Mϕ) and in particularly dendritic cells (DC) are specialized in antigen uptake and presentation and thereby function as activator cells for the adaptive immune response. The other cell type is natural killer (NK) cells, which apart from their innate functions also can eliminate infected cells in an antibody-dependent manner as well as providing a cytokine milieu, which can influence the balance between different subtypes of adaptive immune responses (see later).

In recent years DCs have attracted significant interest. In the activation of the naïve host cells the DCs are mandatory; the adaptive host response will not be sufficiently activated without the action of the DCs (Banchereau and Steinman 1998). However, DCs have been shown to consist of different subtypes and, furthermore, they are highly plastic.

Based on the observation that human DCs could be divided into myeloid (mDCs) or lymphoid (later plasmacytoid DCs, pDCs) depending on their surface markers and cytokine response, that those DC subtypes depended on different cytokines in their surroundings, and that they induced distinct T-helper cell responses initiated a study on inflammatory cytokines and T-helper cell responses (Moser et al. 2005). mDCs were characterized as IL-12 producers and dependent on GM-CSF and were designated DC1 cells due to their Th1 inducing ability. Lymphoid DCs (later pDCs) were weak IL-12 producers and dependent on IL-3. Furthermore, they were shown to be induced by G-CSF and were designated as DC2 cells due to their ability to induce Th2 responses. Therefore, the speculation was that G-CSF in CF patients not only recruited PMNs from the bone marrow but in addition also induced DC2 cells and Th2 response. Indeed, a positive correlation was observed between the GM-CSF/G-CSF ratio and the IFN-γ response as well as the lung function in CF patients with chronic *P. aeruginosa* lung infection (Moser et al. 2000). In addition, an inverse correlation between IL-3 and the IFN-γ response was also observed (Moser et al. 2005). This lead to a GM-CSF treatment study in mice using the alginate embedded *P. aeruginosa* model. Treating the infected mice with GM-CSF from day 3 until day 6 with 0.2 mg day s.c. changed the response from a Th2-dominated to a Th1-dominated response, confirming the Th1 inducing ability of GM-CSF (own observation). This is an area, which we find promising, and is currently under further investigations in our research group.

However, the nomenclature of DCs is currently undergoing adjustments. DCs are only present in very limited numbers, which together with their plasticity makes them very difficult to investigate. However, at least three subtypes of mDCs have been reported, and in addition to the pDCs there is also reported monocyte-derived DCs as well as langerhans cell-derived DCs. The subtype of the adaptive immune response the DCs elicit is depending on a number of factors in a highly complex manner at least involving (1) subtype of DC that is recognizing the invading pathogen, (2) the cytokine milieu, (3) the pathogen itself, (4) antigen amount, and (5) site of infection. In this context the type of DC is difficult to assess, since DCs undergo maturation and phenotypic changes upon antigen uptake and migration to secondary lymphoid tissue. Besides maturation changes DCs also express different Toll-like receptors and they differ in cytokine production and cytokine receptors (Jarrossay et al. 2001, Kadowaki et al. 2001, Penna et al. 2002, Piccioli et al. 2007, Demedts et al. 2006, Ito et al. 2006).

12.4 Activation of the Adaptive Immune Response – Diagnostic Tool

The use of serological tests in diagnosing chronic (e.g. syphilis, borreliosis and Q-fever) and sometimes even acute infections (e.g. legionnaires disease or mycoplasma) as well as during disentangle of hyper reactive phenomena initiated by infectious diseases (e.g. reactive arthritis and Reiter's syndrome) is well established. However, serological tests can also be involved in biofilm infections. The majority of patients with CF acquire chronic *P. aeruginosa* lung infections with time. The infection cannot be completely eradicated from the lungs due to the biofilm mode of growth involving hyper production of exopolysaccharide. A significant characteristic of the disease is a development of a pronounced humoral response against *P. aeruginosa* when the patients become chronically infected (Høiby et al. 1977). In fact, several CF centres use detection of a specific antibody response as a marker of chronic infection per se in contrast to harmless colonization (Pressler et al. 2009). Similarly, antibody responses in chronic or lente endocarditis can be used as a diagnostic tool, either by ELISA technique or as precipitating antibodies, a useful diagnostic tool since infective endocarditis can be difficult to diagnose (Kjerulf et al. 1998). This is in contrast to acute endocarditis where an adaptive immune response is of limited value due to the combination of the aggressive infection and inerty in the activation of the adaptive immune response.

Patients suffering from spinal cord lesions are in high risk of acquiring recurrent or chronic urinary tract infections due to their impaired ability to empty the urinary bladder. Many of those patients empty their bladder by use of catheters either by intermittent catheterization or as catheter a demeure. To distinguish whether those patients have developed a chronic infection antibody response to the most prevalent pathogens of the urinary tract can be estimated (Moser et al. 1998). Actually, the finding of precipitating antibodies to cultured pathogens of the urinary tract in a subgroup of those patients (patients with myelomeningocele) correlates to levels of serum creatinine indicating impaired renal function, probably due to immune complex disease (Moser et al. 1998).

T-cells response as a diagnostic tool is best known in tuberculosis as delayed type hypersensitivity response in the skin after injection of purified TB antigen (the Mantoux test). Another newer TB test is the so-called quantiferon test where peripheral blood cells are exposed to a specific TB antigen and the release of IFN–γ is being measured. A T- cell response has to our knowledge not been used as a diagnostic tool, although specific changes may happen when the infection changes from intermittent colonization to chronic infection (see Th1/Th2 responses and CF).

12.5 Antibody Responses and Biofilm Infections

Antibody responses and biofilm infections have primarily been studied in CF patients with chronic *P. aeruginosa* lung infections. Primarily a humoral response in CF was investigated to reveal whether *P. aeruginosa* could be considered as a pathogen in CF or whether it was a harmless colonizer which was selected for due to antibiotic treatment of *Staphylococcus aureus* and pneumococci lung infections in CF. Since then a number of studies have looked at the classes and subclasses of antibody responses against biofilm growing *P. aeruginosa* in CF. Almost all classes have been shown to be involved except for IgD antibodies. Since classes and subclasses of antibodies have distinct functions the levels of those have been correlated to the course of the chronic *P. aeruginosa* lung infections in CF patients. Interestingly both specific subclasses IgG2 and IgG3 correlated to a poor lung function and poor clinical condition in CF. The mechanism behind this correlation was believed to be the ability of IgG3 antibodies to activate complement, and thereby contribute to inflammation (Pressler et al. 1988, 1990). A typical example on how the two immune responses can act in synergy – although not to the benefit of the patient in this case.

Another interesting study has looked at the maturation of anti-pseudomonas antibodies during different time periods of the infection. In contrast to what is usually reported during the course of an infection there was no maturation in avidity (binding strength between antibody and antigen) of antibodies directed against chromosomal β-lactamase of *P. aeruginosa* or the *P. aeruginosa* heat shock protein Gro-EL during an 11-year follow-up period of the chronic *P. aeruginosa* lung infection in CF (Ciofu et al. 1999). Other investigations on antibody responses in infected CF patients have also reported development of reduced opsonic killing by antibody responses directed against *P. aeruginosa* exopolysaccharide (Meluleni et al. 1995). Such failure in maturation probably results in a reduced ability of the humoral response to control the infection and increase the tendency to immune complex disease (Devey et al. 1984). However, studies investigating the T cell responses have also observed a reduced mitogenic response of T cells during the chronic lung infection. The significance of this observation still needs to be clarified.

Antibody responses may not only be harmful and generate immune complex disease. Antibody responses against specific antigens/virulence factors have actually been shown to be somehow protective during biofilm infections. In infected CF patients specific antibodies directed against protease and elastase were able to neutralize the enzymatic activity of those virulence factors and measurements of those exoenzymes (including exotoxin A) was negative in the sputum from those patients (Döring et al. 1985). An important reason for the antibiotic resistance observed in *P. aeruginosa* and CF is due to increased production of chromosomal β-lactamases. The observations that β-lactamases were secreted outside *P. aeruginosa* in small blebs (vesicles) and that high avidity anti-β-lactamase antibody response correlated to better lung function initiated a study to investigate whether the CF patients generated a humoral response directed against β-lactamases. This was not only the case; actually the production of high avidity anti-β-lactamase antibodies correlated to a better lung function (Ciofu et al. 2002). Furthermore, an animal study where β-lactamase vaccination was performed in rats, later infected with a *P. aeruginosa* strain producing high levels of β-lactamase, showed that vaccinated rats that responded to the vaccine had a more beneficial lung inflammation and reduced quantitative lung bacteriology as non-vaccinated and non-responding vaccinated rats when the vaccination was combined with ceftazidim treatment (Ciofu et al. 2002).

An interesting example of the consequence of simultaneous activity of both the innate and the adaptive immune response is the development of autoantibodies to parts of the immune system. This is probably best described as anti-neutrophil cytoplasmic auto-antibodies (ANCA) directed against bacterial/permeability increasing protein (BPI) in patients suffering from chronic biofilm infections with *P. aeruginosa*, e.g. in patients with diffuse panbronchiolitis (DPB) (Ohtami et al. 2001). The authors reported that serum BPI-ANCA correlated to severity of clinical symptoms and that the titer was reduced with improvements in clinical status (Ohtami et al. 2001). Later on this has also been shown in CF patients, since BPI-ANCA was inversely correlated to the lung function in CF patients chronically infected with *P. aeruginosa* (Carlsson et al. 2003). The mechanism was suggested to be inhibition of the phagocytic activity since BPI-ANCA reduced this activity in a dose-dependent manner (Ohtami et al. 2001).

12.6 Marker for Appropriate Treatment

Since specific antibody responses have been shown useful in distinguishing between infection and colonization in a number of chronic infections it would be obvious to include the specific antibody response as a marker for treatment response. If an appropriate antibiotic treatment has been initiated, the burden of the pathogen should presumably be reduced, and the antigenic induction similarly disappears. It is, however, important to consider the method used to measure the antibody response because some methods are designed to measure a number of clonal antibodies (e.g. crossed immunoelectrophoresis), whereas other methods are designed to estimate the concentration of antibodies against a pool of bacterial antigens (e.g. sonicates) or a specific microbial component (e.g. elastase). Enzyme Linked Immuno Sorbent Assay (ELISA) is an example of a method measuring the latter. If crossed immunoelectrophoresis is used to determine the number of precipitating antibodies against a pathogen there is a significant inertia and the number of precipitating antibodies will only be reduced very slowly over months to years, although the height of the precipitating curves can be used semi-quantitatively. However, if a technique like ELISA is being used, the fluctuations of the levels of antibodies presented as an optical density seems to follow the microbial burden in the infected lung (Pressler et al. 2009). The most valuable marker initially is probably a stop in further increase in humoral response, and then later maybe a reduction. It is important though, to underline that the clinical state of the patient is the most important indicator for appropriate treatment.

12.7 Types of Adaptive Immune Response

Beginning with the era of organ transplantations in attempts to avoid rejection of grafts adaptive immune responses attracted substantial attention. Reports on how acceptance or rejection of the organ transplants could be correlated to whether the organ recipients had a predominantly humoral or cellular immune response initiated speculations in researchers working with chronic infections. A dominant feature of the adaptive immune response for some years has been the correlation between distinct courses of chronic infections and the balance between subtypes of the adaptive immune response. Inspiration for revealing the significance of the adaptive immune response on the course of the chronic *P. aeruginosa* biofilm infection in the lungs of CF patients came from observations on the course of lepra. In patients suffering from this disease a lepramatoid course of the disease with rapid dissemination and organ involvement of the infection was correlated to a poor cellular immune response and a pronounced antibody response. In contrast, other patients had a so-called tuberculoid course of the infection where dissemination was not observed. Patients with this milder course of the infection had a strong cellular immune response and low antibody production.

Later on similar dichotomized courses of Leishmania infections were reported, and since various inbred mouse strains in animal models of Leishmania infections had distinct courses of the infection intense investigations on the immune response were performed.

Mosmann and Coffman first reported the background for division of the Th-response into subtypes in 1986 (Mosmann et al. 1986, Locksley et al. 1987). In their famous work they described how murine Th-cells differed in their ability to release cytokines in two distinct patterns, although the cells were identical in the expression of classical surface markers. The two subsets were designated Th1 and Th2 cells, where Th1 cells produced IFN-γ and Th2 cells produced IL-4 and -5. Moreover, cytokines from one subset could down-regulate the other subset and anti-IL-4 or anti-IFN-γ could induce the opposite Th-response especially in the early stages. Different other cytokines have been ascribed to the two subsets and today IL-9 and -13 are also considered Th2 cytokines. With time more subsets have been suggested and some disappeared again but a Th17 subtype (producing IL-17 and IL-22) and a regulatory T-cell subset (Treg1) producing IL-10 and TGF-β (as well as IFN-γ and IL-5) seems to be widely accepted. Although the subset concept in the beginning seemed to polarize any adaptive response completely into one or the other, it is widely accepted today that more subsets, if not all are involved, and the responses are characterized by a balance of the involved subsets, which however can be an inappropriate balance.

With the presentation of the subtypes and correlation of those to the course of leishmaniasis in mouse models, the next question was of course whether the response could be manipulated into the opposite response, and whether the changed response influenced on the course of the infection and change of the responses from Th1 to Th2 or vice versa also inversed the course of the leishmnania infection in mice. In addition, the two Th-cell subsets were also shown to influence differently on major parts of the immune system; Th1 responses are thus related to activation of Mϕ and the cellular immune response, whereas Th2 responses are stimulating the humoral immune response and mast cells.

The role of the IL-17 from Th17 cells seems to be acting as an inducer of chemo attractants for PMNs, and therefore, those cells can significantly act as proinflammatory cells. The Treg subset is activated Th- or Tc cells, which produce high levels of IL-10 and TGF-β. The Treg cells both inhibit the development of Th1 cells as well as antibody production. Finally a subset of CD25+CD4+ cells appears to be accepted as a T suppressor cell type. This cell type however does not act through cytokine production; instead the CD25+CD4+ subtype acts through cell contact to inhibit IL-2 and IL-2 receptor expression, and induce apoptosis. Moreover, this subset seems to be independent of antigen stimulation. To our best knowledge the possible role of a Th17 response or Treg cells have not been published. Similarly with respect to T suppressor cells. Clarification of the involvement of those subsets in biofilm infections is going to be interesting to follow, especially concerning Th17 as a PMN inducing subset.

12.8 Adaptive Immune Response and CF

Before modern aggressive antibiotic treatments of the chronic *P. aeruginosa* lung infection in CF it was observed that there seemed to be a dichotomized course of the infection. Either a deteriorating course with a poor prognosis in the majority of the patients where the antibody response was pronounced or rapidly increasing, or a more beneficial course of the chronic infection in a minority of the patients where the antibody response remained low, was observed (Høiby et al. 1977). This strongly indicated a decisive role of the adaptive immune response during the chronic *P. aeruginosa* lung infection and initiated a number of studies trying to reveal the mechanisms behind this observation.

In a study where peripheral blood mononuclear cells (PBMCs) from CF patients with chronic *P. aeruginosa* lung infection were re-stimulated with Pseudomonas-antigens a significantly reduced release of IFN-γ (Th1 marker) was observed, as compared to CF patients who were not yet chronically infected (Moser et al. 2000). IL-4 release (Th2 marker) from the re-stimulated PBMCs was almost exclusively seen in cells from the chronically infected CF patients indicating a skewing of the Th1/Th2 balance to a Th2-dominated response in CF patients with chronic *P. aeruginosa* lung infection. Moreover, IFN-γ release from PBMCs correlated to the lung function of the chronically infected CF patients indicating a possible beneficial effect if the Th1/Th2 balance could be tipped in favour of a more Th1-dominated response (Moser et al. 2000). Figure 12.1 shows the results in more details. A skewing of the Th1/Th2 balance in CF has been confirmed by findings in other groups (Moss et al. 2000, Brazova et al. 2005, Hartl et al. 2006).

Using the chronic lung infection model in rats infected with *P. aeruginosa* embedded in seaweed alginate IFN-γ treatment was shown to render the lung inflammation from an acute type dominated by PMNs to an inflammation dominated by MN cells (Johansen et al. 1996). Using the strategy of infecting two different inbred mouse strains with alginate embedded *P. aeruginosa* in the lungs revealed that the C3H/HeN mouse strain had a Th1-dominated response and a beneficial course of the infection, in contrast to the BALB/c mouse strain which had a Th2-dominated response and a more serious course of the infection (Moser et al. 1997, 1999). However, if the susceptible BALB/c mice were re-infected with alginate-embedded *P. aeruginosa* the mice became resistant and the immune response changed to a Th1-dominated response resembling the course of a primary infection in the resistant C3H/HeN mice (Moser et al. 2002).

The mechanism behind an improved course of the chronic *P. aeruginosa* lung infection and a more Th1-dominated response is mainly believed to be an increased stimulation of the alveolar Mϕ. An increase in number and activation of the alveolar Mϕ may improve the resolving of the pulmonary inflammation by phagocytization of apoptotic PMNs and cell debris from necrotic PMNs (Ware and Matthay

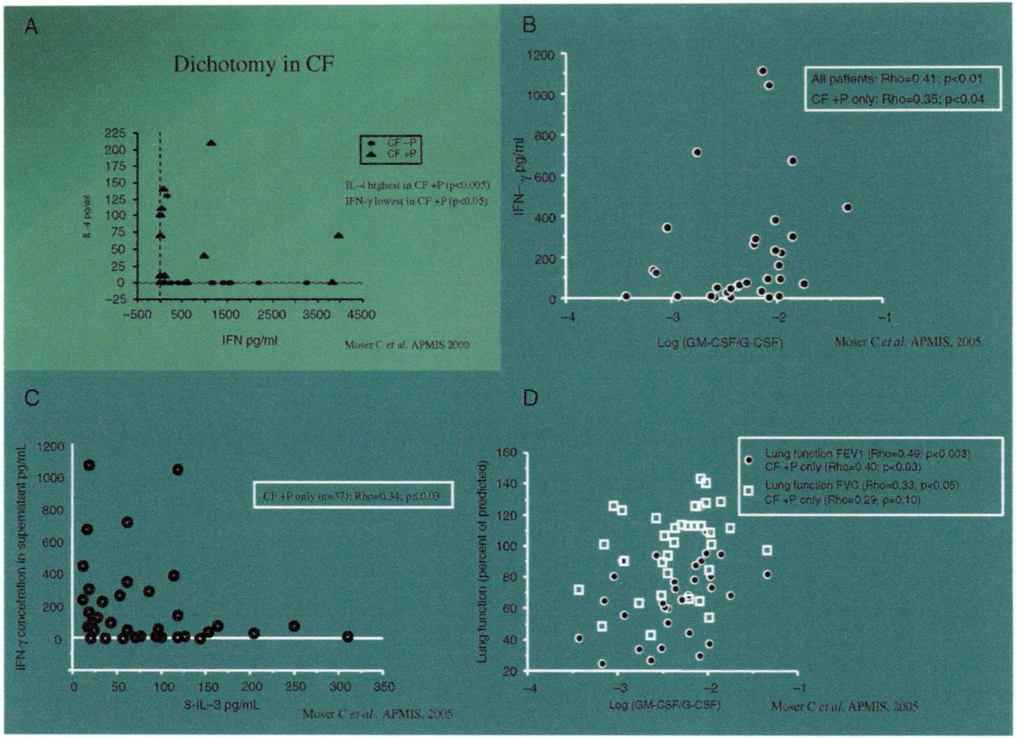

Fig. 12.1 Measurements of cytokines characteristic for the T-helper cell response and factors that influence the T-helper cell response. Panel A presents release of IL-4 and IFN-γ from peripheral mononuclear cells (PBMC) isolated from cystic fibrosis (CF) patients with chronic *Pseudomonas aeruginosa* lung infection as compared to a group of CF patients without chronic infection. IL-4 and IFN-γ was significantly increased and decreased in the chronically infected group. Panel **b**, **c** and **d** are from the same study where cytokines (GM-CSF, G-CSF and IL-3) influencing the T-helper cell response were measured. The ratio between GM-CSF and G-CSF was found to correlate to the IFN-γ release from PBMC (Panel **b**) and the lung function (Panel **d**) in CF patients with chronic *P. aeruginosa* lung infection. In contrast, an inverse correlation between IL-3 and the IFN-γ release from PBMC from the same group of patients (Panel **c**)

2000). Especially the removal of apoptotic PMNs before they proceed into necrosis, and thereby further increase the inflammation, is thought to be important. In addition, a more Th1-dominated response may also result in reduced production of IL-8 (Cassatella et al. 1993, Schnyder-Candrian et al. 1995) and thereby reduced chemo attraction of PMNs.

A more Th1-dominated response would presumably also result in a down regulated Th2 response and thereby reduced B-cell stimulation. This would result in reduced antibody response and reduced formation of immune complexes and therefore reduced tissue damage. The Th2 cytokine IL-13 has also been shown to stimulate pulmonary mucus production, an important factor in CF lung disease, and presumably reduced IL-13 production could also result in diminished mucus production, which might reduce the tendency for aspirated pathogens to be captured by the copiously mucus in the CF lung. However, any relationship between such mechanisms and the Th1/Th2 balance in CF remains to be investigated. Figure 12.2 sums up our current view on the relationship between the adaptive immune response and the inflammation in CF.

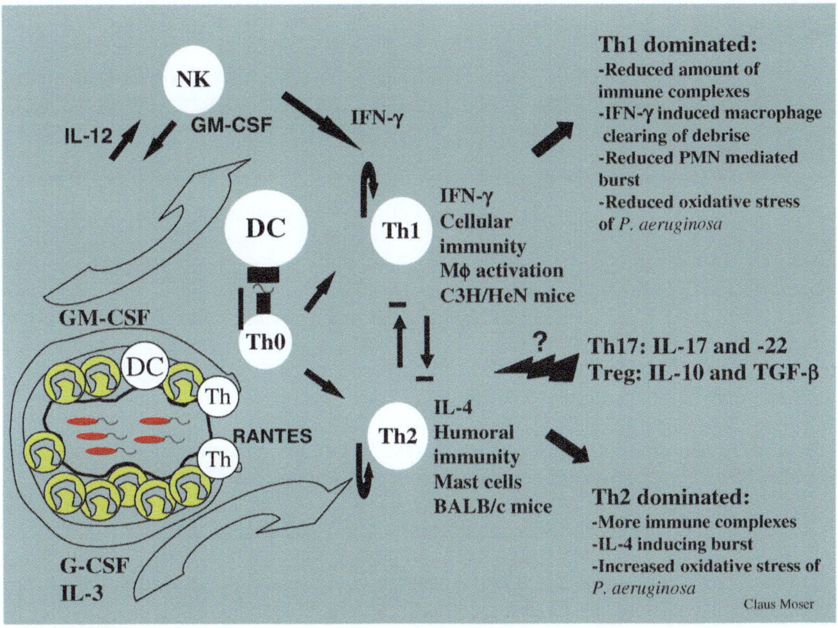

Fig. 12.2 Schematic presentation of the different factors influencing activation of the T-helper cell response during chronic *Pseudomonas aeruginosa* lung infection. The antigen (*P. aeruginosa*) is phagocytozed in the alveoles and processed and presented to uncommitted T-helper cells in the secondary lymphoid tissue. Those cells subsequently differentiate into Th1 or Th2 cells, and probably also Th17 and Treg; the role of the two latter has not been revealed yet, however. The balance of this system is influenced by both NK cells and DCs, their cytokines and the local cytokine milieu

In contrast, a direct anti-microbial effect on the *P. aeruginosa* biofilms by Mϕ seems not to be the mechanism; e.g. Leid and colleagues have observed an increased phagocytation of young *P. aeruginosa* biofilm after activation with IFN-γ. However, when exogenous alginate was added to the biofilms the increased killing of IFN-γ activated Mϕ was impaired (Leid et al. 2005). This observation further supports that the beneficial effect of a more Th1-dominated response in CF patients with chronic biofilm infections is probably mediated through modulation of the host responses, and not by a direct antibiofilm mode of action per se.

In the case of osteomyelitis *S. aureus* biofilm infection there also seems to be a skewing of the T-helper cell response. However, in this case the mechanisms may be quite different. This subject is reviewed in the osteomyelitis chapter of this book to where we refer.

12.9 Conclusion

In conclusion, the adaptive immune response and biofilm infections are important. Not only because the adaptive response is part of the immunopathology in biofilm infections, but it may also provide an important treatment tool during the otherwise immunotolerant biofilm infections.

References

Banchereau J, Steinman RM (1998) Dendritic cells and the control of immunity. Nature 392: 245–252

Brady RA, Leid JG, Cathoun JH et al (2008) Osteomylitis and the role of biofilms in chronic infection. FEMS Immunol Med Microbiol 52:13–22

Brazova J, Sediva A, Pospisilo vaD et al (2005) Differential cytokine profile in children with cystic fibrosis. Clin Immunol 115:210–215

Carlsson M, Eriksson L, Erwander I et al (2003) Pseudomonas-induced lung damage in cystic fibrosis correlates to bactericidal-permeability increasing protein (BPI)-autoantibodies. Clin Exp Immunol 21(suppl 32):S95–100

Cassatella MA, Guasparri I, Ceska M et al (1993) Interferon-gamma inhibits interleukin-8 production by human polymorphonuclear leukocytes. Immunology 78:177–184

Ciofu O, Bagge N, Høiby N (2002) Antibodies against beta-lactamase can improve ceftazidime treatment of lung infection with beta-lactam-resistant Pseudomonas aeruginosa in a rat model of chronic lung infection. APMIS 110:881–891

Ciofu O, Petersen TD, Jensen P et al (1999) Avidity of anti-P. aeruginosa antibodies during chronic infection in patients with cystic fibrosis. Thorax 54:141–144

Demedts IK, Bracke KR, Maes T et al (2006) Different roles for human lung dendritic cell subsets in pulmonary immune defense mechanisms. Am J Respir Cell Mol Biol 35:387–393

Devey ME, Bleasdale K, Stanley C et al (1984) Failure of maturation leads to increased susceptibility to immune complex glomerulonephritis. Immunology 52:377–383

Döring G, Høiby N (1983) Longitudinal study of immune response to Pseudomonas aeruginosa antigens in cystic fibrosis. Infect Immun 42:197–201

Döring G, Goldstein W, Röll A et al (1985) Role of Pseudomonas aeruginosa exoenzymes in lung infections of patients with cystic fibrosis. Infect Immun 49:557–562

Hartl D, Griese M, Kappler M et al (2006) Pulmonary T(H)2 response in Pseudomonas aeruginosa infected patients with cystic fibrosis. J Allergy Clin Immunol 117:204–211

Høiby N, Flensborg EW, Beck B et al (1977) Pseudomonas aeruginosa infection in cystic fibrosis. Diagnostic and prognostic significance of Pseudomonas aeruginosa precipitins determined by means of crossed immunoelectrophoresis. Scan J Respir Dis 58:65–79

Høiby N, Johansen HK, Moser C et al (2001) *Pseudomonas aeruginosa* and the *in vitro* and *in vivo* biofilm mode of growth. Microbes Infect 3:23–35

Horvat RT, Parmely MJ (1988) Pseudomonas aeruginosa alkaline protease degades human gamma interferon and inhibits its bioactivity. Infect Immun 56:2925–2932

Ito T, Kanzler H, Duramad O et al (2006) Specialization, kinetics and repetoire of type1 interferon responses by human plasmacytoid predendritic cells. Blood 107:2423–2431

Janeway CA, Travers P (1997) Immunobiology, 3rd edn. Current Biology ltd. Churchill Livingstone. Garland, New York

Jarrossay D, Napolitani G, Colonna M et al (2001) Specialization and complementarity in microbial molecule recognition by human myeloid and plasmacytoid dendritic cells. Eur J Immunol 31:3388–3393

Johansen HK, Hougen HP, Rygaard J et al (1996) Interferon-gamma treatment decreases the inflammatory response in chronic Pseudomonas aeruginosa pneumonia in rats. Clin Exp Immunol 103:212–218

Kadowaki N, Ho S, Antonenko S et al (2001) Subsets of human dendritic cell precursors express different toll-like receptors and respond to different microbial antigens. J Exp Med 194: 863–869

Kharazmi A, Nielsen H (1991) Inhibition of human monocyte chemotaxis and chemiluminescence by Pseudomonas aeruginosa elastase. APMIS 99: 93–95

Kharazmi A, Döring G, Høiby N et al (1984) Interaction of *Pseudomonas aeruginosa* alkaline protease and elastase with human polymorphonuclear leukocytes *in vitro*. Infect Immun 43:161–165

Kjerulf A, Tvede M, Aldershvile J et al (1998) Bacterial endocarditis at a tertiary hospital – how do we improve diagnosis and delay of treatment? A retrospective study of 140 patients. Cardiology 89:79–86

Koch C, Høiby N (1993) Pathogenesis of cystic fibrosis. Lancet 341:1065–1069

Leid JG, Willson CJ, Shirtliff ME et al (2005) The exopolysaccharide alginate protects Pseudomonas aeruginosa biofilm bacteria from IFN-g-mediated macrophage killing. J Immunol 175:7512–7518

Locksley RM, Heinzel FP, Sadick MH et al (1987) Murine cutaneous leishmaniasis: susceptibility correlates with different expansion of helper T cell subsets. Annales de l'Institut Pasteur de Paris/Immunology 138:744–749

McCormick LL, Karulin AY, Schreiber JR et al (1997) Bispecific antibodies overcome the opsonin-receptor mismatch of cystic fibrosis in vitro: restoration of neutrophil-mediated phagocytosis and killing of Pseudomonas aeruginosa. J Immunol 158:3474–3482

Meluleni GJ, Grout M, Evans DJ et al (1995) Mucoid Pseudomonas aeruginosa growing in a biofilm in vitro are killed by opsonic antibodies to the mucoid exoploysaccharide capsule but not by antibodies produced during chronic lung infection in cystic fibrosis patients. J Immunol 155:2029–2038

Moser C, Hougen HP, Song Z et al (1999) Early immune response in susceptible and resistant mice strains with chronic Pseudomonas aeruginosa lung infection determines the type of T-helper cell response. APMIS 107:1093–1100

Moser C, Jensen PØ, Kobayashi O et al (2002) Improved outcome of chronic Pseudomonas aeruginosa lung infection is associated with induction of a Th1-dominated cytokine response. Clin Exp Immunol 127:206–213

Moser C, Jensen PØ, Pressler T et al (2005) Serum concentrations of GM-CSF and G-CSF correlate with the Th1/Th2 cytokine response in cystic fibrosis patients with chronic Pseudomonas aeruginosa lung infection. APMIS 113:400–409

Moser C, Johansen HK, Song Z et al (1997) Chronic Pseudomonas aeruginosa lung infection is more severe in Th2 responding BALB/c mice compared to Th1 responding C3H/HeN mice. APMIS 105:838–842

Moser C, Kjaergaard S, Pressler T et al (2000) The immune response to chronic Pseudomonas aeruginosa lung infection in cystic fibrosis patients is predominantly of the Th2 type. APMIS 108:329–335

Moser C, Kriegbaum NJ, Larsen SO et al (1998) Antibodies to urinary tract pathogens in patients with spinal cord injuries. Spinal cord 36:613–616

Mosmann TR, Cherwinski H, Bond MW et al (1986) Two types of murine helper T cell clone. I. Definition according to profiles of lymphokine activities and secreted proteins. J Immunol 136:2348–2357

Moss RB, Hsu YP, Olds L (2000) Cytokine dysregulation in activated cystic fibrosis (CF) peripheral lymphocytes. Clin Exp Immunol 120:518–525

Ohtami S, Kobayashi O, Ohtami H (2001) Analysis of intractable factors in chronic airway infections: role of the autoimmunity induced by BPI-ANCA. J Infect Chemother 7:228–238

Penna G, Vulcano M, Roncari A et al (2002) Cutting edge: differential chemokine production by myeloid and plasmacytoid dendritic cells. J Immunol 169:6673–6676

Petersen TD, Ciofu O, Pressler T et al (1996) Quantitative analysis of the IgG and IgG subclasses immune responses to chromosomal Pseudomonas aeruginosa beta-lactamase in serum from patients with cystic fibrosis by western blotting and laser scanning densitometry. Thorax 51:733–738

Piccioli D, Tavarini S, Borgogni E et al (2007) Functional specialization of human circulating CD16 and CD1c myeloid dendritic cell subsets. Blood 109:5371–5379

Pressler T, Karpati F, Granström M et al (2009) Diagnostic significance of measurements of specific IgG antibodies to Pseudomonas aeruginosa by three different serological methods. J Cyst Fibros 8:37–42

Pressler T, Mansa B, Jensen T et al (1988) Increased IgG2 and IgG3 concentration is associated with advanced Pseudomonas aeruginosa infection and poor pulmonary function in cystic fibrosis. Acta Paediatr Scand 77:576–582

Pressler T, Pedersen SS, Espersen F et al (1990) IgG subclass antibodies to Pseudomonas aeruginosa in sera from patients with chronic Pseudomonas aeruginosa infection investigated by ELISA. Clin Exp Immunol 81:428–434

Roitt I, Brostoff J, Male D (2006) Immunology, 6th edn. Mosby, London

Schaudinn C, Gorur A, Keller D et al (2009)Periodontitis: an archetypical biofilm disease. J Am Dent Assoc 140:978–986

Schnyder-Candrian S, Strieter RM, Kunkel SL et al (1995) Interferon-alpha and interferon-gamma down-regulate the production of interleukin-8 and ENA-78 in human monocytes. J Leukoc Biol 57:929–935

Theander TG, Kharazmi A, Pedersen BK et al (1988) Inhibition of human lymphocyte proliferation and cleavage of interleukin-2 by Pseudomonas aeruginosa proteases. Infect Immun 56: 1673–1677

Ware LB, Matthay MA (2000) The acute respiratory distress syndrome. N Engl J Med 342: 1334–1349

Chapter 13
Antibiotic Tolerance and Resistance in Biofilms

Oana Ciofu and Tim Tolker-Nielsen

13.1 Introduction

One of the most important features of microbial biofilms is their tolerance to antimicrobial agents and components of the host immune system. The difficulty of treating biofilm infections with antibiotics is a major clinical problem. Although antibiotics may decrease the number of bacteria in biofilms, they will not completely eradicate the bacteria in vivo which may have important clinical consequences in form of relapses of the infection. Therefore, common antibiotic regimes for treating biofilm-associated infections imply removal of infected tissues or implanted devices (if possible) associated with long-term anti-microbial therapy. In some cases chronic suppressive therapy with antibiotics may be necessary.

The difficulty of treatment of biofilm infections involves the inefficacy of the immune system to eradicate the biofilm-embedded microorganisms as well the recalcitrance of biofilms to antimicrobial therapy. The inefficacy of the immune system to eradicate biofilms is covered in Chapters 11 and 12. In the present chapter we describe factors involved in antibiotic tolerance in *P. aeruginosa* biofilms, as well as factors involved in the occurrence and selection of antibiotic resistant mutants in *P. aeruginosa* biofilms.

The chronic *P. aeruginosa* infection of the airways of patients with cystic fibrosis (see Chapter 10) gives a unique opportunity to study antibiotic resistance in biofilms. In these patients bacteria survive in biofilms for more than 30 years and are exposed to repeated courses of antibiotics which represent an enormous selective pressure for emergence of resistance (Bjarnsholt et al. 2009). The antipseudomonal treatment consists of combination therapy of intravenous beta-lactams and aminoglycosides supplemented with aerosolized colistin or tobramycin and orally ciprofloxacin and azithromycin (Doring et al. 2000). The response of these *P. aeruginosa* biofilms to some of the clinically relevant drugs will be presented.

T. Tolker-Nielsen (✉)
Department of International Health, Immunology and Microbiology, Faculty of Health Sciences, University of Copenhagen, Blegdamsvej 3B, DK-2200 Copenhagen, Denmark
e-mail: ttn@sund.ku.dk

T. Bjarnsholt et al. (eds.), *Biofilm Infections*, DOI 10.1007/978-1-4419-6084-9_13,

13.2 Antibiotic Tolerance in Biofilms

When bacteria that survived antibiotic treatment of biofilms are susceptible to the antibiotics during planktonic growth, the recalcitrance to antibiotic treatment displayed by the biofilm is often referred to as tolerance. Antibiotic tolerance is a physiological condition that does not involve mutation and allows the bacteria to survive, but not necessarily grow, in the presence of antibiotic concentrations above their planktonic minimal inhibitory concentration. Research in the last decade has shown that biofilm tolerance is multifactorial. The mechanisms underlying the phenomenon can be categorised in the five groups described below.

13.2.1 Restricted Penetration of Antimicrobials

The bacteria in biofilms are embedded in a matrix consisting of exopolysaccharide, extracellular DNA, and protein. When antibiotics are used in the attempt to combat biofilm infections they must traverse through this extracellular matrix to reach the encased bacteria. The available evidence suggests that biofilm matrices do not inhibit diffusion of antibiotics in general, but restricted penetration of antibiotics through biofilms may occur in cases where the antibiotic binds to components of the biofilm matrix. Walters, III et al. (2003) analysed the roles of antimicrobial penetration, oxygen limitation, and low metabolic activity in the tolerance of *P. aeruginosa* biofilms to ciprofloxacin and tobramycin. The authors found that although penetration of tobramycin into the biofilms was slower than penetration of ciprofloxacin, both antibiotics eventually penetrated the biofilms but were able to kill only metabolically active bacteria located in zones with high oxygen concentration, suggesting that low metabolic activity is more relevant to biofilm tolerance than restricted antimicrobial penetration. Although oxygen limitation and low metabolic activity appear to be more important factors for the tolerance of biofilms to aminoglycosides, delayed penetration of the aminoglycosides might play a role in the tolerance of biofilms to aminoglycosides in zones of the CF lung with poor access to aminoglycoside aerosols, where the antibiotic concentration is too low to saturate the exopolysaccharide matrix. Recently, it has been shown that administration of DNase and alginate lyase, which are believed to break down *P. aeruginosa* biofilm matrix components, enhanced the activity of tobramycin in biofilms (Alipour et al. 2009).

13.2.2 Differential Physiological Activity

In a study of flow-chamber (see Chapter 15) grown *P. aeruginosa* biofilms Pamp et al. (2008) provided evidence that the metabolic/physiological activity was high in a subpopulation of cells located in the upper part of the biofilm close to the bulk medium, and low in the inner part of the biofilm more distant to the bulk medium. Because similar observations have been obtained for *P. aeruginosa* biofilms grown as colonies or established in capillary glass tubes (Werner et al. 2004) (see Chapter 15) the observed spatial distribution of active and non-active cells might be a general characteristic of *P. aeruginosa* in vitro biofilms. The available evidence suggests that the differential physiological activity seen in biofilms is caused by limited oxygen and nutrient penetration through the biofilm due to bacterial consumption (e.g. Walters et al. 2003). Studies of flow-chamber-grown *P. aeruginosa* biofilms have provided evidence that the antibiotics tobramycin, ciprofloxacin, and tetracycline preferentially kill the physiologically active bacteria located in the upper part of the biofilm, whereas the antibiotic colistin, the detergent SDS, and the chelator EDTA preferentially kill the physiologically inactive bacteria located in the lower part of the biofilm (e.g. see Fig. 13.1a–c) (Bjarnsholt et al. 2005, Kim et al. 2009). The existence of physiologically distinct subpopulations in heterogeneous biofilms therefore appears to function as an insurance mechanism against environmental insult. Combination treatment with colistin and ciprofloxacin or colistin and

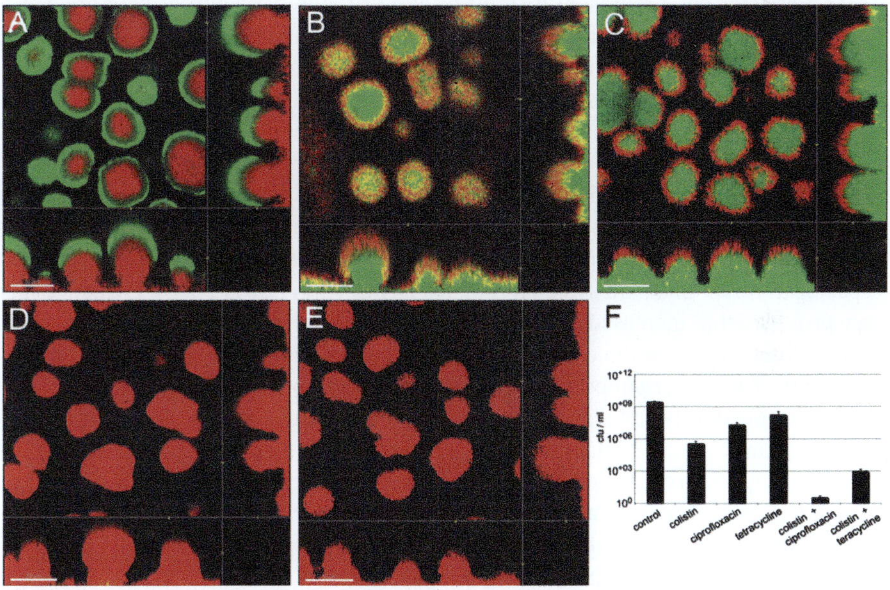

Fig. 13.1 Targeting of distinct subpopulations in *P. aeruginosa* biofilms by single and combined antimicrobial treatment. Biofilms of *P. aeruginosa* PAO1 Gfp were grown for 4 days and then continuously exposed to propidium iodide (dead cell indicator) and either 25 μg/ml colistin (**a**), 60 μg/ml ciprofloxacin (**b**), 200 μg/ml tetracycline (**c**), 25 μg/ml colistin and 200 μg/ml tetracycline (**d**), 25 μg/ml colistin and 60 μg/ml ciprofloxacin (**e**) for 24 h. The confocal laser scanning micrographs show horizontal sections with flanking images representing sagittal sections. Live cells appear green due to expression of Gfp and dead cells appear red due to staining with the dead-cell indicator propidium iodide. The number of cells which survived the antimicrobial treatment (**f**) was determined by plating and counting of cells harvested from biofilms. Reproduced from Pamp et al. (2008) with permission from John Wiley & Sons, Inc

tetracyclin was able to kill almost all the cells in biofilms of *P. aeruginosa* PAO1 in flow-chambers (Pamp et al. 2008) (Fig. 13.1). This indicated that by combined antimicrobial treatment using compounds that target separate physiological subpopulations within biofilms, it is possible to kill the majority of the cells in a biofilm. Because various niches are present in the CF lung, it is reasonable to assume that physiologically distinct subpopulations of *P. aeruginosa* are present in such infected lungs. In agreement, combination therapy with colistin and ciprofloxacin is used successfully for early eradication treatment of *P. aeruginosa* in CF patients in the Copenhagen CF Center (Hansen et al. 2008).

13.2.3 Persisters and Phenotypic Variants

Another type of bacteria that contribute to antibiotic tolerance in biofilms are the so-called persister cells (Brooun et al. 2000, De et al. 2009, Spoering and Lewis 2001). Persister cells represent a small fraction of slowly or non-dividing bacteria that are less vulnerable to antibiotics than the bulk of the population. They are generally believed to be the result of a small subpopulation of bacteria that differentiates into a dormant state. Because of reduced metabolism the persister cells are thought to escape the activity of antibiotics that target fundamental cellular processes such as DNA replication, translation, and cell wall synthesis, but they also in some cases show tolerance to antibiotics that kill non-growing cells (Brooun et al. 2000, De et al. 2009, Spoering and Lewis 2001). The pathways leading to the formation of persister cells are not known, but it has been proposed that they are connected to bacterial toxin/antitoxin systems in *Escherichia coli* (Lewis 2000). Evidence has been provided that increased expression of toxins may block cell metabolism, so that random fluctuations in the expression of toxin and antitoxin genes can lead to the formation of persister subpopulations (Vazquez-Laslop et al. 2006). Thereby bacteria with transiently unbalanced toxin/antitoxin systems are dormant and escape killing by antibiotics. Mutant screens have provided evidence that a number of genes (e.g. *rpoS*, *spoT*, *relA*, *dksA*, *dinG*, *spuC*, *algR*, *pilH*, *ycgM*, and *pheA*) are involved in persister formation in *P. aeruginosa* (Murakami et al. 2005, Viducic et al. 2006, De Groote et al. 2009) suggesting that the persister phenotype can be reached through multiple pathways.

Tolerance to antimicrobial compounds can also arise by other phenotypic variance mechanisms. In contrast to the persister cells that generally do not grow, Drenkard and Ausubel isolated (from CF lungs) *P. aeruginosa* phenotypic variants that were able to form colonies on plates with antibiotics (Drenkard and Ausubel 2002) (see also Chapter 15). In addition to antibiotic tolerance these phenotypic variants also displayed high surface hydrophobicity and adhesiveness and formed biofilms with increased biomass. Through mutant analysis a regulatory protein (PvrR) was identified, that controls the conversion between the parental form and tolerant variants, and also regulates biofilm formation (Drenkard and Ausubel 2002).

13.2.4 Specific Tolerance Mechanisms Connected to the Biofilm Mode of Growth

In addition to tolerance caused by the mechanisms described above, certain gene products that are produced specifically in biofilms may exert unique functions that enhance the antibiotic tolerance of the biofilm. One example of a biofilm specific factor is the *ndvB* gene of *P. aeruginosa* PA14. This gene evidently encodes an enzyme involved in the synthesis of periplasmic glucans that binds tobramycin and prevents cell death most likely by sequestering the antibiotic (Mah et al. 2003). Biofilms formed by a *P. aeruginosa ndvB* mutant were much more sensitive to tobramycin than wild type biofilms. In contrast, the *ndvB* mutant and wild type showed no difference in tobramycin sensitivity when grown in planktonic culture. Reverse transcriptase PCR provided evidence that the *ndvB* gene is expressed specifically in *P. aeruginosa* PA14 biofilms and not in planktonic cells (Mah et al. 2003). However, microarray analysis has provided evidence that *ndvB* is expressed at the same level in biofilm and planktonic cells of *P. aeruginosa* PAO1 (Hentzer et al. 2005), and therefore the *ndvB*-mediated mechanism appears to be restricted to specific *P. aeruginosa* strains. A novel *P. aeruginosa* efflux pump that is expressed only during the biofilm mode of growth and mediates resistance to tobramycin, gentamicin, and ciprofloxacin in *P. aeruginosa* PA14 has been reported (Zhang and Mah 2008). This efflux-pump, encoded by *P. aeruginosa* genes PA1874-1877, is apparently acting in synergy with the *ndvB* gene required for the synthesis of periplasmic glucans (Zhang and Mah 2008).

As described above, Pamp et al. (2008) provided evidence that a spatially distinct subpopulation of metabolically active cells in *P. aeruginosa* biofilms is able to develop tolerance to colistin treatment. On the contrary, biofilm cells exhibiting low metabolic activity are killed by the colistin treatment. It was demonstrated that the subpopulation of metabolically active cells was able to adapt to colistin by inducing a specific adaptation mechanism mediated by the *pmr*-operon, as well as an unspecific adaptation mechanism mediated by the *mexAB-oprM*-genes. The *pmr*-operon encodes a system that adds aminoarabinose to LPS thereby changing the charge and preventing the interaction between colistin and the surface of the bacteria (Trent et al. 2001, McPhee et al. 2003). The *mexAB-oprM* genes encode an RND-efflux-pump which can extrude antimicrobials across the outer membrane of the bacteria (Li et al. 1994). Mutants defective in either *pmr*-mediated LPS-modification or in *mexAB-oprM*-mediated antimicrobial efflux were not able to develop a tolerant subpopulation in biofilms. In contrast to the findings in biofilms, it was found that planktonic exponential-phase cells (exhibiting high metabolic activity) and planktonic stationary-phase cells (exhibiting low metabolic activity) were equally sensitive to colistin, suggesting that the mechanisms resulting in the observed colistin tolerance are biofilm-specific.

The involvement of quorum-sensing in antibiotic tolerance in biofilms may be another example of biofilm-specific tolerance mechanisms. Quorum-sensing (QS) has been reported to be involved in tolerance of *P. aeruginosa* biofilms to tobramycin, kanamycin, and hydrogen peroxide (Hassett et al. 1999, Shih and Huang 2002, Bjarnsholt et al. 2005). However, QS has not been reported to be involved in antibiotic tolerance of planktonic cells, suggesting that this kind of regulation of tolerance genes occurs only in biofilms.

13.2.5 Specific Tolerance Mechanisms Not Connected to the Biofilm Mode of Growth

It is expected that antibiotic treatment can be a harsh stress even for bacteria within biofilms, and that some of the conventional inducible antibiotic resistance mechanisms, that are not specific to biofilm cells, will be induced in biofilms. Below we will describe a role in biofilm tolerance of mechanisms involving antibiotic degrading enzymes and efflux pumps.

The antibiotic degrading enzymes that play a role in the resistance of *P. aeruginosa* to antibiotics are the aminoglycoside-degrading enzymes and the beta-lactamases. While the role of aminoglycoside-degrading enzymes is less important in aminoglycoside resistance compared to impermeability (efflux pumps), beta-lactamases play a crucial role in the development of resistance to beta-lactams. Rapid development of resistance to beta-lactam antibiotics due to beta-lactamase production has been shown in in vitro and in vivo animal models of chronic *P. aeruginosa* lung infection (Bagge et al. 2000) (see also Chapter 16) as well as in CF patients (Giwercman et al. 1990). In biofilms of *P. aeruginosa* that overproduce beta-lactamase, the presence in the biofilm matrix of beta-lactamases will lead to the hydrolysis of the beta-lactam antibiotics before they reach the bacterial cells. Nichols (Nichols et al. 1989) predicted from mathematical models that bacteria expressing high levels of chromosomal beta-lactamase growing in biofilms would be exposed to reduced concentrations of beta-lactam antibiotics due to accumulation of the enzyme in the polysaccharide matrix. Giwercman et al. (1991) showed that imipenem and piperacillin were able to induce beta-lactamase production in *P. aeruginosa* biofilms.

The source of beta-lactamase in biofilms has been considered to be from a sacrificial layer of bacteria exposed to an antibiotic, with release of defensive enzymes into the extracellular space. Bagge et al. (2004) showed that strong inducers like imipenem will induce the beta-lactamase through all the bacterial layers while poorer inducers like ceftazidime will influence just the superficial layers of the biofilm, probably due to the inactivation of the antibiotic by beta-lactamase (Fig. 13.2) (Bagge et al. 2004a). Ciofu et al. (2000) provided evidence that the source of beta-lactamase in biofilms may also be membrane vesicles (MVs) containing beta-lactamase liberated by resistant *P. aeruginosa* bacteria (Ciofu et al. 2000). The concentration of antibiotic might be a limiting factor in the treatment of biofilms with beta-lactams changing the pharmocodynamic parameters for the otherwise time-dependent killing exerted by this group of antibiotics.

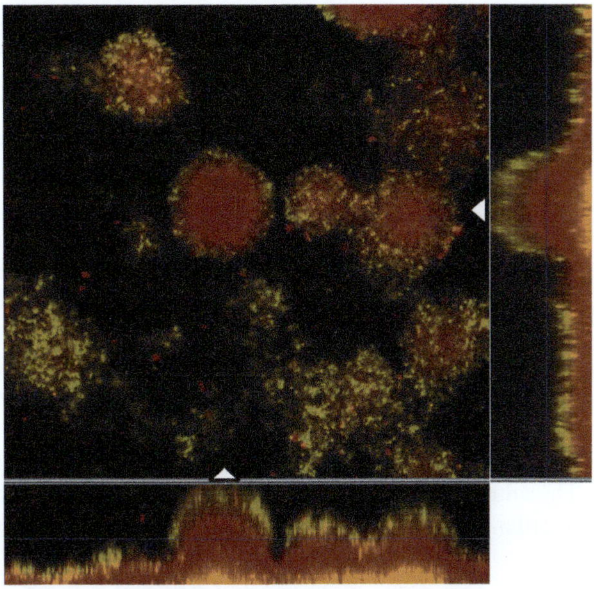

Fig. 13.2 Induction of beta-lactamase in a subpopulation in *P. aeruginosa* biofilm. A biofilm of a *P. aeruginosa* PAO1(P*ampC-gfp*) reporter strain, expressing green fluorescent protein when the promoter of the AmpC beta-lactamase is induced, was grown for 6 days and exposed to 100 μg/ml ceftazidime for 4 h. Bacteria that are not expressing the P*ampC-gfp* fusion appear red due to staining with SYTO62. Reproduced from Bagge et al. (2004a) with permission from the American Society for Microbiology (ASM)

A potential role of beta-lactamase inhibitors in the treatment of biofilms of bacteria hyperproducing beta-lactamase has been suggested by the improved clearance of the bacteria and the milder pathological changes observed in a rat model of chronic lung infection in animals who mounted a protective humoral immune response after vaccination with beta-lactamase. The protective humoral immune response was represented by anti-beta-lactamase antibodies with neutralizing activity (Ciofu et al. 2002). These data were also supported by clinical studies which showed that CF patients with good lung function had higher titers of anti-beta-lactamase antibody of high affinity than patients with poor lung function (Ciofu et al. 1999).

In addition, meropenem, a beta-lactamase-stable beta-lactam, showed good efficacy in the treatment of *P. aeruginosa* biofilms (Hill et al. 2005, Moskowitz et al. 2004).

Together, the data described above show that beta-lactamases play an important role in the tolerance of biofilms to beta-lactam antibiotics and suggest the potential use of beta-lactamase inhibitors together with beta-lactams for the treatment of beta-lactam tolerant biofilms.

Evidence has been provided that MexAB-OprM, MexCD-OprJ, MexEF-OprN, and MexXY efflux pumps, which play an important role in the intrinsic resistance to antibiotics in planktonic *P. aeruginosa*, had no impact on biofilm tolerance when mature biofilms were tested (De Kievit et al. 2001). However, recent studies are starting to modify this perception as it has been suggested that MexAB-OprM and MexCD-oprJ are involved in biofilm tolerance to the macrolide azithromycin (Gillis et al. 2005) and that MexAB-OprM is involved in biofilm tolerance to colistin (Pamp et al. 2008). The MexEF-OprN and MexXY-oprM efflux systems of *P. aeruginosa* were shown to be up-regulated in response to reactive oxygen species (ROS), and it was proposed that this efflux system exports cellular constituents damaged by ROS (Poole 2008). This is especially interesting because bacteria in biofilms, as addressed below, experience increased oxidative stress which might promote up-regulation of these pumps. Thus, in contrast to earlier reported results, it seems that the conventional efflux pumps may play a role in antibiotic tolerance in *P. aeruginosa* biofilms. Similar results have been reported in biofilms formed by *E. coli* isolates from urinary tract infection where many of the efflux pumps involved in removal of toxic substances, including many antibiotics, were highly up-regulated during biofilm growth (Kvist et al. 2008).

In this context of increasing evidence for a role of efflux pumps in the tolerance of biofilms to antibiotics it is obvious that the use of efflux-pumps inhibitors might improve the efficacy of antibiotic treatment. Interestingly, it has been shown that inactivation of efflux pumps abolished *E. coli* biofilm formation (Kvist et al. 2008). The authors speculated that efflux pump activity might be required in the biofilms in order to remove waste products from the bacterial cells.

13.3 Antibiotic Resistance in Biofilms

The described tolerance mechanisms all contribute to the persistence of biofilms, which provide a fertile ground for the emergence of antibiotic resistant mutants. Estimates suggest that 25–45% of adult CF patients are chronically infected with multidrug resistant (MDR) bacteria (Lechtzin et al. 2006), defined as resistance of an organism to two out of three different classes of antibiotics, namely beta-lactams, aminoglycosides, or quinolones (as defined by the CF Foundation http://www.cff.org). The genetic basis for MDR *P. aeruginosa* is often simultaneous mutations in a single bacterial cell of several genes causing antibiotic resistance. Thus, one bacterial cell might produce enzymes that degrade antibiotics have antibiotic targets of low-affinity and overexpress efflux pumps which have a broad spectrum of substrates. Each of the necessary mutations arises at frequencies of 10^{-7} to 10^{-9} per cell per generation. Achievement of bacteria with multiple mutations in a bacterial population size of 10^{8}–10^{10}/ml sputum as is attained under infection of the CF lung (Hoiby 2001) suggests the presence of hypermutable subpopulations, and accordingly the presence of high percentages of hypermutable *P. aeruginosa* isolates associated with antibiotic resistance has been demonstrated in CF patients in several cases (Ciofu et al. 2005, Ferroni et al. 2009, Hoboth et al. 2009, Oliver et al. 2000).

Evidence has been provided that the hypermutable phenotype of CF *P. aeruginosa* isolates is due to alterations in the genes of the DNA repair systems of either the mismatch repair system (MMR) which involves *mutS*, *mutL*, and *uvrD* or the DNA oxidative lesions repair system (GO) which involves *mutT*, *mutY*, and *mutM* (Mandsberg et al. 2009, Oliver et al. 2000, 2002). It has been shown that mutations in either of the two systems promote emergence of antibiotic resistant isolates, especially due to selection of isolates expressing multidrug efflux-pumps (Macia et al. 2005, Mandsberg et al. 2009). Ciofu and Høiby (2007) proposed that the biofilm mode of growth is one of the main conditions which together with the antibiotic selective pressure can predispose to bacterial hypermutability in the CF lung (Ciofu and Høiby 2007). This hypothesis has been recently confirmed by elegant studies of the role of mutation in *P. aeruginosa* biofilm development which show that bacteria within microcolonies exhibit enhanced mutation frequencies (Conibear et al. 2009). These recent data suggest that microcolony structures are specific sites within biofilms for enhanced genetic adaptation and evolutionary change during chronic infections.

An increased production of endogenous ROS (Barraud et al. 2006, Mai-Prochnow et al. 2008, Park et al. 2005) and a deficient anti-oxidant system (Driffield et al. 2008, Hassett et al. 1999) determine an imbalance between oxidative burden and defences leading to oxidative stress in biofilms (Boles and Singh 2008). The oxidative stress is considered to cause enhanced mutability in biofilms (Driffield et al. 2008, Conibear et al. 2009) as well as extensive diversity and adaptability in biofilm communities (Boles and Singh 2008). We have found increased levels of the oxidative DNA lesions (measured as the oxidized form of guanosine 8-oxodG per 10^6 dG molecules) in PAO1 after biofilm growth compared to planktonic growth (unpublished results) confirming the hypothesis of increased oxidative stress in biofilms. In addition, Boles and Singh (2008) showed that the endogenous oxidative stress in biofilms promote antibiotic resistance and that addition of anti-oxidants reduced the occurrence of diversity in biofilms. The notion of oxidative stress in biofilms might be paradoxical given the low oxygen tension detected deep within biofilm structures; however, recent work shows that respiration can produce enough oxidative stress to produce DNA damage (Park et al. 2005). Ciofu et al. (2005) have previously shown that oxidative stress is linked to the occurrence of hypermutable *P. aeruginosa* strains in CF patients (Ciofu et al. 2005). In addition to the endogenous oxidative stress the biofilm-growing bacteria in CF airways are exposed to ROS from the activated immune cells which surround the biofilm. Kolpen et al. (2009) recently showed that the hypoxic environment in the CF sputum is due to the consumption of oxygen by PMNs which liberate reactive oxygen species that can react with the biofilm-embedded bacterial cells thus creating a unique environment with low-oxygen tension filled with reactive oxygen species.

The hypermutability of biofilms promotes the emergence of mutations conferring antibiotic resistance, and antibiotic resistant mutants will be selected for by the repeated antibiotic courses administered in order to maintain the lung function of CF patients. Development of resistance to all classes of antibiotics during the chronic lung infection in CF have been documented (Ciofu et al. 1994). Recently it has been shown that azithromycin might select in *P. aeruginosa* biofilms for mutants overexpressing MexCD-OprJ leading to cross-resistance to ciprofloxacin (Mulet et al. 2009). As the ROS burden is high in the lungs of chronically infected CF patients, and because the MexEF-OprN and MexXY-oprM efflux systems of *P. aeruginosa* are up-regulated in response to ROS (Poole 2008), a selection of mutants overexpressing these pumps might occur. This may explain that MexXY-mediated aminoglycoside resistance is disproportionately represented among strains of *P. aeruginosa* recovered from the CF lung (Islam et al. 2009). In a recent study Mandsberg et al. (2009) found that mutants overexpressing MexCD-OprJ were selected by ciprofloxacin treatment of hypermutable *P. aeruginosa mutT* and *mutY* strains (Mandsberg et al. 2009). Resistance to beta-lactam antibiotics occurs due to mutations in regulatory genes of beta-lactamase production leading to the occurrence of isolates with stable or partially stable derepressed production of AmpC beta-lactamase (Ciofu 2003). Resistance to ciprofloxacin and tobramycin of CF *P. aeruginosa* isolates was shown to be mediated by overexpression of multi-drug efflux pumps due to mutations in the genes involved in the regulation of pump expression (Islam et al. 2009, Jalal et al. 2000). Finally, resistance to colistin was shown to occur due to mutations in the *pmr* system involved in modification of LPS (Johansen et al. 2008).

13.4 Prevention of Tolerance and Resistance in Biofilms in Clinical Settings

The presence of antibiotic tolerant and antibiotic resistant subpopulations in biofilms is the cause of persistent infections. It has been shown that use of antibiotics that proved to be effective in in vitro biofilm susceptibility testing had better clinical effect in CF patients than treatment with antimicrobial drugs based on planktonically determined susceptibility (Keays et al. 2009), emphasizing that antibiotic efficacy should be tested on biofilms as opposed to on planktonic cells (see also Chapter 15). Biofilm susceptibility testing of 100 CF isolates demonstrated diminished activity of several antipseudomonal antibiotics compared to standard in vitro susceptibility testing, and suggested that the use of standard drug dosages result in sub-optimal drug concentrations at the site of infection (Moskowitz et al. 2004). The negative effects of biofilm subinhibitory concentration are multiple: lack of bacterial killing, development of antibiotic resistance due to exposure of bacterial cells to concentrations lower than the mutant-preventing concentration, and enhancement of biofilm formation. It has been shown that sub-MIC concentrations of aminoglycosides (Hoffman et al. 2005), beta-lactam antibiotics (Bagge et al. 2004b) and quinolones (Takahashi et al. 1995) up-regulate genes involved in biofilm formation. To eliminate different antibiotic tolerant subpopulations and to circumvent occurrence of mutations causing antibiotic resistance in biofilm bacteria, high dosages of antibiotics and combination therapy should be used in the treatment of biofilm infections. For *P. aeruginosa* biofilm in the conductive zones of the CF airways high antibiotic concentrations can be achieved by inhalations while for the biofilm in respiratory zones of the airway high antibiotic concentrations can be achieved by intravenous treatment. Combination therapy was shown to efficiently eliminate the bacteria in in vitro *P. aeruginosa* biofilms (Pamp et al. 2008), and the increased efficacy of combination therapy for treatment of hypermutable populations was shown in an animal model of biofilm infection (Macia et al. 2006). As in vitro studies showed a beneficial effect of anti-oxidants in term of decreased mutability and decreased resistance of biofilms to antibiotics (Boles and Singh 2008), the use of anti-oxidants as drugs that prevent resistance development in biofilms may be feasible. Due to the presence of hypermutable subpopulations, mutants resistant to virtually all single antipseudomonal agents are anticipated to be already present in a high proportion of CF patients with *P. aeruginosa* lung infection prior to treatment (Oliver et al. 2004). The direct consequence is that the use of combination therapy with pairs of antibiotics of different classes with synergistic activities should be applied at all stages of the infection, starting with early aggressive treatment as recommended in the European Consensus Document on Early Intervention and Prevention of Lung Disease in Cystic Fibrosis (Doring and Hoiby 2004).

13.5 Conclusions

The persistence of biofilm-growing bacteria in spite of antibiotic treatment is multifactorial involving biofilm-related tolerance mechanisms and accelerated development of antibiotic resistance due to increased mutagenesis and selection by repeated antibiotic courses. The biofilm-related tolerance mechanisms involve restricted penetration of antimicrobials, differential physiological activity, the presence of phenotypic variants and persisters, and specific cellular tolerance mechanisms connected to the biofilm mode of growth. To eliminate different antibiotic tolerant subpopulations and to circumvent occurrence of mutations causing antibiotic resistance in biofilm bacteria, high dosages of antibiotics and early combination therapy should be used in the treatment of biofilm infections

References

Alipour M, Suntres ZE, Omri A (2009) Importance of DNase and alginate lyase for enhancing free and liposome encapsulated aminoglycoside activity against Pseudomonas aeruginosa. J Antimicrob Chemother 64:317–325

Bagge N, Ciofu O, Skovgaard LT, Hoiby N (2000) Rapid development in vitro and in vivo of resistance to ceftazidime in biofilm-growing Pseudomonas aeruginosa due to chromosomal beta-lactamase. Apmis 108:589–600

Bagge N, Hentzer M, Andersen JB, Ciofu O, Givskov M, Hoiby N (2004a) Dynamics and spatial distribution of beta-lactamase expression in Pseudomonas aeruginosa biofilms. Antimicrob Agents Chemother 48:1168–1174

Bagge N, Schuster M, Hentzer M, Ciofu O, Givskov M, Greenberg EP et al (2004b) Pseudomonas aeruginosa biofilms exposed to imipenem exhibit changes in global gene expression and beta-lactamase and alginate production. Antimicrob Agents Chemother 48:1175–1187

Barraud N, Hassett DJ, Hwang SH, Rice SA, Kjelleberg S, Webb JS (2006) Involvement of nitric oxide in biofilm dispersal of Pseudomonas aeruginosa. J Bacteriol 188:7344–7353

Bjarnsholt T, Jensen PO, Burmolle M, Hentzer M, Haagensen JA, Hougen HP et al (2005) Pseudomonas aeruginosa tolerance to tobramycin, hydrogen peroxide and polymorphonuclear leukocytes is quorum-sensing dependent. Microbiology 151:373–383

Bjarnsholt T, Jensen PO, Fiandaca MJ, Pedersen J, Hansen CR, Andersen CB et al (2009) Pseudomonas aeruginosa biofilms in the respiratory tract of cystic fibrosis patients. Pediatr Pulmonol 44:547–558

Boles BR, Singh PK (2008) Endogenous oxidative stress produces diversity and adaptability in biofilm communities. Proc Natl Acad Sci USA 105:12503–12508

Brooun A, Liu S, Lewis K (2000) A dose-response study of antibiotic resistance in Pseudomonas aeruginosa biofilms. Antimicrob Agents Chemother 44:640–646

Ciofu O (2003) Pseudomonas aeruginosa chromosomal beta-lactamase in patients with cystic fibrosis and chronic lung infection. Mechanism of antibiotic resistance and target of the humoral immune response. APMIS Suppl 1–47

Ciofu O, Bagge N, Hoiby N (2002) Antibodies against beta-lactamase can improve ceftazidime treatment of lung infection with beta-lactam-resistant Pseudomonas aeruginosa in a rat model of chronic lung infection. Apmis 110:881–891

Ciofu O, Beveridge TJ, Kadurugamuwa J, Walther-Rasmussen J, Hoiby N (2000) Chromosomal beta-lactamase is packaged into membrane vesicles and secreted from Pseudomonas aeruginosa. J Antimicrob Chemother 45:9–13

Ciofu O, Giwercman B, Pedersen SS, Hoiby N (1994) Development of antibiotic resistance in Pseudomonas aeruginosa during two decades of antipseudomonal treatment at the Danish CF Center. Apmis 102:674–680

Ciofu O, Høiby N (2007) Cystic Fibrosis-coping with resistance. In: Gould, van der Meer (eds) Antibiotic policies: fighting resistance. Springer, New York, pp 149–174

Ciofu O, Petersen TD, Jensen P, Hoiby N (1999) Avidity of anti-P aeruginosa antibodies during chronic infection in patients with cystic fibrosis. Thorax 54:141–144

Ciofu O, Riis B, Pressler T, Poulsen HE, Hoiby N (2005) Occurrence of hypermutable Pseudomonas aeruginosa in cystic fibrosis patients is associated with the oxidative stress caused by chronic lung inflammation. Antimicrob Agents Chemother 49: 2276–2282

Conibear TC, Collins SL, Webb JS (2009) Role of mutation in Pseudomonas aeruginosa biofilm development. PLoS One 4:e6289

De Kievit TR, Parkins MD, Gillis RJ, Srikumar R, Ceri H, Poole K et al (2001) Multidrug efflux pumps: expression patterns and contribution to antibiotic resistance in Pseudomonas aeruginosa biofilms. Antimicrob Agents Chemother 45:1761–1770

De Groote VN, Verstraeten N, Fauvart M, Kint CI, Verbeeck AM, Beullens S et al (2009) Novel persistence genes in Pseudomonas aeruginosa identified by high-throughput screening. FEMS Microbiol Lett 297:73–77

Doring G, Conway SP, Heijerman HG, Hodson ME, Hoiby N, Smyth A et al (2000) Antibiotic therapy against Pseudomonas aeruginosa in cystic fibrosis: a European consensus. Eur Respir J 16:749–767

Doring G, Hoiby N (2004) Early intervention and prevention of lung disease in cystic fibrosis: a European consensus. J Cyst Fibros 3:67–91

Drenkard E, Ausubel FM (2002) Pseudomonas biofilm formation and antibiotic resistance are linked to phenotypic variation. Nature 416:740–743

Driffield K, Miller K, Bostock JM, O'Neill AJ, Chopra I (2008) Increased mutability of Pseudomonas aeruginosa in biofilms. J Antimicrob Chemother 61:1053–1056

Ferroni A, Guillemot D, Moumile K, Bernede C, Le BM, Waernessyckle S et al (2009) Effect of mutator P. aeruginosa on antibiotic resistance acquisition and respiratory function in cystic fibrosis. Pediatr Pulmonol 44:820–825

Gillis RJ, White KG, Choi KH, Wagner VE, Schweizer HP, Iglewski BH (2005) Molecular basis of azithromycin-resistant Pseudomonas aeruginosa biofilms. Antimicrob Agents Chemother 49:3858–3867

Giwercman B, Jensen ET, Hoiby N, Kharazmi A, Costerton JW (1991) Induction of beta-lactamase production in Pseudomonas aeruginosa biofilm. Antimicrob Agents Chemother 35: 1008–1010

Giwercman B, Lambert PA, Rosdahl VT, Shand GH, Hoiby N (1990) Rapid emergence of resistance in Pseudomonas aeruginosa in cystic fibrosis patients due to in-vivo selection of stable partially derepressed beta-lactamase producing strains. J Antimicrob Chemother 26: 247–259

Hansen CR, Pressler T, Hoiby N (2008) Early aggressive eradication therapy for intermittent Pseudomonas aeruginosa airway colonization in cystic fibrosis patients: 15 years experience. J Cyst Fibros 7:523–530

Hassett DJ, Ma JF, Elkins JG, McDermott TR, Ochsner UA, West SE et al (1999) Quorum sensing in Pseudomonas aeruginosa controls expression of catalase and superoxide dismutase genes and mediates biofilm susceptibility to hydrogen peroxide. Mol Microbiol 34: 1082–1093

Hentzer M, Eberl L, Givskov M (2005) Transcriptome analysis of Pseudomonas aeruginosa biofilm development: anaerobic respiration and iron limitation. Biofilms 2:37–61

Hill D, Rose B, Pajkos A, Robinson M, Bye P, Bell S et al (2005) Antibiotic susceptabilities of Pseudomonas aeruginosa isolates derived from patients with cystic fibrosis under aerobic, anaerobic, and biofilm conditions. J Clin Microbiol 43:5085–5090

Hoboth C, Hoffmann R, Eichner A, Henke C, Schmoldt S, Imhof A et al (2009) Dynamics of adaptive microevolution of hypermutable Pseudomonas aeruginosa during chronic pulmonary infection in patients with cystic fibrosis. J Infect Dis 200:118–130

Hoffman LR, D'Argenio DA, MacCoss MJ, Zhang Z, Jones RA, Miller SI (2005) Aminoglycoside antibiotics induce bacterial biofilm formation. Nature 436:1171–1175

Hoiby N (2001) Inflammation and infection in cystic fibrosis – hen or egg? Eur Respir J 17:4–5

Islam S, Oh H, Jalal S, Karpati F, Ciofu O, Hoiby N et al (2009) Chromosomal mechanisms of aminoglycoside resistance in Pseudomonas aeruginosa isolates from cystic fibrosis patients. Clin Microbiol Infect 15:60–66

Jalal S, Ciofu O, Hoiby N, Gotoh N, Wretlind B (2000) Molecular mechanisms of fluoroquinolone resistance in Pseudomonas aeruginosa isolates from cystic fibrosis patients. Antimicrob Agents Chemother 44:710–712

Johansen HK, Moskowitz SM, Ciofu O, Pressler T, Hoiby N (2008) Spread of colistin resistant non-mucoid Pseudomonas aeruginosa among chronically infected Danish cystic fibrosis patients. J Cyst Fibros 7:391–397

Keays T, Ferris W, Vandemheen KL, Chan F, Yeung SW, Mah TF et al (2009) A retrospective analysis of biofilm antibiotic susceptibility testing: a better predictor of clinical response in cystic fibrosis exacerbations. J Cyst Fibros 8:122–127

Kim J, Hahn JS, Franklin MJ, Stewart PS, Yoon J (2009) Tolerance of dormant and active cells in Pseudomonas aeruginosa PA01 biofilm to antimicrobial agents. J Antimicrob Chemother 63:129–135

Kvist M, Hancock V, Klemm P (2008) Inactivation of efflux pumps abolishes bacterial biofilm formation. Appl Environ Microbiol 74:7376–7382

Lechtzin N, John M, Irizarry R, Merlo C, Diette GB, Boyle MP (2006) Outcomes of adults with cystic fibrosis infected with antibiotic-resistant Pseudomonas aeruginosa. Respiration 73:27–33

Lewis K (2000) Programmed death in bacteria. Microbiol Mol Biol Rev 64: 503–514

Li XZ, Ma D, Livermore DM, Nikaido H (1994) Role of efflux pump(s) in intrinsic resistance of Pseudomonas aeruginosa: active efflux as a contributing factor to beta-lactam resistance. Antimicrob Agents Chemother 38:1742–1752

Macia MD, Blanquer D, Togores B, Sauleda J, Perez JL, Oliver A (2005) Hypermutation is a key factor in development of multiple-antimicrobial resistance in Pseudomonas aeruginosa strains causing chronic lung infections. Antimicrob Agents Chemother 49: 3382–3386

Macia MD, Borrell N, Segura M, Gomez C, Perez JL, Oliver A (2006) Efficacy and potential for resistance selection of antipseudomonal treatments in a mouse model of lung infection by hypermutable Pseudomonas aeruginosa. Antimicrob Agents Chemother 50:975–983

Mah TF, Pitts B, Pellock B, Walker GC, Stewart PS, O'Toole GA (2003) A genetic basis for Pseudomonas aeruginosa biofilm antibiotic resistance. Nature 426:306–310

Mai-Prochnow A, Lucas-Elio P, Egan S, Thomas T, Webb JS, Sanchez-Amat A. et al (2008) Hydrogen peroxide linked to lysine oxidase activity facilitates biofilm differentiation and dispersal in several gram-negative bacteria. J Bacteriol 190:5493–5501

Mandsberg LF, Ciofu O, Kirkby N, Christiansen LE, Poulsen HE, Hoiby N (2009) Antibiotic resistance in Pseudomonas aeruginosa strains with increased mutation frequency due to inactivation of the DNA oxidative repair system. Antimicrob Agents Chemother 53:2483–2491

McPhee JB, Lewenza S, Hancock RE (2003) Cationic antimicrobial peptides activate a two-component regulatory system, PmrA-PmrB, that regulates resistance to polymyxin B and cationic antimicrobial peptides in Pseudomonas aeruginosa. Mol Microbiol 50:205–217

Moskowitz SM, Foster JM, Emerson J, Burns JL (2004) Clinically feasible biofilm susceptibility assay for isolates of Pseudomonas aeruginosa from patients with cystic fibrosis. J Clin Microbiol 42:1915–1922

Mulet X, Macia MD, Mena A, Juan C, Perez JL, Oliver A (2009) Azithromycin in Pseudomonas aeruginosa biofilms: bactericidal activity and selection of nfxB mutants. Antimicrob Agents Chemother 53:1552–1560

Murakami K, Ono T, Viducic D, Kayama S, Mori M, Hirota K. et al (2005) Role for rpoS gene of Pseudomonas aeruginosa in antibiotic tolerance. FEMS Microbiol Lett 242:161–167

Nichols WW, Evans MJ, Slack MP, Walmsley HL (1989) The penetration of antibiotics into aggregates of mucoid and non-mucoid Pseudomonas aeruginosa. J Gen Microbiol 135:1291–1303

Oliver A, Canton R, Campo P, Baquero F, Blazquez J (2000) High frequency of hypermutable Pseudomonas aeruginosa in cystic fibrosis lung infection. Science 288:1251–1254

Oliver A, Levin BR, Juan C, Baquero F, Blazquez J (2004) Hypermutation and the preexistence of antibiotic-resistant Pseudomonas aeruginosa mutants: implications for susceptibility testing and treatment of chronic infections. Antimicrob Agents Chemother 48:4226–4233

Oliver A, Sanchez JM, Blazquez J (2002) Characterization of the GO system of Pseudomonas aeruginosa. FEMS Microbiol Lett 217:31–35

Pamp SJ, Gjermansen M, Johansen HK, Tolker-Nielsen T (2008) Tolerance to the antimicrobial peptide colistin in Pseudomonas aeruginosa biofilms is linked to metabolically active cells, and depends on the pmr and mexAB-oprM genes. Mol Microbiol 68:223–240

Park S, You X, Imlay JA (2005) Substantial DNA damage from submicromolar intracellular hydrogen peroxide detected in Hpx-mutants of Escherichia coli. Proc Natl Acad Sci USA 102:9317–9322

Poole K (2008) Bacterial multidrug efflux pumps serve other functions. Microbe 3:179–185. Ref type: generic

Shih PC, Huang CT (2002) Effects of quorum-sensing deficiency on Pseudomonas aeruginosa biofilm formation and antibiotic resistance. J Antimicrob Chemother 49:309–314

Spoering AL, Lewis K (2001) Biofilms and planktonic cells of Pseudomonas aeruginosa have similar resistance to killing by antimicrobials. J Bacteriol 183:6746–6751

Takahashi A, Yomoda S, Ushijima Y, Kobayashi I, Inoue M (1995) Ofloxacin, norfloxacin and ceftazidime increase the production of alginate and promote the formation of biofilm of Pseudomonas aeruginosa in vitro. J Antimicrob Chemother 36:743–745

Trent MS, Ribeiro AA, Lin S, Cotter RJ, Raetz CR (2001) An inner membrane enzyme in Salmonella and Escherichia coli that transfers 4-amino-4-deoxy-L-arabinose to lipid A: induction on polymyxin-resistant mutants and role of a novel lipid-linked donor. J Biol Chem 276:43122–43131

Vazquez-Laslop N, Lee H, Neyfakh AA (2006) Increased persistence in Escherichia coli caused by controlled expression of toxins or other unrelated proteins. J Bacteriol 188: 3494–3497

Viducic D, Ono T, Murakami K, Susilowati H, Kayama S, Hirota K et al (2006) Functional analysis of spoT, relA and dksA genes on quinolone tolerance in Pseudomonas aeruginosa under nongrowing condition. Microbiol Immunol 50:349–357

Walters MC, III, Roe F, Bugnicourt A, Franklin MJ, Stewart PS (2003) Contributions of antibiotic penetration, oxygen limitation, and low metabolic activity to tolerance of Pseudomonas aeruginosa biofilms to ciprofloxacin and tobramycin. Antimicrob Agents Chemother 47:317–323
Werner E, Roe F, Bugnicourt A, Franklin MJ, Heydorn A, Molin S et al (2004) Stratified growth in Pseudomonas aeruginosa biofilms. Appl Environ Microbiol 70:6188–6196
Zhang L, Mah TF (2008) Involvement of a novel efflux system in biofilm-specific resistance to antibiotics. J Bacteriol 190:4447–4452

Chapter 14
Novel and Future Treatment Strategies

Morten Alhede, Tim Holm Jakobsen, and Michael Givskov

14.1 Introduction

The biofilm mode of life is by many scientists acknowledged as a successful, bacterial survival strategy in hostile environments. When bacteria invade their hosts, they encounter harsh conditions such as low levels of iron, oxidative stress, macerating enzymes, phagocytic cells, and host-derived as well as administered antimicrobials. Experimental evidence has accumulated over the years showing that biofilms tolerate antimicrobial properties of the immune system and antibiotics. This multifaceted tolerance relies to a certain extend on general resistance mechanisms including efflux pumps and enzymatic modifications in addition to innate tolerances offered by integral structure-functions of the biofilm. Among the features contributing to this are the decreased penetration of phagocytes and antibiotics through the biofilm matrix and reduced nutrient availability ensued by lowered metabolic activity (Anderson and O'Toole 2008).

Bacteria living within in vitro biofilms (see also Chapter 15) display highly heterogeneous physiology. For example, in vitro biofilms of the model organism *Pseudomonas aeruginosa* have been shown to accommodate several subpopulations with different susceptibility to antimicrobials (Haagensen et al. 2007, Parsek and Tolker-Nielsen 2008). Most conventional antimicrobials interfere with fundamental and basal life processes of the bacteria. Such antimicrobials preferentially kill biofilm cells with high metabolic activity in particular those active in cell division and multiplication. Consequently, the top layers of exposed biofilms (which are maintaining high metabolic activities) are killed leaving less active or even dormant deeper layers alive (Hentzer et al. 2003, Pamp et al. 2009, Walters et al. 2003). On the other hand, the stalks of the characteristic mushroom structures can be selectively killed with the antimicrobial peptide colistin and compounds, such as SDS and EDTA, all of which affects membrane integrity (Haagensen et al. 2007, Pamp

M. Givskov (✉)
Department of International Health, Immunology and Microbiology, Faculty of Health Sciences, University of Copenhagen, DK-2200, Copenhagen N, Denmark
e-mail: mgivskov@sund.ku.dk

T. Bjarnsholt et al. (eds.), *Biofilm Infections*, DOI 10.1007/978-1-4419-6084-9_14,

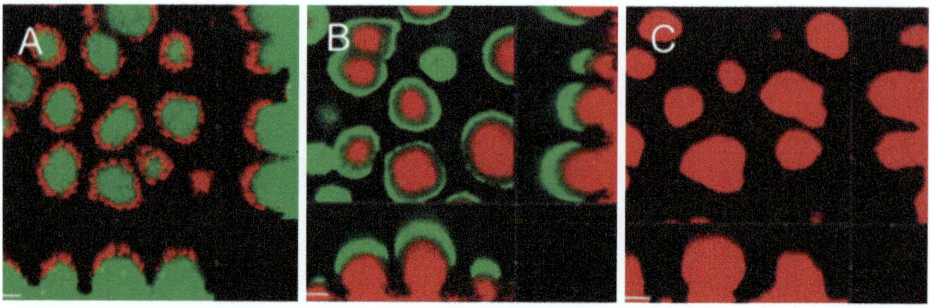

Fig. 14.1 Subpopulations in *P. aeruginosa* in vitro biofilms have different susceptibility for antimicrobials. A 4-day-old biofilms grown in glucose medium were treated with tetracycline (**a**) colistin (**b**) and tetracycline + colistin (**c**), upon which they were stained with propidium iodide (viable cells are *green* and dead cells are *red*). Reproduced from Pamp et al. (2008) with permission from Blackwell Publishing Ltd

et al. 2008). Figure 14.1 demonstrates how the various subpopulations can be killed by antibiotic combination treatments.

The bulk of evidence strongly suggests that a single compound is not sufficient to eradicate or protect against biofilm infections, since such treatment will leave invulnerable bacteria left to colonize encore (Drenkard 2003). The combination of antimicrobials targeting several biofilm subpopulations has shown its potential as successful treatment strategy. The blends of tetracycline + colistin and ciprofloxacin + colistin have been shown to kill most bacteria in *P. aeruginosa* biofilms in vitro (Pamp et al. 2008). The latter combination has furthermore shown great results, with respect to prevention of chronic colonisation, in vivo and is recommended by a European Consensus report for treatment of early *P. aeruginosa* infection in Cystic Fibrosis patients (Doring and Høiby 2004, Høiby et al. 2005, Valerius et al. 1991).

Since the importance of biofilms in infections has just recently been acknowledged by the pharmaceutical industry there are at the present no antimicrobial agents that specifically target the biofilm mode of growth. Furthermore, the increase in bacteria with resistance to available antibiotics and the lack of new approved antimicrobials as well as preclinical leads underlines the urgent need for development of antibacterial drugs with novel targets, hereunder anti-biofilm drugs (Boucher et al. 2009, Chopra et al. 2008, Nnis System 2004). Furthermore, history has taught us the importance of developing drugs that minimize development of resistance. The development of biofilm specific drugs focuses on three major strategies:

- Prevention of biofilm formation
- Removal/killing of biofilm
- Weakening of biofilm

Below we will outline and discuss each of these strategies and their significance for future drugs development. Unless otherwise stated, none of the presented drugs have yet been approved for treatments of humans.

14.2 Prevention of Biofilm Formation

A first and preferable strategy would be to prevent invading bacteria from forming biofilm. Since biofilms show increased tolerance to both antibiotics and the immune system (Bjarnsholt et al. 2005b, Donlan and Costerton 2002, Mah et al. 2003, Stewart and Costerton 2001), development of drugs that impede surface attachment or other specific events in the early stages of biofilm formation may keep infecting bacteria in a planktonic, susceptible state. This strategy will most likely entail prophylactic administration, thus incoming bacteria are prevented to settle on vacant surfaces in particular those present in the form of introduced medical equipment. It may also show potential in immuno-compromised patients such as those suffering from cystic fibrosis (see also Chapter 10), HIV or burn wounds. In the western world, the group of susceptible patients will increase dramatically in the coming years due to the growing fraction of elderly patients and patients being subjected to surgery, transplantation and chemotherapy (Boucher et al. 2009, Chopra et al. 2008). What speaks in favour of this is that prophylactic administration of today's antimicrobials clearly leads to fewer infections (Gibson et al. 2003, Pallasch and Slots 1996, Smyth and Walters 2003). However this kind of treatment has been reported to select for resistant bacterial variants or adverse secondary effects such as colonization with more pathogenic species (Gibson et al. 2003, Smyth and Walters 2003). When designing future drugs for prophylaxis, special attention should be given to these and other possible long-term effects.

Several novel methods, able to prevent biofilm formation, have already been presented in the literature.

14.2.1 Antibodies

Antibodies (delivered as a vaccine) against bacterial components facilitating adhesion and accumulation on surfaces, such as the polysaccharide intercellular adhesin (PIA) and the accumulation-associated protein (Aap) in *Staphylococcus aureus* (Maira-Litran et al. 2004, Sun et al. 2005) and Opr86 in *P. aeruginosa* (Tashiro et al. 2008), have shown promising results in preventing biofilm formation in vitro. By using antibodies, which block bacterial attachment to surfaces, the authors have prevented formation of biofilms. However, the setback to this approach lies in the fact that antibodies have to be tailor-made against specific species of invading bacteria. If however a universal, biofilm specific antigen should be identified this would at least in theory circumvent the problem. Until such has been identified, one could benefit from using antibodies that would weaken the biofilm by for example neutralizing important virulence factors (Cegelski et al. 2008).

14.2.2 Pilicides

Extracellular fibres such as pili allow for adherence and colonization of epithelial cells (Dodson et al. 2001, Mulvey et al. 1998, Roberts et al. 1994). These extracellular fibres are often assembled via a conserved mechanism termed chaperone-usher pathway (Sauer et al. 2004). Impeding this conserved mechanism is a novel strategy that seems feasible due to its potential broad-spectrum capabilities. Several research groups have designed small synthetic compounds termed pilicides which inhibits the biogenesis of both type 1 and P pili (Åberg et al. 2007, Pinkner et al. 2006, Sauer et al. 2004, Svensson et al. 2001). Ring fused 2-pyridones are pilicides which interfere with the chaperone part of the chaperone-usher pathway. The chaperone is essential for the folding, stabilization and transport of the pili subunits (Klemm 1992, Kuehn et al. 1993, 1994). The effect of the 2-pyridones seems promising since an uropathogenic *E. coli* strain grown with pilicides-reduced biofilm formation as well as the adherence to cultured bladder cells by 90% compared to untreated. Pilicides do not affect growth and are therefore not likely to impose selective pressure for resistance development (Pinkner et al. 2006).

14.3 Removal/Killing of Biofilm

Killing infecting bacteria has for long been the preferred strategy. This has been achieved by conventional antimicrobials targeting basal life processes of the bacteria. But the dissemination of resistance and the lack of new antimicrobials have forced the world to look for new strategies. It is generally accepted that the application of lethal or growth inhibiting compounds will impose a selective pressure upon resistance genes and hence purify such in the population (Boucher et al. 2009, Clatworthy et al. 2007, Hawkey 2008, Nnis System 2004, Spellberg et al. 2008). Therefore a constant development of new drugs is essential, as drugs already in use get obsolete. Here we present strategies that can remove or kill established biofilms.

14.3.1 Silver

Silver (in the form of $AgNO_3$) has been used as an agent against bacteria, viruses, algae and fungi for centuries, but it was outcompeted by modern antibiotics. Recently the evolving threat of multiresistent bacteria has given silver a renaissance in wound dressings, etc. (in the form of silver sulfadiazine) (Chopra 2007). The antimicrobial action of silver has been documented to involve binding of silver ions to proteins, DNA and RNA causing fatal changes in the bacterial homeostasis (Lansdown 2002). The lack of standardized breakpoints such including MIC values has complicated the evaluation of silver as an antimicrobial. Furthermore, the diversity of delivery mechanisms makes it even harder to evaluate on the efficacy of silver (Chopra 2007). It is however clear that silver has antimicrobial effects on planktonic cells, but can the ion eradicate biofilms? Recent investigations of the action of silver show that silver in high concentrations eradicate mature in vitro biofilms of *P. aeruginosa* very efficiently. Whereas low concentrations (1 μg/mL) have no effect. These observations strongly indicate that the concentration of silver in currently available wound dressings is much too low for treatment of chronic biofilm wounds (Bjarnsholt et al. 2007). A recent clinical trial showed that silver prevents formation of biofilms. The phase I–II clinical trials were carried out on cardiac surgery patients who were mechanically ventilated for 12–24 h. In these patients it was demonstrated that the lumen of an endotracheal tube coated with silver sulfadiazine had significantly lower bacterial colonization compared to a standard non-coated endotracheal tube (Berra et al. 2008b).

14.3.2 Enzymes

The use of enzymes that degrade the biofilm is another potential strategy. Enzymes that have shown their potential are enzymes that degrade the extracellular polymeric substances (EPS) found in biofilms. Biofilms mainly consist of bacteria and EPS. The EPS have essential roles in defining the physical properties of biofilm; hence degrading EPS has been shown to promote biofilm detachment (Xavier et al. 2005). By disintegrating the biofilm and subsequent release of susceptible planktonic cells will lead to an easier clearance by immune cells and antibiotics.

14.3.2.1 Dispersin B

Dipersin (DspB) is an enzyme that hydrolyses the crucial extracellular polysaccharide poly-*N*-acetylglucosamine, which is needed for biofilm formation and integrity. Exogenously added DspB was found to disrupt and reduce the biomass of biofilms in several bacterial species including *Escherichia coli*, *Staphylococcus epidermidis*, *Yersinia pestis* and *Pseudomonas fluorescens* (Itoh et al. 2005, Kaplan et al. 2004). As previously discussed, the heterogeneity of the biofilm raises the need for combination treatments to target all subpopulations in the biofilm. Hence, it has been shown that treatment with DspB followed by treatment with a protease (proteinase K or trypsin) is capable of eradicating biofilms more effectively in a variety of staphylococcal strains (Chaignon et al. 2007).

14.3.2.2 DNase

Another interesting enzyme is DNase. Recently it was shown that *P. aeruginosa* produces extracellular DNA, which functions as an interconnecting matrix component stabilizing the stalks of the in vitro biofilm. An anti-biofilm strategy could involve destabilisation of the biofilm by removing the extracellular DNA by enzymes degrading DNA called DNases (Allesen-Holm et al. 2006). In vitro studies have shown that treatment of already existing biofilms with DNase only has effect on young biofilms; older biofilms seemed to be independent of stability offered by DNA. Interestingly, the same study showed that biofilm formation was prevented by the presence of DNase in the growth medium (Whitchurch et al. 2002). DNase is already in use as an antibiofilm strategy by the human body. Both living layers of epidermis and the stratum corneum contain DNase (DNase1-like2). This DNase was shown to suppress in vitro biofilm formation in *P. aeruginosa* and *S. aureus* (Eckhart et al. 2007). DNase is a promising tool against biofilm formation and it is administered to chronically infected CF patients. DNAases are administered to degrade extracellular DNA released from dead leukocytes in purulent sputum, which is lowering the viscosity of sputum and thereby facilitating clearance (Ramsey et al. 1993).

14.3.3 Bacteriophages

A novel method to eradicate biofilms is the use of bacteriophages (viruses that kill bacteria) (Merril et al. 2003). In vitro biofilm formation by *S. epidermidis* on silicone catheters is reduced significantly when the catheter is pretreated with a bacteriophage (Curtin and Donlan 2006). The ability to reduce biofilm formation of bacteriophages has also been proved in vivo. Lytic bacteriophages were shown to rescue septicemic mice infected with the multidrug-resistant bacteria *P. aeruginosa* and *Klebsiella pneumonia*, respectively (Vinodkumar et al. 2005, Vinodkumar et al. 2008). A refinement of phages that includes expression of both lytic activity and biofilm-degrading (DspB) enzymes was shown to eradicate biofilm more efficiently than wild type phages (Lu and Collins 2007). Bacteriophages seem to be a promising agent against biofilm infections; however, issues such as specificity of host, efficacy, safety and pharmacokinetics still need to be solved (Merril et al. 2003).

14.3.4 Electrical Currents

Electrical current has been shown to be antibacterial for several species (Del Pozo et al. 2008). Recently, it was shown that low intensity current substantially reduced numbers of viable bacteria in *staphylococcus* and *pseudomonas* biofilms. Alone and in combination with antimicrobials, radiofrequency electrical currents, electromagnetic fields and ultrasound have been shown to have similar effect on biofilms in vitro as well as in vivo (Caubet et al. 2004, Jass et al. 1995, McLeod et al. 1999, Rediske et al. 1999, 2000). Application of currents in combination with antimicrobial chemotherapy in humans could theoretically abrogate the need to remove the device in prosthetic device-related infections (Del Pozo et al. 2008), but the proper administration of these regimens can show to be a huge limitation of their application.

14.3.5 Dispersal Signals

P. aeruginosa produces a small organic compound, *cis*-2-decanoic acid, which induces dispersion of already established biofilms. Davies and colleagues have recently shown that this compound when added exogenously at native concentrations (2.5 nM) was able to induce dispersion in an in vitro *P. aeruginosa* biofilm. Interestingly, *cis*-2-decanoic acid was also shown to induce dispersion of biofilms of *E. coli*, *Klebsiella pneumoniae*, *Proteus mirabilis*, *Streptococcus pyogenes*, *Bacillus subtilis*, *S. aureus* and *Candida albicans* (Davies and Marques 2009).

As previously discussed, this forced transition from biofilm into the more susceptible planktonic phenotype is a very promising strategy. It is suggested that this novel strategy will enhance the activity of conventional treatments through disruption of biofilms. Furthermore, the broad-spectrum activity of this compound makes the use of this compound alone or in combination with antimicrobial agents a very interesting molecule.

14.4 Weakening of Biofilm

Once a biofilm infection has been established, the most obvious alternative to antibiotic-mediated killing would be to attenuate the bacteria with respect to pathogenicity (Bjarnsholt and Givskov 2007). Most pathogenic bacteria including *P. aeruginosa* produce compounds that impair the immune system, e.g. inhibition of antimicrobial production, antimicrobial degradation, inhibition of chemotaxis, induction of apoptosis and necrosis (Allen et al. 2005, Bjarnsholt et al. 2005a, Bortolussi et al. 1987, Jensen et al. 2007, Kharazmi et al. 1984, Kharazmi 1991). Attenuating bacteria by targeting these mechanisms will add an upper hand to the immune system and consequently facilitating eradication (Bjarnsholt et al. 2005a, Hentzer et al. 2003, Wu et al. 2004). It has been put forward (yet not proven) that this strategy imposes a weaker selective pressure with respect to development of resistance compared with conventional antibiotics (Clatworthy et al. 2007).

14.4.1 Quorum-Sensing Inhibitors

Probably the most studied novel strategy in antimicrobials is the quorum-sensing inhibitors. This strategy targets the regulation of virulence expression. The strategy utilizes that bacteria living within the biofilm coordinate a range of social behaviours to maximize their survival potential. Cooperative behaviours are maintained through inter- and extracellular chemical crosstalk comparable to higher organisms (Shapiro 1998). Gram-negative bacteria perform their crosstalk by means of signal molecules such as *N*-acyl homoserine lactones (AHL) (Withers et al. 2001). These molecules are employed in the process of synchronizing activities and pooling the efforts of large groups of cells. Among those synchronized activities are the expression of an impressive arsenal of virulence factors (Davies et al. 1998, Smith and Iglewski 2003). This type of bacterial communication was by Fuqua et al. (1994) termed Quorum Sensing (QS). QS systems allow bacteria to "sense" bacterial density in the environment and respond by gross changes in gene expression. It has been proposed that this mechanism enables arrest in the production of virulence factors until enough bacteria have been amassed to conquer the host defence (Waters and Bassler 2005). Selected virulence factors controlled by QS communication in *P. aeruginosa* are listed in Table 14.1.

QS signals have been detected in sputum as well as lung tissue of infected CF patients, underlining the clinical relevance of this system (Favre-Bonte et al. 2002, Middleton et al. 2002, Wu et al. 2000). In the last decade many researchers have searched for compounds that could block the QS system (Bjarnsholt and Givskov

Table 14.1 Functions and products controlled by the QS systems *las* and *rhl* in *P. aeruginosa* (Juhas et al. 2005)

las controlled	*rhl* controlled
rhl system	*rhl* system
	Rhamnolipids
Alkaline protease	Alkaline protease
Elastase	Elastase
Lipase	Lipase
Hydrogen cyanide	Hydrogen cyanide
Xcp secretion	Xcp secretion
Exotoxin A	Lectins A and B
Neuroaminidase	Pyocyanin
Pvds-regulated endoprotease	Chitinase
Catalase	RpoS
Superoxide dismutase	Exoenzyme S
Aminopeptidase	Iron metabolism
Swarming	Swarming
Twitching	Twitching

2008). By blocking the QS system, one could prohibit that the bacteria produce their harmful cocktail of virulence factors and thus weakening the bacteria. The QS controlled surfactant, rhamnolipid, has recently been identified as an important virulence factor being able to rapidly kill attracted PMN leukocytes. This deadly compound is actively used by *P. aeruginosa* as a shield surrounding the biofilm and thus protecting it against phagocytes (Alhede et al. 2009, Jensen et al. 2007, Van Gennip et al. 2009).

In addition to controlling the production of virulence factors, QS has also been shown to control the biofilms tolerance to antibiotics, such as tobramycin, ciprofloxacin and ceftazidime. QS-deficient biofilm were shown to be more prone to killing by these antibiotics than the QS proficient biofilm (Bjarnsholt et al. 2005a, Bjarnsholt and Givskov 2007). It has been shown that a *P. aeruginosa* wild type biofilm is impermeable to PMN leukocytes and thus it protects the bacteria from phagocytosis (Jesaitis et al. 2003). This mechanism was shown by Bjarnsholt et al. (2005a) to be QS dependent, thus the QS mutant biofilm (*ΔlasRrhlR* mutant) was fully penetrated and grazed by the PMNs (Fig. 14.2a,b). The PMN grazing in the QS deficient biofilm resulted in a reduced biomass whereas the wild type withstood the activity leaving the PMNs paralysed on the top of the biofilm. This tolerance towards phagocytes is probably enhanced by the production of rhamnolipid shielding the biofilm against killing phagocytes (Fig. 14.2c,d).

Since a multitude of virulence factors are controlled by QS, blockage will likely result in many beneficial effects, which could also be synergistic. Both QS-deficient strains as well as treatment with QS inhibitors (QSIs) have been used to generate proof of the antimicrobial concept in animal models (Bjarnsholt et al. 2005a, Hentzer et al. 2003, Pearson et al. 2000, Rumbaugh et al. 1999, Wu et al. 2004) (see also Chapter 16).

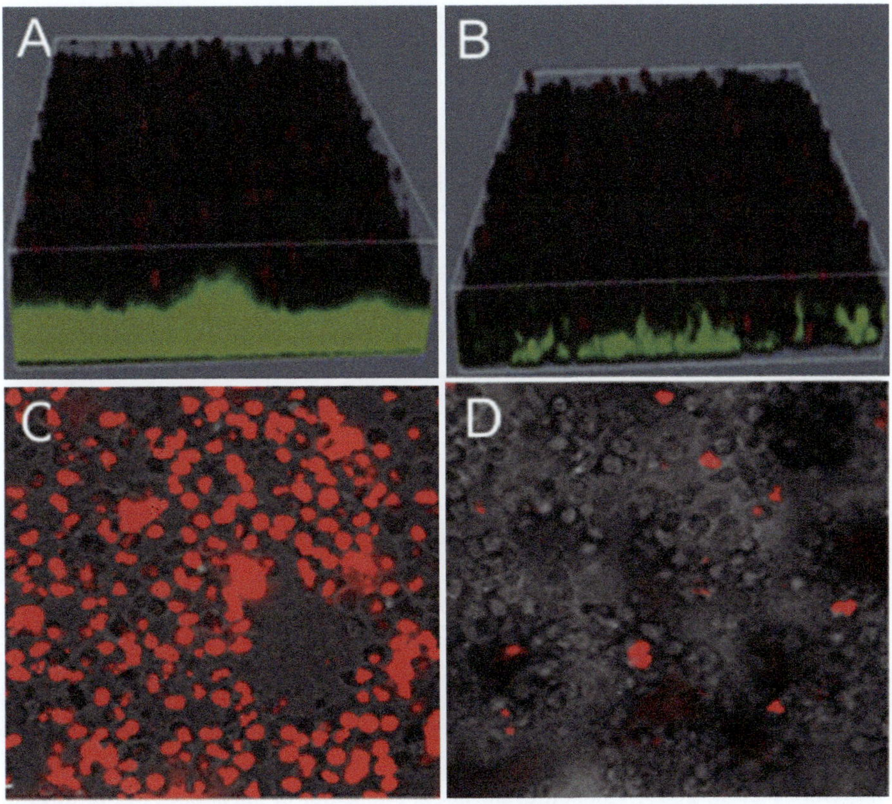

Fig. 14.2 QS dependent tolerance of *P. aeruginosa* biofilms towards phagocytes. Three-day-old in vitro biofilms of wild type *P. aeruginosa* (**a**) and a Δ*lasRrhlR* mutant (**b**), both expressing GFP as a tag, were exposed to PMNs for 2.5 h. The PMNs appear red fluorescent by SYTO-62 staining. The wild type is not penetrated by the PMNs as is the QS mutant. Reproduced from Bjarnsholt et al. (2005a) with permission from Society for General Microbiology. Four-day-old in vitro biofilms were exposed to PMNs. Dead PMNs emit red fluorescent by the DNA stain propidium iodide. A greater number of PMNs became necrotic on the QS-competent wild type biofilm (**c**) compared with the Δ*lasRrhlR* mutant (**d**). Reproduced from Jensen et al. (2007) with permission from Society for General Microbiology

To breakthrough, a general QSI with activity against several QS-containing pathogens should be identified. So far, more than 70 Gram-negative species have been reported to contain functional QS-regulatory circuits (producing AHL signal molecules) (Taga and Bassler 2003). Thus, it seems that at least in theory, the concept of QSI is extremely promising; only the right compound has to be identified. Despite, the principles of QS in both Gram-negative and Gram-positive bacteria are the same; the molecular mechanisms and signal molecules differ. QSI peptides which block QS in Gram-positive bacteria have also shown proof of concept in animal models (Balaban et al. 2007). It should be noted that a controversy about the QSI effect on biofilm formation has evolved over the years. In *Staphylococcus*, inhibition of QS has been found to enhance biofilm formation by some groups but to be reduced by others (Balaban et al. 2005, Otto 2004, Vuong et al. 2003, 2004). Therefore, QSIs should be carefully tested, in each organism, for adverse effects that might worsen the infection.

QS deficiency leads to reduced tolerance to a variety of conventional antibiotics. In particular, QSI compounds block production of the rhamnolipid shield, which will make the biofilm more prone to eradication by the immune system (Alhede et al. 2009, Bjarnsholt et al. 2005a, Rasmussen et al. 2005). Consequently, prophylactic administration of QSIs or administration of QSIs in combination with antibiotic or other antimicrobials may become a useful strategy in the treatment of biofilm infections.

14.4.2 Inhibition of Type III Secretion

In addition to target the virulence expression, disrupting the delivery of virulence factors would also lead to attenuation. The type III secretion system (T3SS) is a bacterial needle-like organelle that is able to deliver proteins such as toxins into eukaryotic cells. This system has been found in a long range of bacterial species thus making T3SS a good ubiquitous target for novel antimicrobials.

Acylated hydrazones of different salicyclaldehydes have been shown to inhibit effector molecule secretion by T3SS, without affecting growth, in various species such as *Yersinia pseudotuberculosis*, *Salmonella enterica* and *Chlamydia trachomatis* (Bailey et al. 2007, Hudson et al. 2007, Muschiol et al. 2006, Nordfelth et al. 2005). The effects of these compounds have been proven in vivo. Preincubation of *S. enterica* serovar Typhimurium with acylated hydrazones of salicyclaldehydes suppressed secretion of T3SS effectors and the inflammatory responses in a bovine intestinal ligated loop model (Hudson et al. 2007). So far, bacteria have to be pretreated with the T3SS-inhibitor for optimal effect in vivo. This underlines the need for additional development of this strategy with respect to identification of compounds that will have an effect on established infections.

14.4.3 C-Di-GMP

Recent research has pointed cyclic diguanylate monophosphate (c-di-GMP) out as a promising target for antibiofilm treatments. C-di-GMP is a bacterial second messenger (relaying signals received at cell surface to target molecules in the cell) used by most bacteria (Tamayo et al. 2007). This ubiquitous messenger molecule seems interesting due to its impact on biofilm formation and dissolution. Its specific mode of action is yet to be elucidated, but since this messenger is found in almost any bacteria it may by first sight be an obvious target for a future universal antibiofilm drug. Compounds that interact or block the conserved domains (GGDEF and EAL) in the enzymes, which regulate the level of c-di-GMP, may be able to affect c-di-GMP production in many different kinds of bacteria (Tal et al. 1998). Despite c-di-GMP has been shown to activate in vitro biofilm formation it also poses an inhibiting effect on motility and virulence (Huang et al. 2003, Simm et al. 2004). Therefore this molecule has opposing ways of action and seems to be a complicated target with possible adverse effects. However, promising results have been delivered in *S. aureus*, where treatment with extracellular c-di-GMP inhibited bacterial colonization in a mouse model (Brouillette et al. 2005). This effect was found to be due to c-di-GMPs effect on virulence and as an immunostimulatory inducing a protective response against the bacteria. The messenger molecule was shown to activate many cells of the immune system including T cells, dendritic cells, monocytes and granulocytes (Karaolis et al. 2007a, Karaolis et al. 2007b).

14.5 Surface Coatings

Biofilm formation on catheters and surgically implanted devices is a considerable clinical problem (see also Chapter 5 and 6). Preventing biofilm formation on these devices is of enormous importance in order to prevent inflammation and ultimately device-related bloodstream infections. It is estimated that infections associated with intravascular catheters or devices stand for up to 20% of all nosocomial infections (Eggimann et al. 2004, Pittet et al. 1994, Rosenthal et al. 2003). The predominant strategy for preventing or reducing development of bacterial biofilm on catheters and endotracheal tubes (ETTs) is coating the device with substances such as antiseptics, antimicrobials or metals (O'Grady et al. 2002). Several common antiseptic substances have been tested for their efficiency in prevention of biofilm formation on catheters. Chlorhexidine-silver sulfaziadine (CH-SS) is one of the most used antiseptic coating materials for short term prevention of biofilm formation on ETT and central venous catheters (Berra et al. 2008a, b, Ramritu et al. 2008). Prophylactic coating with conventional antibiotics have also shown good results. Coating with combinations of minocycline and rifampicin has shown a positive synergistic effect. The use of this combination to prevent catheter colonization is more effective than the CH-SS impregnated catheter (Crnich and Maki 2002). This difference has been speculated to be due to the fact that minocycline–rifampicin catheters were coated both externally and internally whereas CH-SS catheters are only coated externally. The new generation of CH-SS catheters coated both externally and internally prevents colonization and subsequent bloodstream infections equally efficient as the catheters coated with antibiotics (Gastmeier and Geffers 2006, Raad et al. 2008, Tamilvanan et al. 2008).

The duration of insertion has an important impact on the effect of the above-mentioned dressings. Trials with average durations less than 12 days show good results with respect to prevention of infections, where trials with longer average durations did not result in a statistical significant treatment results (Hockenhull et al. 2008).

So far clinical trials have not shown bacterial resistance against minocycline and rifampicin (Chatzinikolaou et al. 2003, Hanna et al. 2003), nevertheless in vitro data indicate that the use of catheters coated with minocycline–rifampicin may increase the development of resistance (Sampath et al. 2001). The reduced effect on long-term insertions and the concerns on resistance clearly illustrate the importance of developing new effective antimicrobials that imposes weaker selective pressure on resistance mechanisms than conventional coatings. The strategies discussed in this chapter have all shown promising effects as anti-biofilm strategies. Many of

Table 14.2 Summery table of novel and future treatment strategies against biofilm infections

Drug	Action	Tested in[a]
Prevention of Biofilm Formation		
Antibodies	Antibodies against bacterial components facilitating adhesion and accumulation on surfaces	*Only in vitro*
Pilicides	Compounds impeding assembly of extracellular fibres such as pili – thus reducing biofilm formation	*Cultured Bladder cells*
Removal/Killing of Biofilm		
Silver	Silver ions binds to macromolecules which in turn causes fatal changes in bacterial homeostasis	*Human wounds, Endotracheal tubes and various animal models*
Enzymes	Enzymes degrading biofilm components	*Only in vitro*
DNase	Breaks down the stabilizing extracellular DNA in the biofilm	*Human Lungs /in vitro*
Bacteriophages	Bacteriophages infects and kills the bacteria living within the biofilm	*Murine Sepsis*
Electrical currents	Electrical current can disrupt and kill biofilms	*Rabbit implant*
Dispersal signals	Compounds that induce dispersion of already established biofilms	*only in vitro*
Weakening of Biofilm		
Quorum-sensing inhibitors	Compounds that prevent the production of virulence factors by interfering with intercellular communication and hence promoting clearance by the immune defence	*Murine Lungs and Grafts*
T3SS inhibitors	Inhibition of Type III secretion system. Inhibiting this effector delivery mechanism the bacteria living within the biofilm will be unable to deliver its virulence factors	*Bovine intestinal ligated loop model*
Cyclic-Di-GMP	A conserved second messenger that is highly involved in biofilm formation and virulence factor production. It inhibits virulence and activates cells of the immune system	*Murine Mastitis*

[a]Examples of test models

these, such as QSI compounds and bacteriophages, could be used as future coating materials.

14.6 Future Perspectives

This chapter has discussed the recent progression in the field of novel therapies against biofilm infections (Table 14.2). Biofilm infections have proven to be hard to prevent and treat. Recent findings of biofilm heterogeneity have opened a window of novel treatment strategies. Research has shown that distinct subpopulations have different susceptibility to antimicrobials and therefore the biofilm should preferably be eradicated with more than one regimen. Combinations of already approved antimicrobials have shown good results in vivo, but novel combinations of compounds such as those weakening the biofilms could ultimately be the end of chronic infections. The synergistic use of novel drugs in combination with conventional antibiotics could improve the efficacy of treatment by attenuating the renitent biofilm and target several subpopulations within it. In fact, some of the most active antibiotics (azithromycin, ceftazidime and ciprofloxacin) have recently been shown to have dual activities. In addition to working as conventional antibiotics they were also shown to function as QSIs (Hoffmann et al. 2007, Mizukane et al. 1994, Nalca et al. 2006, Skindersoe et al. 2008). These compounds are excellent epitomes of the future drug against biofilm infections since they combine several targets in one drug. In addition to the discovery of new drug scaffolds and designing effective combinations of existing compounds, another important task is to identify drug targets that are ubiquitous in many bacteria. Consequently, a drug that would hit a ubiquitous target could then be used to eradicate many different pathogenic organisms and hopefully attract the interest of the pharmaceutical industry. The increase in multiresistent bacteria, the sparsity of new approved antibiotics and the desperate lack of leads in the pipeline have raised the need of new strategies against infections. Hopefully, some of the strategies presented in this chapter will soon be transferred from pre-clinical trials to trials in humans.

References

Åberg V, Fallman E, Axner O, Uhlin BE, Hultgren SJ, Almqvist F (2007) Pilicides regulate pili expression in *E. coli* without affecting the functional properties of the pilus rod. Mol Biosyst 3:214–218

Alhede M, Bjarnsholt T, Jensen PO, et al (2009) Pseudomonas aeruginosa recognizes and responds aggressively to the presence of polymorphonuclear leukocytes. Microbiology 155:3500–3508

Allen L, Dockrell DH, Pattery T, Lee DG, Cornelis P, Hellewell PG, Whyte MK (2005) Pyocyanin production by *Pseudomonas aeruginosa* induces neutrophil apoptosis and impairs neutrophil-mediated host defenses in vivo. J Immunol 174:3643–3649

Allesen-Holm M, Barken KB, Yang L, Klausen M, Webb JS, Kjelleberg S, Molin S, Givskov M, Tolker-Nielsen T (2006) A characterization of DNA release in *Pseudomonas aeruginosa* cultures and biofilms. Mol Microbiol 59:1114–1128

Anderson GG, O'Toole GA (2008) Innate and induced resistance mechanisms of bacterial biofilms. Curr Top Microbiol Immunol 322:85–105, 85–105

Bailey L, Gylfe A, Sundin C, et al (2007) Small molecule inhibitors of type III secretion in Yersinia block the Chlamydia pneumoniae infection cycle. FEBS Lett 581: 587–595

Balaban N, Stoodley P, Fux CA, Wilson S, Costerton JW, Dell'Acqua, G (2005) Prevention of staphylococcal biofilm-associated infections by the quorum sensing inhibitor RIP. Clin Orthop Relat Res 437:48–54

Balaban N, Cirioni O, Giacometti A, Ghiselli R, Braunstein JB, Silvestri C, Mocchegiani F, Saba V, Scalise G (2007) Treatment of *Staphylococcus aureus* Biofilm Infection by the Quorum-Sensing Inhibitor RIP. Antimicrob Agents Chemother 51:2226–2229

Berra L, Curto F, Li Bassi G, Laquerriere P, Pitts B, Baccarelli A, Kolobow T (2008a). Antimicrobial-coated endotracheal tubes: an experimental study. Intensive Care Med 34: 1020–1029

Berra L, Kolobow T, Laquerriere P, et al (2008b) Internally coated endotracheal tubes with silver sulfadiazine in polyurethane to prevent bacterial colonization: a clinical trial. Intensive Care Med 34:1030–1037

Bjarnsholt T, Jensen PO, Burmolle M, et al (2005a) Pseudomonas aeruginosa tolerance to tobramycin, hydrogen peroxide and polymorphonuclear leukocytes is quorum-sensing dependent. Microbiology 151:373–383

Bjarnsholt T, Jensen PO, Rasmussen TB, et al (2005b) Garlic blocks quorum sensing and promotes rapid clearing of pulmonary Pseudomonas aeruginosa infections. Microbiology 151:3873–3880

Bjarnsholt T, Givskov M (2007) Quorum-sensing blockade as a strategy for enhancing host defences against bacterial pathogens. Philos Trans R Soc Lond B Biol Sci 362: 1213–1222

Bjarnsholt T, Kirketerp-Moller K, Kristiansen S, Phipps R, Nielsen AK, Jensen PO, Hoiby N, Givskov M (2007) Silver against Pseudomonas aeruginosa biofilms. APMIS 115:921–928

Bjarnsholt T, Givskov M (2008) Quorum sensing inhibitory drugs as next generation antimicrobials: worth the effort? Curr Infect Dis Rep 10:22–28

Bortolussi R, Vandenbroucke-Grauls CM, van Asbeck BS, Verhoef J (1987) Relationship of bacterial growth phase to killing of *Listeria monocytogenes* by oxidative agents generated by neutrophils and enzyme systems. Infect Immun 55:3197–3203

Boucher HW, Talbot GH, Bradley JS, Edwards JE, Gilbert D, Rice LB, Scheld M, Spellberg B, Bartlett J (2009) Bad bugs, no drugs: no ESKAPE! An update from the Infectious Diseases Society of America. Clin Infect Dis 48:1–12

Brouillette E, Hyodo M, Hayakawa Y, Karaolis DKR, Malouin F (2005) 3′,5′-Cyclic Diguanylic Acid Reduces the Virulence of Biofilm-Forming *Staphylococcus aureus* Strains in a Mouse Model of Mastitis Infection. Antimicrob Agents Chemother 49:3109–3113

Caubet R., Pedarros-Caubet F, Chu M, Freye E, de Belem RM, Moreau JM, Ellison WJ (2004) A radio frequency electric current enhances antibiotic efficacy against bacterial biofilms. Antimicrob Agents Chemother 48:4662–4664

Cegelski L, Marshall GR, Eldridge GR, Hultgren SJ (2008) The biology and future prospects of antivirulence therapies. Nat Rev Microbiol 6:17–27

Chaignon P, Sadovskaya I, Ragunah C, Ramasubbu N, Kaplan JB, Jabbouri S (2007) Susceptibility of staphylococcal biofilms to enzymatic treatments depends on their chemical composition. Appl Microbiol Biotechnol 75:125–132

Chatzinikolaou I, Hanna H, Graviss L, et al (2003) Clinical experience with minocycline and rifampin-impregnated central venous catheters in bone marrow transplantation recipients: efficacy and low risk of developing staphylococcal resistance. Infect Control Hosp Epidemiol 24:961–963

Chopra I (2007) The increasing use of silver-based products as antimicrobial agents: a useful development or a cause for concern? J Antimicrob Chemother 59:587–590

Chopra I, Schofield C, Everett M, et al (2008) Treatment of health-care-associated infections caused by Gram-negative bacteria: a consensus statement. Lancet Infect Dis 8:133–139

Clatworthy AE, Pierson E, Hung DT (2007) Targeting virulence: a new paradigm for antimicrobial therapy. Nat Chem Biol 3:541–548

Crnich CJ, Maki DG (2002) The promise of novel technology for the prevention of intravascular device-related bloodstream infection. II. Long-term devices. Clin Infect Dis 34:1362–1368

Curtin JJ, Donlan RM (2006) Using bacteriophages to reduce formation of catheter-associated biofilms by *Staphylococcus epidermidis*. Antimicrob Agents Chemother 50:1268–1275

Davies DG, Parsek MR, Pearson JP, Iglewski BH, Costerton JW, Greenberg EP (1998) The involvement of cell-to-cell signals in the development of a bacterial biofilm. Science 280: 295–298

Davies DG, Marques CN (2009) A fatty acid messenger is responsible for inducing dispersion in microbial biofilms. J Bacteriol 191:1393–1403

Del Pozo JL, Rouse MS, Patel R (2008) Bioelectric effect and bacterial biofilms. A systematic review. Int J Artif Organs 31:786–795

Dodson KW, Pinkner JS, Rose T, Magnusson G, Hultgren SJ, Waksman G (2001) Structural basis of the interaction of the pyelonephritic *E. coli* adhesin to its human kidney receptor. Cell 105:733–743

Donlan RM, Costerton JW (2002) Biofilms: survival mechanisms of clinically relevant microorganisms. Clin Microbiol Rev 15:167–193

Doring G, Høiby N (2004) Early intervention and prevention of lung disease in cystic fibrosis: a European consensus. J Cyst Fibros 3:67–91

Drenkard E (2003) Antimicrobial resistance of *Pseudomonas aeruginosa* biofilms. Microbes Infect 5:1213–1219

Eckhart L, Fischer H, Barken KB, Tolker-Nielsen T, Tschachler E (2007) DNase1L2 suppresses biofilm formation by *Pseudomonas aeruginosa* and *Staphylococcus aureus*. Br J Dermatol 156:1342–1345

Eggimann P, Sax H, Pittet D (2004) Catheter-related infections. Microbes Infect 6:1033–1042

Favre-Bonte S, Pache JC, Robert J, Blanc D, Pechere JC, Van DC (2002) Detection of *Pseudomonas aeruginosa* cell-to-cell signals in lung tissue of cystic fibrosis patients. Microb Pathog 2:143–147

Fuqua WC, Winans SC, Greenberg EP (1994) Quorum sensing in bacteria: the LuxR-LuxI family of cell density-responsive transcriptional regulators. J Bacteriol 176:269–275

Gastmeier P, Geffers C (2006) Prevention of catheter-related bloodstream infections: analysis of studies published between 2002 and 2005. J Hosp Infect 64:326–335

Gibson RL, Burns JL, Ramsey BW (2003) Pathophysiology and Management of Pulmonary Infections in Cystic Fibrosis. Am J Respir Crit Care Med 168:918–951

Haagensen JAJ, Klausen M, Ernst RK, Miller SI, Folkesson A, Tolker-Nielsen T, Molin S (2007) Differentiation and distribution of colistin- and sodium dodecyl sulfate-tolerant cells in *Pseudomonas aeruginosa* biofilms. J Bacteriol 189:28–37

Hanna HA, Raad II, Hackett B, Wallace SK, Price KJ, Coyle DE, Parmley CL (2003) Antibiotic-impregnated catheters associated with significant decrease in nosocomial and multidrug-resistant bacteremias in critically ill patients. Chest 124:1030–1038

Hawkey PM (2008) The growing burden of antimicrobial resistance. J Antimicrob Chemother 62(Suppl 1): i1–9

Hentzer M, Wu H, Andersen JB, et al (2003) Attenuation of *Pseudomonas aeruginosa* virulence by quorum sensing inhibitors. EMBO J 22:3803–3815

Hockenhull JC, Dwan K, Boland A, et al (2008) The clinical effectiveness and cost-effectiveness of central venous catheters treated with anti-infective agents in preventing bloodstream infections: a systematic review and economic evaluation. Health Technol Assess 12:iii–iv, xi–xii, 1–154

Hoffmann N, Lee B, Hentzer M, Rasmussen TB, Song Z, Johansen HK, Givskov M, Hoiby N (2007) Azithromycin blocks quorum sensing and alginate polymer formation and increases the sensitivity to serum and stationary-growth-phase killing of Pseudomonas aeruginosa and attenuates chronic P. aeruginosa lung infection in Cftr(–/–) mice. Antimicrob Agents Chemother 51:3677–3687

Høiby N, Frederiksen B, Pressler T (2005) Eradication of early *Pseudomonas aeruginosa* infection. J Cyst Fibros 4(Suppl 2):49–54

Huang B, Whitchurch CB, Mattick JS (2003) FimX, a multidomain protein connecting environmental signals to twitching motility in *Pseudomonas aeruginosa*. J Bacteriol 185:7068–7076

Hudson DL, Layton AN, Field TR, Bowen AJ, Wolf-Watz H, Elofsson M, Stevens MP, Galyov EE (2007) Inhibition of type III secretion in Salmonella enterica serovar Typhimurium by small-molecule inhibitors. Antimicrob Agents Chemother 51:2631–2635

Itoh Y, Wang X, Hinnebusch BJ, Preston JF, III, Romeo T (2005) Depolymerization of {beta}-1,6-N-Acetyl-D-Glucosamine Disrupts the Integrity of Diverse Bacterial Biofilms. J Bacteriol 187:382–387

Jass J, Costerton JW, Lappin-Scott HM (1995) The effect of electrical currents and tobramycin on *Pseudomonas aeruginosa* biofilms. J Ind Microbiol 15:234–242

Jensen PO, Bjarnsholt T, Phipps R, et al (2007) Rapid necrotic killing of polymorphonuclear leukocytes is caused by quorum-sensing-controlled production of rhamnolipid by Pseudomonas aeruginosa. Microbiology 153:1329–1338

Jesaitis AJ, Franklin MJ, Berglund D, Sasaki M, Lord CI, Bleazard JB, Duffy JE, Beyenal H, Lewandowski Z (2003) Compromised host defense on *Pseudomonas aeruginosa* biofilms: characterization of neutrophil and biofilm interactions. J Immunol 171:4329–4339

Juhas M, Eberl L, Tummler B (2005) Quorum sensing: the power of cooperation in the world of *Pseudomonas*. Environ Microbiol 7:459–471

Kaplan JB, Ragunath C, Velliyagounder K, Fine DH, Ramasubbu N (2004) Enzymatic detachment of *Staphylococcus epidermidis* biofilms. Antimicrob Agents Chemother 48:2633–2636

Karaolis DK, Means TK, Yang D, et al (2007a). Bacterial c-di-GMP is an immunostimulatory molecule. J Immunol 178:2171–2181

Karaolis DK, Newstead MW, Zeng X, Hyodo M, Hayakawa Y, Bhan U, Liang H, Standiford TJ (2007b) Cyclic di-GMP stimulates protective innate immunity in bacterial pneumonia. Infect Immun 75:4942–4950

Kharazmi A, Doring G, Høiby N, Valerius NH (1984) Interaction of *Pseudomonas aeruginosa* alkaline protease and elastase with human polymorphonuclear leukocytes *in vitro*. Infect Immun 43:161–165

Kharazmi A (1991) Mechanisms involved in the evasion of the host defence by *Pseudomonas aeruginosa*. Immunol Lett 30:201–205

Klemm P (1992) *FimC*, a chaperone-like periplasmic protein of *Escherichia coli* involved in biogenesis of type 1 fimbriae. Res Microbiol 143:831–838

Kuehn MJ, Ogg DJ, Kihlberg J, Slonim LN, Flemmer K, Bergfors T, Hultgren SJ (1993) Structural basis of pilus subunit recognition by the *PapD* chaperone. Science 262:1234–1241

Kuehn MJ, Jacob-Dubuisson F, Dodson K, Slonim L, Striker R, Hultgren SJ (1994) Genetic, biochemical, and structural studies of biogenesis of adhesive pili in bacteria. Methods Enzymol 236:282–306

Lansdown AB (2002) Silver. I: Its antibacterial properties and mechanism of action. J Wound Care 11:125–130

Lu TK, Collins JJ (2007) Dispersing biofilms with engineered enzymatic bacteriophage. Proc Natl Acad Sci USA 104:11197–11202

Mah TF, Pitts B, Pellock B, Walker GC, Stewart PS, O'Toole GA (2003) A genetic basis for *Pseudomonas aeruginosa* biofilm antibiotic resistance. Nature 426:306–310

Maira-Litran T, Kropec A, Goldmann D, Pier GB (2004) Biologic properties and vaccine potential of the *staphylococcal* poly-N-acetyl glucosamine surface polysaccharide. Vaccine 22: 872–879

McLeod BR, Fortun S, Costerton JW, Stewart PS (1999) Enhanced bacterial biofilm control using electromagnetic fields in combination with antibiotics. Methods Enzymol 310:656–670

Merril CR, Scholl D, Adhya SL (2003) The prospect for bacteriophage therapy in Western medicine. Nat Rev Drug Discov 2:489–497

Middleton B, Rodgers HC, Camara M, Knox AJ, Williams P, Hardman A (2002) Direct detection of N-acylhomoserine lactones in cystic fibrosis sputum. FEMS Microbiol Lett 207:1–7

Mizukane R, Hirakata Y, Kaku M, Ishii Y, Furuya N, Ishida K, Koga H, Kohno S, Yamaguchi K (1994) Comparative in vitro exoenzyme-suppressing activities of azithromycin and other macrolide antibiotics against *Pseudomonas aeruginosa*. Antimicrob Agents Chemother 38:528–533

Mulvey MA, Lopez-Boado YS, Wilson CL, Roth R, Parks WC, Heuser J, Hultgren SJ (1998) Induction and evasion of host defenses by type 1-piliated uropathogenic *Escherichia coli*. Science 282:1494–1497

Muschiol S, Bailey L, Gylfe A, et al (2006) A small-molecule inhibitor of type III secretion inhibits different stages of the infectious cycle of Chlamydia trachomatis. Proc Natl Acad Sci USA 103:14566–14571

Nalca Y, Jansch L, Bredenbruch F, Geffers R, Buer J, Haussler S (2006) Quorum-Sensing Antagonistic Activities of Azithromycin in *Pseudomonas aeruginosa* PAO1: a Global Approach. Antimicrob Agents Chemother 50:1680–1688

Nnis System Arf t (2004) National Nosocomial Infections Surveillance (NNIS) System Report, data summary from January 1992 through June 2004, issued October 2004. Am J Infect Control 32:470–485

Nordfelth R, Kauppi AM, Norberg HA, Wolf-Watz H, Elofsson M (2005) Small-molecule inhibitors specifically targeting type III secretion. Infect Immun 73:3104–3114

O'Grady NP, Alexander M, Dellinger EP, et al (2002) Guidelines for the prevention of intravascular catheter-related infections. Centers for Disease Control and Prevention. MMWR Recomm Rep 51:1–29

Otto M (2004) Quorum-sensing control in *Staphylococci* – a target for antimicrobial drug therapy? FEMS Microbiol Lett 241:135–141

Pallasch TJ, Slots J (1996) Antibiotic prophylaxis and the medically compromised patient. Periodontol 2000 10:107–138

Pamp SJ, Gjermansen M, Johansen HK, Tolker-Nielsen T (2008) Tolerance to the antimicrobial peptide colistin in *Pseudomonas aeruginosa* biofilms is linked to metabolically active cells, and depends on the *pmr* and *mexAB-oprM* genes. Mol Microbiol 68:223–240

Pamp SJ, Sternberg C, Tolker-Nielsen T (2009) Insight into the microbial multicellular lifestyle via flow-cell technology and confocal microscopy. Cytometry A 75:90–103

Parsek MR, Tolker-Nielsen T (2008) Pattern formation in *Pseudomonas aeruginosa* biofilms. Curr Opin Microbiol 11:560–566

Pearson JP, Feldman M, Iglewski BH, Prince A (2000) *Pseudomonas aeruginosa* cell-to-cell signaling is required for virulence in a model of acute pulmonary infection. Infect Immun 68:4331–4334

Pinkner JS, Remaut H, Buelens F, et al. (2006) Rationally designed small compounds inhibit pilus biogenesis in uropathogenic bacteria. Proc Natl Acad Sci USA 103:17897–17902

Pittet D, Tarara D, Wenzel RP (1994) Nosocomial bloodstream infection in critically ill patients. Excess length of stay, extra costs, and attributable mortality. JAMA 271: 1598–1601

Raad II, Fang X, Keutgen XM, Jiang Y, Sherertz R, Hachem R (2008) The role of chelators in preventing biofilm formation and catheter-related bloodstream infections. Curr Opin Infect Dis 21:385–392

Ramritu P, Halton K, Collignon P, Cook D, Fraenkel D, Battistutta D, Whitby M, Graves N (2008) A systematic review comparing the relative effectiveness of antimicrobial-coated catheters in intensive care units. Am J Infect Control 36:104–117

Ramsey BW, Astley SJ, Aitken ML, et al (1993) Efficacy and safety of short-term administration of aerosolized recombinant human deoxyribonuclease in patients with cystic fibrosis. Am Rev Respir Dis 148:145–151

Rasmussen TB, Skindersoe ME, Bjarnsholt T, et al (2005) Identity and effects of quorum-sensing inhibitors produced by *Penicillium* species. Microbiology 151:1325–1340

Rediske AM, Roeder BL, Brown MK, Nelson JL, Robison RL, Draper DO, Schaalje GB, Robison RA, Pitt WG (1999) Ultrasonic enhancement of antibiotic action on *Escherichia coli* biofilms: an in vivo model. Antimicrob Agents Chemother 43:1211–1214

Rediske AM, Roeder BL, Nelson JL, Robison RL, Schaalje GB, Robison RA, Pitt WG (2000) Pulsed ultrasound enhances the killing of *Escherichia coli* biofilms by aminoglycoside antibiotics in vivo. Antimicrob Agents Chemother 44:771–772

Roberts JA, Marklund BI, Ilver D, et al (1994) The Gal(alpha 1-4) Gal-specific tip adhesin of *Escherichia coli* P-fimbriae is needed for pyelonephritis to occur in the normal urinary tract. Proc Natl Acad Sci USA 91:11889–11893

Rosenthal VD, Guzman S, Orellano PW (2003) Nosocomial infections in medical-surgical intensive care units in Argentina: attributable mortality and length of stay. Am J Infect Control 31:291–295

Rumbaugh KP, Griswold JA, Iglewski BH, Hamood AN (1999) Contribution of quorum sensing to the virulence of *Pseudomonas aeruginosa* in burn wound infections. Infect Immun 67: 5854–5862

Sampath LA, Tambe SM, Modak SM (2001) In vitro and in vivo efficacy of catheters impregnated with antiseptics or antibiotics: evaluation of the risk of bacterial resistance to the antimicrobials in the catheters. Infect Control Hosp Epidemiol 22:640–646

Sauer FG, Remaut H, Hultgren SJ, Waksman G (2004) Fiber assembly by the chaperone-usher pathway. Biochim Biophys Acta 1694:259–267

Shapiro JA (1998) Thinking about bacterial populations as multicellular organisms. Ann Rev Microbiol 52:81–104

Simm R, Morr M, Kader A, Nimtz M, Romling U (2004) GGDEF and EAL domains inversely regulate cyclic di-GMP levels and transition from sessility to motility. Mol Microbiol 53: 1123–1134

Skindersoe ME, Alhede M, Phipps R, et al (2008) Effects of antibiotics on quorum sensing in Pseudomonas aeruginosa. Antimicrob Agents Chemother 52:3648–3663

Smith RS, Iglewski BH (2003) *P. aeruginosa* quorum-sensing systems and virulence. Curr Opin Microbiol 6:56–60

Smyth A, Walters S (2003) Prophylactic antibiotics for cystic fibrosis. Cochrane Database Syst Rev 3:CD001912

Spellberg B, Guidos R, Gilbert D, Bradley J, Boucher HW, Scheld WM, Bartlett JG, Edwards J Jr (2008) The epidemic of antibiotic-resistant infections: a call to action for the medical community from the Infectious Diseases Society of America. Clin Infect Dis 46:155–164

Stewart PS, Costerton JW (2001) Antibiotic resistance of bacteria in biofilms. Lancet 358:135–138

Sun D, Accavitti MA, Bryers JD (2005) Inhibition of biofilm formation by monoclonal antibodies against *Staphylococcus epidermidis* RP62A accumulation-associated protein. Clin Diagn Lab Immunol 12:93–100

Svensson A, Larsson A, Emtenas H, Hedenstrom M, Fex T, Hultgren SJ, Pinkner JS, Almqvist F, Kihlberg J (2001) Design and evaluation of pilicides: potential novel antibacterial agents directed against uropathogenic *Escherichia coli*. Chembiochem 2:915–918

Taga ME, Bassler BL (2003) Chemical communication among bacteria. Proc Natl Acad Sci USA 100 Suppl 2:14549–14554

Tal R, Wong HC, Calhoon R, et al (1998) Three cdg operons control cellular turnover of cyclic di-GMP in *Acetobacter xylinum*: genetic organization and occurrence of conserved domains in isoenzymes. J Bacteriol 180:4416–4425

Tamayo R, Pratt JT, Camilli A (2007) Roles of cyclic diguanylate in the regulation of bacterial pathogenesis. Annu Rev Microbiol 61:131–148

Tamilvanan S, Venkateshan N, Ludwig A (2008) The potential of lipid- and polymer-based drug delivery carriers for eradicating biofilm consortia on device-related nosocomial infections. J Controlled Release 128:2–22

Tashiro Y, Nomura N, Nakao R, et al (2008) *Opr86* is essential for viability and is a potential candidate for a protective antigen against biofilm formation by *Pseudomonas aeruginosa*. J Bacteriol 190:3969–3978

Valerius NH, Koch C, Høiby N (1991) Prevention of chronic *Pseudomonas aeruginosa* colonisation in cystic fibrosis by early treatment. Lancet 338:725–726

Van Gennip M, Christensen LD, Alhede M, et al (2009) Inactivation of the rhlA gene in Pseudomonas aeruginosa prevents rhamnolipid production, disabling the protection against polymorphonuclear leukocytes. APMIS 117:537–546

Vinodkumar CS, Neelagund YF, Kalsurmath S (2005) Bacteriophage in the treatment of experimental septicemic mice from a clinical isolate of multidrug resistant *Klebsiella pneumoniae*. J Commun Dis 37:18–29

Vinodkumar CS, Kalsurmath S, Neelagund YF (2008) Utility of lytic bacteriophage in the treatment of multidrug-resistant *Pseudomonas aeruginosa* septicemia in mice. Indian J Pathol Microbiol 51:360–366

Vuong C, Gerke C, Somerville GA, Fischer ER, Otto M (2003) Quorum-sensing control of biofilm factors in *Staphylococcus epidermidis*. J Infect Dis 188:706–718

Vuong C, Kocianova S, Yao Y, Carmody AB, Otto M (2004) Increased colonization of indwelling medical devices by quorum-sensing mutants of *Staphylococcus epidermidis* in vivo. J Infect Dis 190:1498–1505

Walters MC III, Roe F, Bugnicourt A, Franklin MJ, Stewart PS (2003) Contributions of antibiotic penetration, oxygen limitation, and low metabolic activity to tolerance of *Pseudomonas aeruginosa* biofilms to ciprofloxacin and tobramycin. Antimicrob Agents Chemother 47:317–323

Waters CM, Bassler BL (2005) Quorum sensing: cell-to-cell communication in bacteria. Annu Rev Cell Dev Biol 21:319–346

Whitchurch CB, Tolker-Nielsen T, Ragas PC, Mattick JS (2002) Extracellular DNA required for bacterial biofilm formation. Science 295:1487

Withers H, Swift S, Williams P (2001) Quorum sensing as an integral component of gene regulatory networks in Gram-negative bacteria. Curr Opin Microbiol 4:186–193

Wu H, Song Z, Hentzer M, et al (2000) Detection of N-acylhomoserine lactones in lung tissues of mice infected with *Pseudomonas aeruginosa*. Microbiology 146 (Pt 10):2481–2493

Wu H, Song Z, Hentzer M, Andersen JB, Molin S, Givskov M, Høiby N (2004) Synthetic furanones inhibit quorum-sensing and enhance bacterial clearance in *Pseudomonas aeruginosa* lung infection in mice. J Antimicrob Chemother 53:1054–1061

Xavier JB, Picioreanu C, Rani SA, van Loosdrecht MC, Stewart PS (2005) Biofilm-control strategies based on enzymic disruption of the extracellular polymeric substance matrix – a modelling study. Microbiology 151:3817–3832

Chapter 15
Different Methods for Culturing Biofilms In Vitro

S. Brook Peterson, Yasuhiko Irie, Bradley R. Borlee, Keiji Murakami, Joe J. Harrison, Kelly M. Colvin, and Matthew R. Parsek

15.1 Introduction

The field of biofilm microbiology, while by no means new, has been experiencing significant "growing pains" as more and more researchers become involved. One of the underlying reasons is the lack of standardized methods for culturing biofilm communities. Many times, the culturing format will be unique to the study in question, resulting in difficulties when other labs attempt to confirm results produced by another lab. Another issue has been the limited utility of different culturing methods for the specific research questions being asked. For example, culturing formats designed to be accessible to microscopy are not always suited for other types of analyses, such as harvesting biofilm biomass for biochemical measurements.

This chapter is intended to introduce the reader to some of the more commonly employed culturing methodologies. A common feature of biofilm culturing methods is a mechanism for separating adherent bacteria from those growing planktonically. In many cases these culturing methods can be adapted to mimic certain aspects of the system in which the researcher is interested. For example, abiotic attachment surfaces can be replaced with pathogenically relevant biotic surfaces. This chapter will also discuss the advantages and disadvantages of each system, in an attempt to help the reader select a culturing format that is most suitable for the research questions being asked.

15.2 Static Microtiter Plate Assays

One of the first methods standardized for quantifying biofilm formation is the microtiter plate static biofilm model. This method is designed to quantify the extent to which a microbe will attach to an abiotic surface. This method was developed

M.R. Parsek (✉)
Department of Microbiology - Box 357242, School of Medicine, University of Washington, 1959 NE Pacific Street, Seattle, WA 98195-7242, USA
e-mail: parsem@u.washington.edu

T. Bjarnsholt et al. (eds.), *Biofilm Infections*, DOI 10.1007/978-1-4419-6084-9_15,

in the context of studying coagulase-negative *Staphylococcus aureus* attachment to plastic (Christensen et al. 1985), but has been since adapted for use in many other species. Using short incubation times (1–2 h), initial attachment to a surface can be assessed, while longer incubation times (~20 h) allow for a measure of biofilm formation.

The primary advantage of this method is that it is relatively high-throughput, enabling screens for mutants defective in attachment or evaluation of the effects of different treatments or compounds on attachment or biofilm formation. For example, O'Toole and Kolter used static microtiter well assays to identify genetic requirements for surface attachment by *Pseudomonas fluorescens* and *Pseudomonas aeruginosa* (O'Toole and Kolter 1998a, b). Other organisms for which microtiter assays have been used to identify attachment-deficient mutants include *S. aureus* (Valle et al. 2003, Tu Quoc et al. 2007), *Bacillus subtilis* (Hamon and Lazazzera 2001), and *Vibrio cholerae* (Watnick and Kolter 1999). Many studies have also used microtiter dish assays to evaluate the effects of environmental parameters on the ability of bacteria to attach and form biofilms. A few examples include testing the effects of nutrients in the medium on attachment of *P. fluorescens* to polyvinyl chloride (O'Toole and Kolter 1998a), measuring *Escherichia coli* attachment promotion and disruption by indolic compounds (Lee et al. 2007), and evaluating the role of salt and ethanol stress on biofilm formation in *Staphylococcus epidermidis* (Knobloch et al. 2001).

The effects of surface properties on attachment can also be studied in microtiter dishes. For example, one study characterized *S. aureus* attachment in wells coated with the immobilized host proteins fibrinogen and fibronectin (Cucarella et al. 2002). Microtiter dish assays are less well suited to studies of biofilm structure or of antimicrobial resistance properties. This is due in part to the difficulty in microscopically visualizing the biofilms, and in distinguishing between live cells and matrix material stained by the dye.

15.2.1 Basic Microtiter Dish Biofilm Assay Protocol

1. Preparing inoculating culture. Subculture overnight cultures of the strain of interest and grow to midlog phase (OD_{600} ~0.5). Dilute subcultures again to $OD_{600} = 0.2$.
2. Inoculation and incubation. Distribute 95 μl per well of chosen medium into 96-well microtiter plates. Add 5 μl per well of the culture prepared in step 1. Include wells containing media-alone as negative controls, and avoid use of wells on the periphery of the plate (fill with media alone to prevent desiccation). Seal with parafilm and incubate at the desired temperature.
3. Removing planktonic cells. After the desired incubation time (see above), gently pour-off the media remaining in the wells containing unattached cells and gently submerge the plate in deionized water. Remove the plate, again pour off contents gently, and repeat wash. Depending on desired stringency, further washing steps may also be included.

4. Staining of adherent cells. Add 150 μl 0.1% W/V Crystal violet to each well, and stain for 10 min at room temperature. Gently wash the plate three times, as described in Step 3. Allow the plate to air-dry.
5. Measuring dye absorbed by adherent cells and matrix. Add 200 μl 95% ethanol [or other solubilizing agent, depending on bacterial species (6)] to each well, incubate 10 min at RT. Pipette this solution containing resolublized dye up and down twice to mix, transfer to a clean 96-well plate, and read absorbance at 595 nm.

15.3 Flow Cells

The basic flow cell system (Fig. 15.1) consists of a chamber where biofilm bacteria are grown, usually attached to a coverslip overlying the chamber. Fresh,

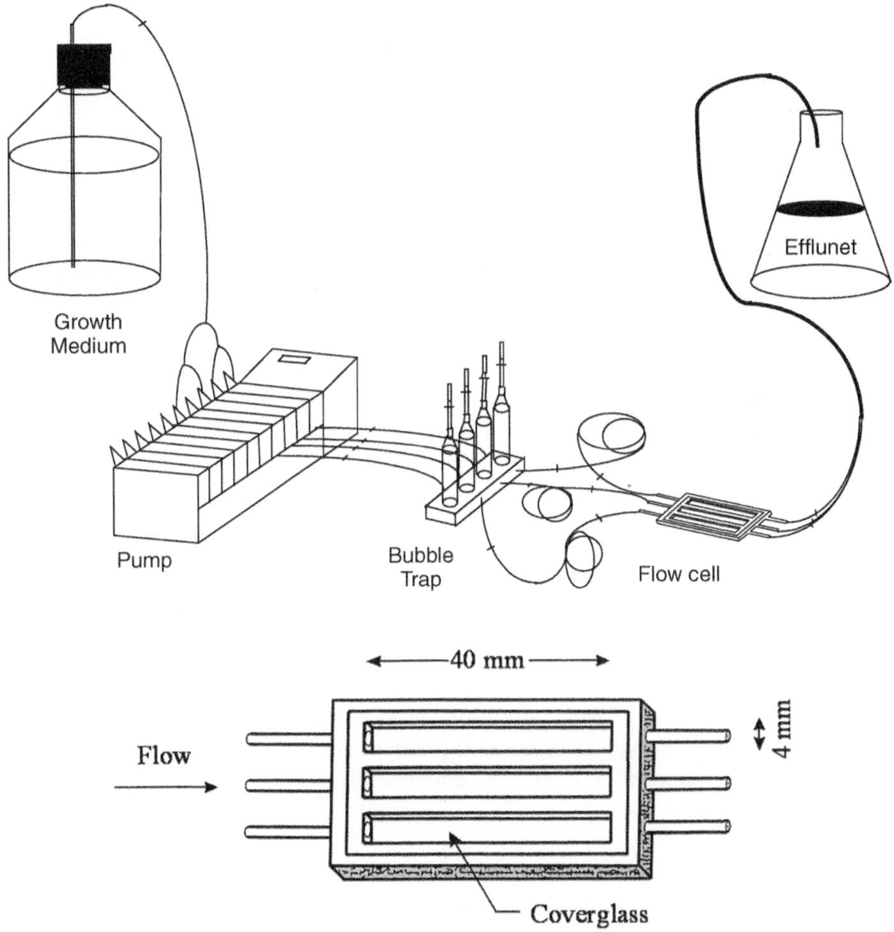

Fig. 15.1 Diagram of a flow cell apparatus

sterile medium is continuously pumped through the chamber, promoting growth of the surface-attached bacteria and washing away unattached planktonic cells. The introduction of various fluorescent proteins and dyes, as well as improvements on microscopy technologies, enables the analysis of the spatial distribution and growth of individual cells in real time. Additionally, flow cells can be used in conjunction with fluorescent reporter genes to study spatial patterns of gene expression, a feature of sessile microbial community becoming increasingly recognized (reviewed in Stewart and Franklin 2008).

Various flow cell systems have been developed by several groups studying biofilms. Several representative models are presented on the website maintained by the Center for Biofilm Engineering at Montana State University (http://www.erc.montana.edu/CBEssentials-SW/research/Biofilm%20Mechanics/flow_cells.htm). Several manufacturers also sell flow cell systems, including Stovall Life Science (Greensboro, NC, USA), Bioptechs Inc. (Butler, PA, USA), and BioSurface Technologies Corp. (Bozeman, MT, USA). Microscopy has traditionally been useful for qualitative analyses, but not for quantitative interpretations. However, for fluorescent imaging of biofilms in a flow cell, the computer software COMSTAT has been a particularly powerful tool in determining biofilm structures on a quantitative basis (Heydorn et al. 2000).

In recent years, flow cell systems have been further adapted to experimentally examine biofilm formation on more biologically relevant surfaces, especially in the context of host–pathogen interactions. Biological molecules can be used to coat the surface of the flow cells to test their effects on bacterial attachment. Mucin-coated glass surfaces were discovered to enhance *P. aeruginosa* adherence using this method (Landry et al. 2006). George O'Toole and Bruce Stanton's groups have described an in vitro continuous flow model in which *Pseudomonas aeruginosa* biofilms are grown on polarized epithelial cells and visualized by live high-resolution imaging (Moreau-Marquis et al. 2008). Using this system, they observed that cystic fibrosis epithelial cells promote *P. aeruginosa* biofilm formation. Such modifications of flow cell applications to biofilm development have only recently begun to be exploited and will be exciting tools for future biofilm research.

Despite the powerful experimental potential of live imaging as a tool for studying microbial biofilm development, two major concerns must be addressed when applying its use for interpretations of pathogenesis. First and foremost, flow cell experiments are in vitro systems and caution must be exercised when tying data generated from in vitro into an in vivo pathogenesis systems. Most microbes have a variety of environmental adaptation mechanisms that may drastically alter their behavior depending on subtle changes in their surroundings. Flow cells are capable of creating a uniform and constant environment for in vitro purposes, but our current state of technologies is a distant representation from the dynamics of host environment in vivo.

A second consideration for using flow cells is the difficulty collecting biofilm biomass. Many models of flow cells do not allow for easy access to the biofilm growing in the chambers. Even for those that are optimized for biofilm collection, collecting sufficient biomass for subsequent analyses is a challenge. For optimized biomass collection, spin-disc reactors and tube biofilms are recommended. Furthermore, flow cells are ideal for visualization of biofilms at real-time, but these experiments are not available for high-throughput assays. Microtiter-well crystal violet attachment assays and peg biofilm methods are more ideal for quantitative measurements and genetic screens.

15.3.1 Basic Flow Cell Protocol

Of course, each organism is generally different in respect to biofilm formation. This is most likely the consequence of multiple factors, including growth rate, EPS matrix production, and the relative tendency to stick to a surface. Therefore, a variety of culturing conditions (e.g., flow rates, growth media, etc.) must be empirically tested to determine what is best suited for the organism of interest. Some representative conditions that are often variable from one system to another are summarized in Table 15.1.

Table 15.1 Parameters of flow cell biofilm systems which may be varied and phenotypic traits that will influence the choice of flow cell parameters

Variable conditions	Some major phenotypic traits to consider
Growth medium Incubation temperature	Organism doubling time, motility effects, and gene expression
Flow rate	Can impact growth, some adhesins enhance binding in higher sheer force (Sokurenko et al. 2008)
Cover slip material	Some organisms have poor attachment to certain surfaces. Acid-washing glass cover slips may help prevent non-specific interactions to contaminants on the surface
Inoculum size	May be dependent on the biofilm development stage of interest. Can affect initial distribution of cells on a surface, and thus downstream effects on structure

15.3.2 Sample Protocol

1. Autoclave flow components (some flow cells do not withstand autoclaves).
2. Further sterilize by flowing H_2O_2 (3% vol/vol, at least 200 ml) through the tubing and flow cells for 4+ h.
3. Wash the flow components with sterile H_2O, followed by media overnight.
4. Inject bacteria directly into the flow cells, taking care not to introduce air bubbles. Many researchers inoculate diluted bacteria from log phase cultures, but some researchers studying biofilm detachment have inoculated at a higher bacterial density (Boles, Thoendel and Singh 2005).
5. Flip the flow cell cover slip side down so as to allow bacterial attachment and incubate for 30 min to 1 h.
6. Flip back the flow cell upright, and start the flow.

7. Observe biofilm development at time points of interest via microscopy. Biofilms will be growing attached to the cover slip.
8. For researchers interested in introducing foreign materials to the developing biofilms (e.g., antibiotics and staining materials), it is generally recommended to do this upstream of the flow cell. When staining biofilms, it may be advantageous to inject directly into the tubing. In this scenario, flow should be stopped, clamps should be applied downstream of the flow cell, and the bubble traps opened. Inject materials into the tubing upstream of the flow cell in the direction of the bubble traps. Close bubble traps, unclamp the tubing, and start the flow. Depending on the staining conditions, researchers may have to stop the flow to allow for prolonged incubation of the staining materials with the biofilm. These steps are precautionary methods for minimizing physiological changes to the biofilm, which may change the culturing conditions and therefore affect the experiment.
9. Re-usable flow cells may be sterilized with H_2O_2, followed by autoclaving for subsequent experiments.

For a more comprehensive description of flow cell apparatus system, readers may refer to Palmer (1999).

15.4 Tube Biofilms

Like flow cells, tube biofilm reactors provide a platform for studying biofilms that develop under flow. The primary difference between the two methods is in the tube reactor; biofilms are grown on the interior surface of silicone tubing, rather than on a cover slip attached to a chamber. The advantage of the tube biofilm is that it enables the accumulation of a large biofilm mass which can be collected easily by scraping from the tubing. One potential use of the system is to quantitatively evaluate the effect of antibacterial agents on biofilms by counting colony forming unit recovered from tubes before and after exposure to treatment with the agent of choice. Biochemical and gene expression studies can also be conducted on biomass collected from tubes. Finally, tube reactors can also be used to study the phenotypic diversification which can accumulate as a result of biofilm growth (Kirisits and Parsek 2006). However, unlike biofilms grown in flow cells, tube biofilms are not amenable to microscopy. A limitation of this system shared with flow cells is the difficulty in using the methods for high-throughput analysis.

15.4.1 Basic Protocol for Growing Tube Biofilms

1. Assembling the apparatus. The tube biofilm system consists of a growth medium reservoir, a peristaltic pump (Watson-Marlow PumpPro MPL), a bubble trap, a connector silicone tubing (3.18 mm-inner–diameter, NALGEN), a polypropylene connector, a silicone biofilm reactor tubing (6.35 mm-inner–diameter, Cole

Parmer), and a waste bottle (Schaefer et al. 2001). The medium flows from the reservoir through silicone tubing into the silicone biofilm reactor tubing on which the biofilms are grown. After all devices are sterilized, the system is assembled and media is run through the system at a low flow rate.

2. Inoculation and attachment. Cells (generally in log phase) are inoculated and diluted to OD 600 0.1 with medium. Before inoculation to the tubing, flow in the system is stopped and the connector tubing is clamped directly upstream of the biofilm tubing. The samples are slowly injected to the biofilm reactor tubing by syringe and needle. After the injection, the injection hole is immediately sealed with silicone. The bacteria are allowed to incubate for 30 min before flow is reinitiated, allowing bacterial attachment to the inside surface of the tubing. Biofilms are grown at a flow rate of approximately 50 ml/h.
3. Sampling and recovery of biomass. At selected time points, flow is stopped and the biofilm tubing is separated from the connecting tubing. The tubing is washed once with a phosphate-buffered saline (PBS, pH 7). The biofilm can then be collected using cell scrapers and suspended in PBS.

15.5 Colony Biofilms

A colony biofilm is grown on a semi-permeable polycarbonate filter (GE) placed onto an appropriate solid medium. The filter provides a convenient surface that is easily maneuverable allowing transfer of the surface-grown biofilm from one media source to another. This technique analyzes biofilm development at the air–surface interface, without submersing the biofilm in liquid. Due to the nature of growth, colony biofilms have the ability to become quite large in a short period of time. After 48 h of growth, a *P. aeruginosa* colony biofilm is approximately 150–300 μm thick while *S. epidermidis* produces a biofilm that is approximately 100 μm thick (Borriello et al. 2004, Werner et al. 2004, Stewart et al. 2007). Colony biofilms display a typical stratified profile similar to flow cell biofilms with anaerobic conditions and lysed cells predominating within the interior colony biofilm, while exponential phase cells are found at the colony–air interface where there is more oxygen (Werner et al. 2004).

A colony biofilm is a particularly useful method to analyze bacterial survival in the face of antimicrobial treatment. The filter biofilm technique has the benefit of analyzing antibiotic susceptibility by complementary measurements. First, the biofilm susceptibility can be quantified by CFU. Second, relative antibiotic penetration through the biofilm can be assayed and finally, the spatial pattern of antibiotic activity, oxygen content, extracellular matrix components, and protein synthesis can be determined using electrodes or microscopy. Additionally, one caveat to analyzing antimicrobial susceptibility in other biofilm systems, such as the PEG biofilm and flow cell biofilm, is the starting inoculum will only be similar when comparing strains with equal ability to attach and develop into a mature biofilm. In contrast, colony biofilm are simply relocated to an agar medium with or without antibiotic treatment, which does not select for cell attachment since the filter is not submersed in liquid to remove the antibiotic. An additional benefit to the using colony biofilms is that changes in viable CFU are attributed to cell death rather than detachment due to the stable and spatially restricted nature of this system.

Colony biofilms are not suited to perform live, non-destructive imaging of biofilm development. However, bacterial cell morphology, DNA replication, and protein expression can be assessed by transmission electron microscopy and confocal scanning laser microscopy as described (Borriello et al. 2004, Werner et al. 2004, Rani et al. 2007). In these methods colonies must be fixed with gluteraldehyde or frozen and sectioned for analysis. Stratification of oxygen can also be measured in colony biofilms using a microelectrode (Borriello et al. 2004, Werner et al. 2004, Rani et al. 2007). Colony biofilms have become a useful model to begin to visualize the heterogeneity of nutrients, extracellular components, and gene expression profiles unique to biofilm growth.

One drawback to this method is after complete bacterial coverage of the membranes, the membranes are difficult to handle- becoming flimsy and periodically folding on top of themselves, thus destroying the biofilm. An additional consideration to this technique is that bacterial strains that differ in surface motility will spread across the filter at variable rates. To counter this problem, the polycarbonate filter can be placed into a cartridge (similar to mating filters) and the bacterial suspension can be injected into the cartridge allowing complete coverage of the polycarbonate filter.

15.5.1 Colony Biofilm Preparation Protocol

1. Bacteria are grown overnight in liquid culture in the same medium that will be used for biofilm growth.
2. The 25 mm polycarbonate filters are UV sterilized for 10 min, glossy side up. To prevent the filters from curling, place them onto an agar plate with a maximum of seven filters per plate prior to UV treatment.
3. Using two sterile forceps, transfer the polycarbonate filter to a fresh media source and dispose of the UV-treated agar plates from step 2.
4. Dilute overnight cultures to an OD_{600} 0.05 in the same medium used for biofilm growth and spot 5 μl of culture in the center of each polycarbonate filter.
5. Incubate at 37°C for 24 h. Using two sterile forceps transfer the colony biofilm to fresh media.
6. After 48 h of growth, biofilm susceptibility can be assessed by transferring the colony biofilm to fresh media with or without the antibiotic of interest present.
7. Optional antibiotic penetration assay: in addition to assessing antimicrobial susceptibility of different biofilm strains and/or conditions, a slightly modified version allows the measurement of relative antimicrobial penetration. Prior to moving the colony biofilm to an antibiotic treatment, a smaller 13 mm polycarbonate filter is gently placed on top of the biofilm followed by a 6 mm moistened paper disc (Fig. 15.2). Antibiotics will travel from the agar through the first polycarbonate filter and the biofilm, and then past the second filter into the

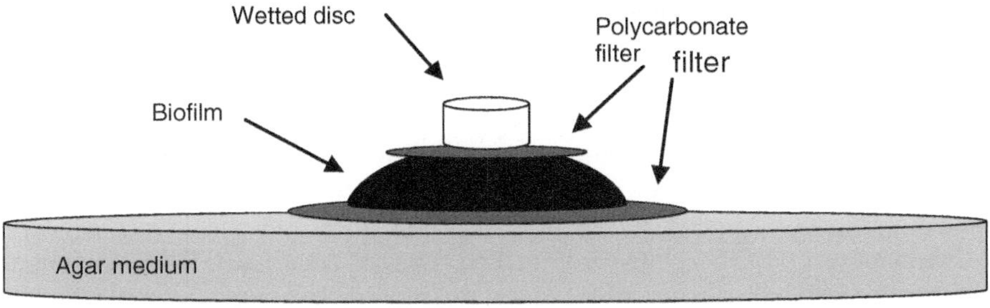

Fig. 15.2 Side view of the colony biofilm technique for determining relative antibiotic penetration

moistened disc. The relative concentration of the antibiotic in the wetted disk is determined by a well-described Kirby–Bauer disc diffusion assay (Morley 1945).

15.6 Biofilm Growth on Peg Lids

A disadvantage shared by several of the biofilm cultivation methods discussed in this chapter – such as flow cells, rotating-disk, or tube reactors – is the inability to produce more than a few biofilm samples at one time. Moreover, these systems depend on continuous flow and are somewhat prone to contamination and leakage. To enable high-throughput experimental designs, one possible approach might be to grow biofilms on a peg lid. An example of this is the Calgary Biofilm Device, which consists of a polystyrene lid with 96 pegs that can be fit into a standard 96-well microtiter plate. Originally described by Ceri et al. (1999) and akin to the microtiter plate method popularized by O'Toole and Kolter (O'Toole and Kolter 1998a) (see above), the primary purpose of growing biofilms on pegs is phenotypic screening. Batch culture apparatuses, such as the Calgary Biofilm Devices, are neither prone to contamination, since they are manipulated in a laminar flow hood, nor prone to leakage, since they are sealed in a fashion similar to a Petri dish. Biofilm cells are typically recovered from pegs by using low frequency sonication to disrupt the surface-adherent cells into a recovery medium.

Several modifications of this technique have been reported in the literature, including growth of biofilms on lids containing 384 pins (Musk et al. 2005), as well as growth of biofilms on chemically modified plastic pegs, such as those used for solid-surface enzyme-linked immunosorbant assays (Spoering and Lewis 2001). Although certain peg lids, including that of the Calgary Biofilm Device, are made of untreated polystyrene and bear an overall neutral electrostatic charge, it is possible to coat the plastic with a conditioning film to facilitate the adhesion of fastidious microorganisms that might not otherwise stick to the surface. Such coatings might alter surface hydrophobicity or might provide specific attachment sites for various human or plant pathogens, and include hydroxyapetite, human serum, methylcellulose, or metal alloys. For example, *Candida tropicalis* forms robust biofilms on polystyrene pegs coated with L-lysine or bovine serum albumin, although not on the uncoated surface itself (Harrison et al. 2007b). This corresponds with well-established protocols for growing biofilms of *Candida* spp. in microtiter plates manufactured from cell culture treated plastics (Pierce et al. 2008).

Reproducible growth of biofilms on pegs follows a standard five-step protocol which serves as the starting point for a variety of downstream applications, including biomass staining, metabolic assays and viable cell counting (Fig. 15.3). The foremost use of this cultivation method remains quantitative determination of bacterial growth and survival after exposure of biofilms to gradient arrays of antibiotics. A correctly implemented experimental protocol is able to distinguish between biofilm resistance and tolerance toward the tested antimicrobial agent (Harrison et al. 2005a). The transfer of pegs between microtiter plates is easy and enables simultaneous comparisons of multiple biofilm growth and exposure conditions. As a basic research tool, assay systems such as the Calgary Biofilm Device have been used for checkerboard susceptibility testing to screen combinations of biocides for synergistic anti-biofilm activity (Harrison et al. 2008), as well as to assess the role

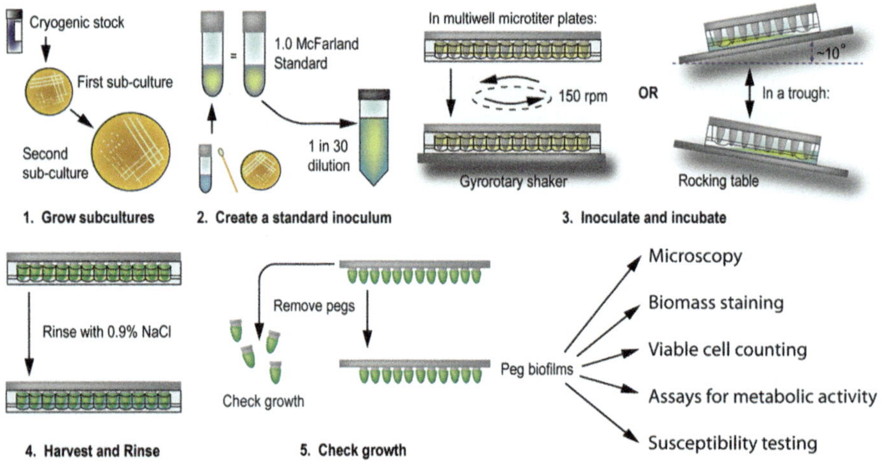

Fig. 15.3 A five-step standard procedure for cultivation of biofilms on peg lids serves as a starting point for versatile downstream workflows. (1) Microbial strains recovered from a cryogenic stock are streaked out twice in succession on agar media. (2) Using a protocol set by the Clinical Laboratory Standards Institute for creating standardized bacterial inocula, colonies from a second agar subculture are collected with a sterile cotton swab and suspended in saline to match an optical McFarland standard. This is then typically diluted 30-fold in growth medium, which serves as the inoculum for biofilm apparatus. (3) Depending on the format of the peg lid apparatus, inoculum is either transferred into microtiter plate wells or into a trough, into which the peg lid is then inserted. The assembled apparatus is then placed on a gyrorotary shaker or rocking platform, respectively, and incubated at an appropriate temperature and for a suitable time. (4) The cultivated biofilms are rinsed, and (5) a series of pegs are broken from the lid in order to assess starting biofilm cell number by viable cell counting. The harvested biofilms may then be used in one of many downstream applications. Adapted from Harrison et al. (2006) by permission of the authors

of different chromosomal genetic elements in biofilm formation and multidrug tolerance (Parkins et al. 2001, Harrison et al. 2009). As a clinical diagnostic tool, the Calgary Biofilm Device was approved in 2008 by Health Canada for selecting antibiotics to treat *Pseudomonas aeruginosa* infections.

A key advantage of peg lids over microtiter plates is the ability to detach pegs for control measurements and for microscopy. Although biofilms cultivated in this fashion are subject to complex fluid dynamics, gross morphological changes in biofilm structure may still be determined in situ by SEM or CLSM (Harrison et al. 2006, 2007b). It is also possible to discern patterns of cell death in antibiotic-exposed biofilms that are similar to those reported in the literature for flow cells (J. J. Harrison and H. Ceri, unpublished data). As a *caveat*, strain-to-strain comparisons of biofilm formation and antimicrobial susceptibility require optimized growth conditions prior to high-throughput screening (Harrison et al. 2005b, 2006). Lastly, these methods are ill-suited for downstream biochemistry because of the limited amounts of biomass produced on pegs. Nonetheless, devices such as the Calgary Biofilm Device are among some of the most versatile cultivation tools available in biofilm research.

15.6.1 Basic Protocol

See Fig. 15.3.

15.7 Rotating Disk and Concentric Cylinder Reactors

Both rotating disk reactors (RDR) and concentric cylinder reactors (CCR) have been engineered for the growth of biofilms under shear stress for the reproducible evaluation of biocide efficacy. The RDR (Fig. 15.4) is comprised of a circular disk that is engineered to allow for the incorporation of removable coupons or chips that are flush with the overall surface of the disk and can be made from a variety of materials depending upon the question under investigation (Zelver et al. 1999, 2001, Ramey and Parsek 2005). The disk is attached to a star-head magnet which is then placed in a 1 L glass side-arm reactor vessel. The vessel is placed on top of magnetic stirrer which provides for adjustable rotational speed of the disk to generate a liquid shear force across the surface of the disk. Rather than on a rotating disk, the biofilm in the CCR forms on the surface of rotating stainless steel cylinders inside concentric chambers (Willcock et al. 2000). The speed of rotation and the diameter of the cylinders can be varied to create different shear strength. In both reactors, the flow of medium through the reactor is maintained as a constant drip regulated by pump. Medium is pumped from a reservoir and drips into the reactor vessel that contains approximately 180–300 ml of medium depending upon the location of the outlet for the drain which leads to a waste reservoir.

Growth or viability can be monitored aseptically by removing coupons from the RDR or cylinders from the CCR and plating bacteria for relative viable cell counts at various intervals of growth. In the evaluation of therapeutic efficacy, molecules

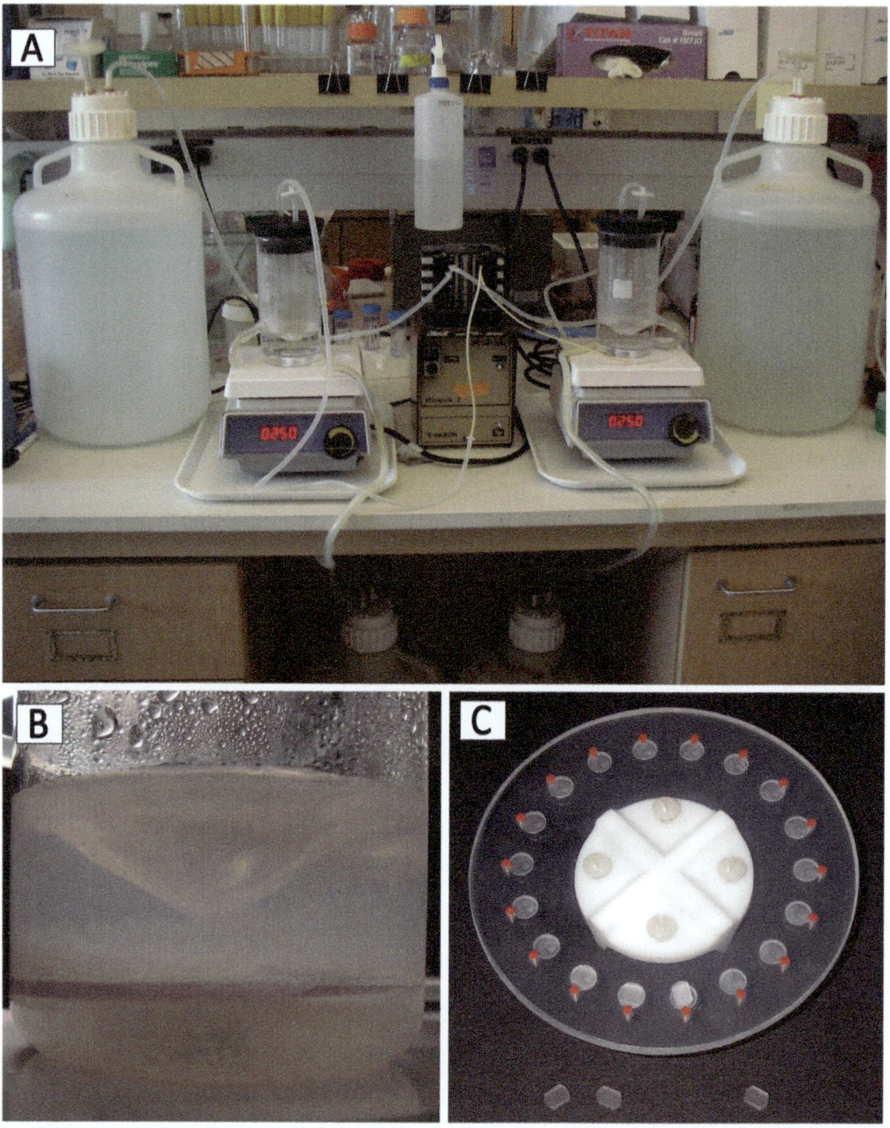

Fig. 15.4 A rotating disk reactor. (**a**) The assembled reactor, depicting influent and effluent media reservoirs, and two reactor vessels housing rotating disks, placed on stir plates. (**b**) The reactor vessel, containing medium and a rotating disk. (**c**) A rotating disk with removable coupons

or chemicals can be introduced into the reaction vessel or delivered by the continuous flow in the medium and in manner that mimics the relevant parameters associated with treatment regimes such as dose and time of exposure. Alternatively, after biofilms have been developed, coupons can be removed from the RDR and incubated in potential biocides or therapeutics followed by enumeration of surviving bacteria to determine effective doses for eradication or disruption of the biofilm. The use of a rotating disk that accommodates large numbers of coupons or chips allows for dose–response killing relationships to be accurately determined from a single biofilm.

Numerous studies have used the RDR to understand the role of biofilm growth in resistance and tolerance to antibiotics and other chemical compounds (Hentzer et al. 2001, Pitts et al. 2001, Boles et al. 2004, Kirisits et al. 2005, Garo et al. 2007, Banin et al. 2008). The RDR also allows for comparison of various strains or mutants and their ability to form biofilms as measured by biomass accumulation with respect to various stages of biofilm development including maturation mediated by signaling and detachment (Boles et al. 2004, Yarwood et al. 2004). The American Society of testing and materials has approved the RDR for the cultivation and quantification of *P. aeruginosa* biofilms grown with shear and continuous flow (ASTM Standard Method E2196-07) and the system is commercially available from Biosurface Technologies Corp. One example of a use of the CCR is a study of the effects of shear strength on the formation of biofilms by communities of freshwater bacteria (Rickard et al. 2004).

An advantage of this technique for culturing biofilms is the reproducibility in biofilm sampling that can be attained under continuous flow culture conditions that allow for the maintenance of a relatively steady-state chemostat (Ramey and Parsek 2005). Disks or coupons can be coated with biologically relevant substrates, materials that are used in implants (with antibacterial coatings or treatments), or coated with biopolymers that are important at the interface of the biofilm and any surface components that the biofilm may interact with. It is also possible that coupons or chips coated with biofilm could be implanted to mimic a biofilm infection in animal studies of biofilm-associated infections. However the greatest weakness with this biofilm reactor system is the number of strains that can be cultivated and analyzed simultaneously. Only one strain or mutant can be analyzed at a time per reactor. High throughput analysis of large numbers of strains is not possible with this system.

15.7.1 The Basic Protocol for the RDR (The Protocol for the CCR Is Well Described in Willcock et al. 2000)

1. Inoculation. Bacteria are initially inoculated at a low starting density into approximately 250 ml of medium in the reactor vessel. Our method for testing biofilm susceptibility to antibiotics involves operating the reactor in batch mode (no medium flow) with the media stirring at 200–250 rpm until the bacteria have reached mid-log phase of growth as determined by OD_{600}. Alternatively, if a rich medium is used the reactor can be run in chemostat mode during this initial growth phase. In both cases, bacteria are incubated at room temperature for ~24 h or until the desired optical density (mid-exponential phase) is achieved in the growth medium.

2. Reactor operation. Once mid-log phase has been achieved in the starting culture in the reaction vessel, the stir bar used for mixing is replaced aseptically with the rotating disk (with coupons installed, if desired). The reactor should then be operated at chemostat mode at a dilution rate of 0.10 h^{-1}. This flow rate consumes approximately 13 L of medium in 24 h. Biofilm formation is allowed to proceed for 24 h (or longer if desired).
3. Evaluating antimicrobial susceptibility of biofilms. Once biofilms have become established on the rotating disk, antimicrobial agents of interest can be introduced to the reactor vessel or added to the flow, and viability of biofilm bacteria assessed by removing coupons. Adherent bacteria are recovered from coupons by first rinsing in PBS to remove loosely associated planktonic cells, and then vortexting, sonicating, and/or homogenizing, depending on the strength of adherence and accumulation of matrix material. Viable cell count can then be determined by plate counting. Alternatively, antimicrobial susceptibility of coupon-adherent bacteria can be assessed by first removing the coupons, and then incubating them in solutions containing varying concentrations of the antimicrobial prior to adherent cell enumeration as described above. Equal numbers of planktonic cells collected from the reactor can also be treated to determine the difference in susceptibility resulting from biofilm formation.

15.8 Summary

The methods described here are intended to furnish the biofilm researcher with a selection of standardized methods from which to choose to address different research questions. For high throughput analyses, we describe the mirotiter and peg biofilm methods. The flow cell device is described for applications where microscopic observation of biofilm development is desired. The tube biofilm method is provided as a convenient way of recovering biofilm material for downstream biochemical analyses. Several methods are described for assessing biofilm susceptibility to antimicrobials: peg biofilms, colony biofilms and the rotating disk, and concentric cylinder reactors. Finally, we have also provided suggestions for modifications of these methods to enable closer approximation to clinically relevant conditions, such as modification of growth surfaces to incorporate biological substrates.

References

Banin E, Lozinski A, Brady KM et al (2008) The potential of desferrioxamine-gallium as an anti-*Pseudomonas* therapeutic agent. Proc Natl Acad Sci USA 105:16761–16766

Boles BR, Thoendel M, Singh PK (2005) Rhamnolipids mediate detachment of *Pseudomonas aeruginosa* from biofilms. Mol Microbiol 57:1210–1223

Boles BR, Thoendel M, Singh PK (2004) Self-generated diversity produces "insurance effects" in biofilm communities. Proc Natl Acad Sci USA 101:16630–16635

Borriello G, Werner E, Roe F et al (2004) Oxygen limitation contributes to antibiotic tolerance of *Pseudomonas aeruginosa* in biofilms. Antimicrob Agents Chemother 48:2659–2664

Ceri H, Olson ME, Stremick C et al (1999) The Calgary Biofilm Device: new technology for rapid determination of antibiotic susceptibilities of bacterial biofilms. J Clin Microbiol 37:1771–1776

Christensen GD, Simpson WA, Younger JJ et al (1985) Adherence of coagulase-negative staphylococci to plastic tissue culture plates: a quantitative model for the adherence of staphylococci to medical devices. J Clin Microbiol 22:996–1006

Cucarella C, Tormo MA, Knecht E et al (2002) Expression of the biofilm-associated protein interferes with host protein receptors of *Staphylococcus aureus* and alters the infective process. Infect Immun 70:3180–3186

Garo E, Eldridge GR, Goering, MG et al (2007) Asiatic acid and corosolic acid enhance the susceptibility of *Pseudomonas aeruginosa* biofilms to tobramycin. Antimicrob Agents Chemother 51:1813–1817

Hamon MA, Lazazzera BA (2001) The sporulation transcription factor Spo0A is required for biofilm development in *Bacillus subtilis*. Mol Microbiol 42:1199–1209

Harrison JJ, Ceri H, Turner RJ (2007a) Multimetal resistance and tolerance in microbial biofilms. Nat Rev Microbiol 5:928–938

Harrison JJ, Turner RJ, Ceri H (2007b) A subpopulation of *Candida albicans* and *Candida tropicalis* biofilm cells are highly tolerant to chelating agents. FEMS Microbiol Lett 272:172–181

Harrison JJ, Turner RJ, Ceri H (2005a) Persister cells, the biofilm matrix and tolerance to metal cations in biofilm and planktonic *Pseudomonas aeruginosa*. Environ Microbiol 7:981–994

Harrison JJ, Turner RJ, Ceri H (2005b) High-throughput metal susceptibility testing of microbial biofilms. BMC Microbiol 5:53

Harrison JJ, Wade WD, Akierman S et al (2009) The chromosomal toxin *yafQ* is a determinant of multidrug tolerance for *Escherichia coli* growing in a biofilm. Antimicrob Agents Chemother. doi:10.1128/AAC.00043-09

Harrison JJ, Ceri H, Yerly J et al (2006) The use of microscopy and three-dimensional visualization to evaluate the structure of microbial biofilms cultivated in the Calgary Biofilm Device. Biol Proced Online 8:194–215

Harrison JJ, Turner RJ, Joo DA et al (2008) Copper and quaternary ammonium cations exert synergistic bactericidal and antibiofilm activity against *Pseudomonas aeruginosa*. Antimicrob Agents Chemother 52:2870–2881

Hentzer M, Teitzel GM, Balzer GJ et al (2001) Alginate overproduction affects *Pseudomonas aeruginosa* biofilm structure and function. J Bacteriol 183:5395–5401

Heydorn A, Nielsen AT, Hentzer M et al (2000) Quantification of biofilm structures by the novel computer program COMSTAT. Microbiology 146(Pt 10):2395–2407

Kirisits MJ, Parsek MR (2006) Does *Pseudomonas aeruginosa* use intercellular signalling to build biofilm communities? Cell Microbiol 8:1841–1849

Kirisits MJ, Prost L, Starkey M et al (2005) Characterization of colony morphology variants isolated from *Pseudomonas aeruginosa* biofilms. Appl Environ Microbiol 71:4809–4821

Knobloch JK, Bartscht K, Sabottke A et al (2001) Biofilm formation by *Staphylococcus epidermidis* depends on functional RsbU, an activator of the sigB operon: differential activation mechanisms due to ethanol and salt stress. J Bacteriol 183:2624–2633

Landry RM, An D, Hupp JT et al (2006) Mucin-*Pseudomonas aeruginosa* interactions promote biofilm formation and antibiotic resistance. Mol Microbiol 59:142–151

Lee J, Bansal T, Jayaraman A et al (2007) Enterohemorrhagic *Escherichia coli* biofilms are inhibited by 7-hydroxyindole and stimulated by isatin. Appl Environ Microbiol 73:4100–4109

Moreau-Marquis S, Bomberger JM, Anderson GG et al (2008) The DeltaF508-CFTR mutation results in increased biofilm formation by *Pseudomonas aeruginosa* by increasing iron availability. Am J Physiol Lung Cell Mol Physiol 295:L25–37

Morley D (1945) A simple method for testing the sensitivity of wound bacteria to penicillin and sulphathiazole by use of impregnated blotting paper discs. J Pathol Bacteriol 57: 379–382

Musk DJ, Banko DA, Hergenrother PJ (2005) Iron salts perturb biofilm formation and disrupt existing biofilms of *Pseudomonas aeruginosa*. Chem Biol 12:789–796

O'Toole GA, Kolter R (1998a) Initiation of biofilm formation in *Pseudomonas fluorescens* WCS365 proceeds via multiple, convergent signalling pathways: a genetic analysis. Mol Microbiol 28:449–461

O'Toole GA, Kolter R (1998b) Flagellar and twitching motility are necessary for *Pseudomonas aeruginosa* biofilm development. Mol Microbiol 30:295–304

Palmer RJ (1999) Microscopy flowcells: perfusion chambers for real-time study of biofilms. Methods Enzymol 310:160–166

Parkins MD, Ceri H, Storey DG (2001) *Pseudomonas aeruginosa* GacA, a factor in multihost virulence, is also essential for biofilm formation. Mol Microbiol 40:1215–1226

Pierce CG, Uppuluri P, Tristan AR et al (2008) A simple and reproducible 96-well plate-based method for the formation of fungal biofilms and its application to antifungal susceptibility testing. Nat Protoc 3:1494–1500

Pitts B, Willse A, McFeters GA et al (2001) A repeatable laboratory method for testing the efficacy of biocides against toilet bowl biofilms. J Appl Microbiol 91:110–117

Ramey BE, Parsek MR (2005) Growing and analyzing biofilms in fermenters. Curr Protoc Microbiol Chapter 1: Unit 1B.3

Rani SA, Pitts B, Beyenal H et al (2007) Spatial patterns of DNA replication, protein synthesis, and oxygen concentration within bacterial biofilms reveal diverse physiological states. J Bacteriol 189:4223–4233

Rickard AH, McBain AJ, Stead AT et al (2004) Shear rate moderates community diversity in freshwater biofilms. Appl Environ Microbiol 70:7426–7435

Schaefer AL, Greenberg EP, Parsek MR (2001) Acylated homoserine lactone detection in *Pseudomonas aeruginosa biofilms* by radiolabel assay. In: Doyle RJ (ed) Methods in enzymology. Academic, London, pp 41–47

Sokurenko EV, Vogel V, Thomas WE (2008) Catch-bond mechanism of force-enhanced adhesion: counterintuitive, elusive, but ... widespread? Cell Host Microbe 4:314–323

Spoering AL, Lewis K (2001) Biofilms and planktonic cells of *Pseudomonas aeruginosa* have similar resistance to killing by antimicrobials. J Bacteriol 183:6746–6751

Stewart PS, Franklin MJ (2008) Physiological heterogeneity in biofilms. Nat Rev Microbiol 6: 199–210

Stewart PS, Rani SA, Gjersing E et al (2007) Observations of cell cluster hollowing in *Staphylococcus epidermidis* biofilms. Lett Appl Microbiol 44:454–457

Tu Quoc PH, Genevaux P, Pajunen M et al (2007) Isolation and characterization of biofilm formation-defective mutants of *Staphylococcus aureus*. Infect Immun 75:1079–1088

Valle J, Toledo-Arana A, Berasain C et al (2003) SarA and not sigmaB is essential for biofilm development by *Staphylococcus aureus*. Mol Microbiol 48:1075–1087

Watnick PI, Kolter R (1999) Steps in the development of a *Vibrio cholerae* El Tor biofilm. Mol Microbiol 34:586–595

Willcock, L, Gilbert P, Holah J et al (2000) A new technique for the performance evaluation of clean-in-place disinfection of biofilms. J Ind Microbiol Biotechnol 25:235–241

Werner E, Roe F, Bugnicourt A et al (2004) Stratified growth in *Pseudomonas aeruginosa* biofilms. Appl Environ Microbiol 70:6188–6196

Yarwood JM, Bartels DJ. Volper EM et al (2004) Quorum sensing in *Staphylococcus aureus* biofilms. J Bacteriol 186:1838–1850

Zelver N, Hamilton M, Goeres D et al (2001) Development of a standardized antibiofilm test. Methods Enzymol 337:363–376

Zelver N, Hamilton M, Pitts B et al (1999) Measuring antimicrobial effects on biofilm bacteria: from laboratory to field. Methods Enzymol 310:608–628

Chapter 16
In Vivo Models of Biofilm Infection

Kendra P. Rumbaugh and Nancy L. Carty

16.1 Introduction – Diversity of In Vivo Biofilm Models

The in vivo biofilm models that have been described use a variety of animals ranging from goats to chinchillas, and mimic diseases including dental caries, endocarditis, pneumonia, keratitis, and otitis media. Representatives of these in vivo models are listed in Table 16.1. Although not exhaustive, this list provides a survey of the animals and techniques used to examine biofilms in vivo, and the microorganisms that have been studied. Argumentatively the main advantage of using animal models to study medical biofilms is the inclusion of the host milieu, including the components of the host defenses and complex bodily fluids, which are difficult, if not impossible to replicate in vitro. This is especially important when studying the effects of therapeutics on microorganisms in a biofilm, as issues of drug toxicity, bioavailability, clearance, and absorbance are of high concern.

In general, one of two approaches has been used concerning the choice of microorganisms to be studied in vivo. The most commonly used approach is to infect the animal with an organism of choice, which usually reflects a typical human cause of infection. For example, a major cause of otitis media in humans is *Haemophilus influenzae*. Thus a *H. influenzae* strain of choice is inoculated into the middle ear of the experimental animal to initiate an infection. Biofilms are then evaluated at different points during the course of the infection. One major advantage of this approach is that mutant derivatives, that lack putative virulence factors for example, can be used. This approach has been instrumental in determining the contributions of different gene products to biofilm formation (Swords et al. 2004, Nallapareddy et al. 2006). However, the downfall is that it can be difficult to establish bacterial infections in non-human vertebrates, which have the same general sequelae as human infections. For example, the opportunistic bacterium *Pseudomonas aeruginosa* causes debilitating and often fatal chronic infections in humans with the genetic disease

K.P. Rumbaugh (✉)
Department of Surgery, Texas Tech University Health Sciences Center, 3601 4th Street, Lubbock, TX 79430, USA
e-mail: kendra.rumbaugh@ttuhsc.edu

T. Bjarnsholt et al. (eds.), *Biofilm Infections*, DOI 10.1007/978-1-4419-6084-9_16,

Table 16.1 Representative in vivo models of biofilm disease

Clinical indication	Animals used	Organism(s) examined	Infection method	Methods used to evaluate biofilm	References
Periodontal infections	Humans	Unknown plaque flora	Dentin discs worn in oral cavities of volunteers	CSLM (FDA/EB)	Zaura-Arite et al. (2001)
	Rats	*S. gordonii, S. mutans, and S. sobrinus*	Bacteria inoculated into oral cavity	CFU determination (excised teeth), caries scoring	Duarte et al. (2006) and Tanzer et al. (2008)
	Humans	*Veillonella* spp. and *Streptococcus* spp.	Enamel chips worn in oral cavities of volunteers	CSLM (DAPI and bacteria-specific fluorescently-labeled antibodies)	Chalmers et al. (2007)
Foreign body infections	Cats	Unidentified	Biliary stent placed in bile duct	CFU determination, SEM (stent)	Leung et al. (2000)
Stents					
	Rats	*P. aeruginosa*	Urethral stents and bacteria placed in SC pouch	CFU determination (stent)	Minardi et al. (2007)
	Rats	*S. aureus and E. faecalis*	Urethral stents and bacteria placed in bladder	CFU determination (urine and stent)	Orlando et al. (2008)
Catheters UTI Models	Rabbits	*E. coli, P. aeruginosa, Enterococcus, Staphylococcus and Streptococcus spp., and P. vulgaris.*	Sterile catheters inserted into urethra and bladder	Bacterial identification from urine and catheter swab, SEM (catheter)	Cho et al. (2001)
	Rats	*P. aeruginosa*	Polyethylene tube placed into bladder	CFU determination (kidneys, bladders and tubes), LM (H&E, kidneys and bladders), SEM (tubes)	Kurosaka et al. (2001)
	Mice	*P. aeruginosa* and *P. mirabilis*	Biofilm-coated catheters or bacterial suspension with catheters instilled into bladder	CFU determination (urine), BLI (whole animal), SEM (catheter)	Kadurugamuwa et al. (2005)
Other Catheter Models	Mice	*S. aureus and P. aeruginosa*	Infected catheter inserted SC in flank	CFU determination (catheters), BLI (catheters and whole animal)	Kadurugamuwa et al. (2003)

Table 16.1 (continued)

Clinical indication	Animals used	Organism(s) examined	Infection method	Methods used to evaluate biofilm	References
	Rats	*C. albicans*	Central-venous catheters inoculated with fungi inserted into jugular vein	CFU determination (catheter, blood, kidneys), CSLM (FUN-1 and ConA) and SEM (catheters)	Andes et al. (2004)
Osteomyelitis models	Rats	*S. aureus*	Biofilm-coated metallic implant inserted into tibia	X-rays, CFU determination and LM (Gram's stain), ATP-bioluminescence (bones)	Monzon et al. (2001)
	Sheep	*Staphylococcus spp.*	Metal pins inserted into tibia	CFU determination, SEM (pins)	Smith et al. (2006)
	Goats	*S. epidermidis*	Bacteria inoculated with metal pins in tibia	X-ray (bones), CSLM (*Bac*Light, pins)	van der Borden et al. (2007)
	Mice	*S. aureus*	Bacteria-coated pins inserted into tibia	X-ray, micro-computed tomography and LM (H&E and Gram's stain) (bones), BLI (whole animals)	Li et al. (2008)
Other foreign bodies	Rats	*S. aureus*	Silicone implant placed in SC pocket with bacteria	CFU determination, LM (Gomori's trichrome stain, hematoxylin, PAS, silver), and SEM (implants and surrounding tissues)	Nablo et al. (2005)
	Dogs	Unknown rods and fusiform bacteria	Sutures were placed within the mandibular keratinized gingiva	LM (toluidine-blue stain)	Leknes et al. (2005)
	Rats	*S. aureus*	Teflon tissue cages containing coverslips inserted SC then bacteria injected into tissue cage fluid	CFU determination (coverslips and cage fluid)	Murillo et al. (2006)

Table 16.1 (continued)

Clinical indication	Animals used	Organism(s) examined	Infection method	Methods used to evaluate biofilm	References
	Rats	*S. epidermidis*	Bacteria-coated polypropylene mesh graft inserted near abdominal muscle	CFU determination, LM (H&E) and SEM (grafts)	Agalar et al. (2006) and Saygun et al. (2006)
	Mice	*P. aeruginosa*	Infected silicone implant inserted i.p.	CFU determination and CSLM (GFP) (implants)	Christensen et al. (2007)
Other chronic infections					
Lung infections	Mice	*P. aeruginosa*	Bacteria coated intubation tube inserted in brochus	CFU determination (lungs) and SEM (tube)	Yanagihara et al. (2000)
	Rats	*H. influenzae*	Bacteria infected intratracheally suspended in hog gastric mucin	CFU determination (lungs)	Swords et al. (2004)
	Mice	*P. aeruginosa*	Bacteria infected intratracheally	CFU determination, LM (H&E combined with Alcian blue-periodic acid-Schiff stain) (lungs)	Hoffmann et al. (2005)
	Mice	*P. aeruginosa*	Bacteria immobilized in seaweed alginate beads and instilled into lungs	CFU determination, LM (H&E) (lungs)	Pedersen et al. (1990)
	Mice	*Mycobacterium abscessus*	Bacteria suspended in PBS introduced i.n.	CFU determination (lungs and spleens)	Howard et al. (2006)
Endocarditis	Rats	*S. aureus, E. faecalis and S. gordonii*	Bacteria given i.v. to rats with experimental aortic vegetations	Determination of CFU (organs), BLI (whole animal for *S. aureus*)	Xiong et al. (2005), Bizzini et al. (2006), and Nallapareddy et al. (2006)
Keratitis and Endophthalmitis	Mice	*P. aeruginosa*	Bacteria applied to damaged cornea	Determination of CFU and TEM (corneas)	Fleiszig et al. (1994)

Table 16.1 (continued)

Clinical indication	Animals used	Organism(s) examined	Infection method	Methods used to evaluate biofilm	References
	Mice	*S. aureus*	Bacteria injected into vitreous Cavity	TEM (vitreous cavities)	Leid et al. (2002)
Otitis media	Guinea pigs	*S. aureus*	Bacteria injected into middle ear in conjunction with PE tube placement	FM (*Bac*Light), SEM (tubes)	Saidi et al. (1999)
	Chinchillas	*H. influenzae*	Bacteria injected into middle ear	CSLM (*Bac*Light) and SEM (middle-ear mucosa)	Ehrlich et al. (2002)
	Gerbils	*H. influenzae*	Bacteria injected percutaneously into middle ear	CFU determination (middle ear fluid aspirates)	Swords et al. (2004)
	Monkeys	*P. aeruginosa*	Bacteria injected into middle ear	SEM (middle ear mucosa)	Dohar et al. (2005)
Rhinosinusitis	Sheep	*S. aureus*	Bacteria instilled into frontal sinuses	CSLM, SEM, TEM (frontal mucosa)	Ha et al. (2007)
	Rabbits	*P. aeruginosa*	Bacteria instilled into sinuses	CFU determination (nasal fluid), SEM (sinus mucosa)	Chiu et al. (2007)
Wound infections	Mice	*P. aeruginosa*	Bacteria injected into burn wound	CFU determination, CSLM (GFP), SEM, TEM, and FM (alginate antibody and FISH) (hypodermis)	Schaber et al. (2007)
	Pigs	*S. aureus*	Bacteria applied topically to surgical wound	CFU determination, SEM, LM (H&E and gram stain) and FM (calcofluor white and ethidium bromide) (wound beds)	Davis et al. (2008)

cystic fibrosis. However, developing a suitable animal model that displays the same disease sequelae has been extremely challenging (Kukavica-Ibrulj et al. 2008).

A second approach to creating a biofilm infection in vivo is to allow the animal's normal microflora to initiate the infection (Cho et al. 2001, Leknes et al. 2005). Although this can certainly be more reflective of the natural infection process, there can be major differences in the composition of the microflora between some animals and humans (Marsh 1995). Therefore, as with studying any infectious disease in animals, great consideration must be given when choosing the bacterial and animal species and experimental models to use. While making generalizations about human diseases from data collected in animal models can be more realistic than using in vitro methods, care must be taken to ensure the model is as physiologically relevant as possible.

16.2 In Vivo Biofilm Models

16.2.1 Periodontal Biofilm Models

Dentists were among the first clinicians to embrace the concept of biofilms as a *root* cause of dental caries (Nishioka et al. 1988). This is extremely appropriate considering that some of the very first images of microorganisms were illustrated in the journals and notes of Antonie van Leeuwenhoek from the plaque scraped off his own teeth (Leewenhoeck 1684). Although oral biofilms are an intensive area of study, there have been few animal models developed specifically for studying oral biofilms (Marsh 1995). Instead, many techniques use humans as the model. In fact, plaque is the only medical biofilm studied that involves purposefully placing a foreign body into a human with the intention of creating a biofilm. Typically, small discs or chips of various surface composition are fixed into the oral cavity of human volunteers, who then wear the appliance for a period of time (Zaura-Arite et al. 2001, Chalmers et al. 2007). Once the plaque-covered appliance is removed the biofilm can be studied, typically using imaging methods such as confocal laser scanning microscopy (CLSM), as described in the in vivo biofilm detection methods section below.

The human plaque model has significant advantages over other in vitro or in vivo biofilm models. Firstly, instead of a homogenous biofilm infection, initiated by an organism of the investigator's choosing, the biofilm is formed by the actual human oral microflora under physiologically relevant conditions. It is easy to perform and does not pose significant risk to the subject. This model is versatile and can be applied to several areas of biofilm research, including investigating the dynamics of multi-species biofilms (Chalmers et al. 2007) and evaluating the efficacy of anti-biofilm coatings and therapeutics (Zaura-Arite et al. 2001).

16.2.2 Foreign-Body Models of Biofilm Infection

The majority of in vivo biofilm studies have been conducted by placing some type of foreign body into an animal and initiating an infection. As with the disks and chips used in the periodontal models, the foreign-body of choice can be removed at some point during the infection and analyzed for the presence of biofilm. The reasons foreign-body models are so prevalent in biofilm research are twofold. First, biofilm infections involving foreign bodies affect over 7 million patients a year, resulting in over 20 billion dollars in healthcare costs and 100 thousand deaths, just in the United States (Pittet et al. 1994, Darouiche 2001, 2004 Tambyah et al. 2002, Deysine 2004, Blot et al. 2005, Thomas JG 2006, Klevens et al. 2007). Secondly, inserting a foreign body into an animal provides an artificial surface on which a biofilm can form, and it can be easily removed and imaged. Several clinical indications fall under the category of foreign-body infections as listed in Table 16.1; however, the three major indications which are modeled in animals are urinary tract infections (UTI), systemic infections initiated by indwelling catheters, and osteomyelitis (infection of the bone) caused by orthopedic devices.

16.2.2.1 Urinary Tract Infection Models

Catheter-associated UTI affect more than 6 million people annually, cost the US healthcare system over 3.5 billion dollars, and kill over 13,000 people (Tambyah et al. 2002, von Eiff et al. 2005, Klevens et al. 2007). Increasing the urgency of the situation is the recent "Pay for Performance" plan enacted by the US Centers for Medicare and Medicaid Services, which will cease paying hospitals for some of the care made necessary by "preventable complications", which include some catheter-associated infections (Rosenthal 2007). Therefore there is significant need to study this preventable biofilm-related disease. Several different animal species are used to study UTI including cats, rabbits, and rodents (Table 16.1). Typically, a stent or catheter is placed into the urethra or bladder of the animal, either by making a supra-pubic surgical incision through the abdomen or by insertion through the urethra. Different methods are used to create the infection, including pre-coating catheters with established biofilms (Kadurugamuwa et al. 2005), injecting the organisms of interest into the body cavity along with the foreign body (Orlando et al. 2008), or instilling a sterile catheter and allowing the animals normal flora to create the biofilm (Cho et al. 2001). A wide range of organisms have been studied using in vivo UTI biofilm models, including *Proteus spp.*, *Streptococcus spp.*, *Staphylococcus Spp.*, *Enterococcus spp.*, *P. aeruginosa*, and *Escherichia coli* (Table 16.1).

At different stages of the infection, the stent or catheter can be removed from the animal and analyzed for the presence of biofilm using microscopy (see in vivo biofilm detection methods below). In addition to imaging the retrieved foreign body, many investigators also enumerate the bacteria from the infected bladder and kidneys by harvesting tissue, creating a homogenate, and plating on media to determine colony forming units (CFU). CFU from the infected foreign body can also be determined by removing the bacteria from the artificial surface via sonication or other methods of dispersion. This method is often used to quantify and compare the bacteria adhered to catheters made of different materials or coatings (Cho et al. 2001). The same basic approach is used to study biofilm-infected central-venous catheters, except in this case the catheters are usually inserted into the jugular vein of the animal (Andes et al. 2004).

16.2.2.2 Other Foreign Body Models

Several rodent models have been described which study bacterial biofilms involved in pneumonia (Table 16.1). Many models attempt to mimic ventilator-associated pneumonia associated with biofilm laden intubation tubes (Yanagihara et al. 2000). Generally these models are similar to catheter infection models, except intubation tubes are inserted into the trachea or even bronchus of the animal for some period of time. As with the catheter models, the tubes can be removed and imaged for the presence of biofilms (Yanagihara et al. 2000). Other models initiate the infection by introducing bacteria into the lungs by way of the nasal cavity or trachea (Swords et al. 2004, Bjarnsholt et al. 2005, Howard et al. 2006, Pedersen et al. 1990).

Other materials, including silicone (Nablo et al. 2005, Christensen et al. 2007), teflon cages containing coverslips (Murillo et al. 2006), and grafts (Agalar et al. 2006, Saygun et al. 2006) have been implanted into animals to study in vivo biofilm dynamics. These foreign-bodies are typically inserted subcutaneously (SC) or into the peritoneal cavity of rodents. As with the above foreign-body models, the implants and/or tissue surrounding them can be harvested during the infection and analyzed for bacterial growth and biofilm formation. These models are advantageous because a wide array of materials and/or bacteria can be implanted SC or intraperitoneally (IP) with limited surgical knowledge. Implantation into these cavities can also pose less risk to the animal, and the investigator gains the benefit of evaluating biofilms in an in vivo setting.

16.2.2.3 Osteomyelitis Models

Osteomyelitis is another prolific area of in vivo biofilm research, and reasonably so considering that this clinical indication, along with biofilm-infected orthopedic devices, affects over 300,000 patients per year, costing over 2.5 billion in US healthcare dollars and causing over 17,000 deaths (Tice et al. 2003, Darouiche 2004, Partanen et al. 2006, Pollard et al. 2006). Several large-animal species, such as sheep and goats, are commonly used to study biofilms in osteomyelitis, but rodent models are also used (Table 16.1). Regardless of the species used, most infections are initiated by inserting a metal instrument, usually an orthopedic pin, into the animal's tibia. As with the other foreign-body models, bacteria can also be inoculated along with the implant (van der Borden et al. 2007), or the foreign-body is pre-coated with biofilm prior to implantation (Monzon et al. 2001). The majority of osteomyelitis-associated in vivo biofilm studies have been conducted using *Staphylococcus aureus* (Table 16.1). To evaluate the degree of inflammation accompanying the infection,

X-rays of the tibia are often performed during the infection course (Monzon et al. 2001, van der Borden et al. 2007, Li et al. 2008), and the degree of biofilm formed on the implants are typically assessed by various microscopic methods on the harvested implants (Table 16.1).

16.2.3 In Vivo Models of Biofilms Adhered to Host Mucosa or Soft-Tissue

As frequently as biofilms form on artificial surfaces within the host, they also develop on natural tissue. Clinical indications for these biofilm-related infections include lung infections, endocarditis, keratitis, otitis media, sinusitis, and wound infections. Typically, normal host tissue is not susceptible to biofilm infection unless it has sustained damage by disease or trauma. Once infected though, biofilms on soft tissue can be just as debilitating as those on implants. However, unlike foreign-body biofilms where the patient is treated by removing the infected implant, more severe resolutions, such as the amputation of limbs is often necessary with biofilm-infected host tissue. The advantages of using in vivo models of biofilm-associated soft-tissue infections are that there are a wide range of indications and procedures to choose from, which can be performed on a variety of animal species. However, advanced surgical skills and mechanical ventilation are required for many of the models, and the detection of biofilms on host tissue can present unique challenges.

Although excellent animal models exist in which to study these infections, standard biofilm detection techniques can be ambiguous. Imaging bacterial biofilms on artificial surfaces is generally straightforward because any organic material detected is assumed to be bacterial-derived. However, in vivo biofilms may be embedded within host tissue forming a fused complex surrounded by both bacterial-derived and host-derived matrix material, making interpretation more difficult. Issues associated with detecting and imaging bacterial biofilms among host tissue are discussed in the in vivo biofilm detection methods section.

16.2.3.1 Lung Infection Models

As normal healthy animals usually clear bacteria from the lungs relatively rapidly, developing reliable methods that result in a chronic biofilm-associated bacterial lung infection has been a challenge. As mentioned earlier, this problem has been notoriously difficult for *P. aeruginosa* researchers investigating cystic fibrosis-associated pneumonia (Kukavica-Ibrulj et al. 2008). Different methods have been used to encourage bacteria to persist in the lung and prevent mechanical clearing by the host. Typically bacteria are embedded in a matrix, such as mucin or alginate, prior to instillation in the lung (Pedersen et al. 1990, Swords et al. 2004, Bjarnsholt et al. 2005). Another approach, described by Hoffmann et al., was to use a *P. aeruginosa* clinical isolate that overproduced the exopolysaccharide alginate (Hoffmann et al. 2005). While these methods help prevent the bacteria from being mechanically cleared from the lung, and somewhat mimic a chronic pneumonia, the infections they cause are typically limited in time to 2 weeks, after which time the infection becomes fatal or is cleared (Moser et al. 2009). To try to circumvent this limitation Moser et al. serially infected mice with isotypic *P. aeruginosa* strains isolated from one CF patient during different periods of a 20-year infection (Moser et al. 2009). Results of studies using this infection strategy indicate that the immune response and the success of antibiotic treatments varies significantly depending on the infection stage at which the *P. aeruginosa* strain being used was obtained from the human patient (Moser et al. 2009, van Gennip et al. 2009).

Infections in pneumonia models are typically graded based on bacterial load, by determining CFU from lung homogenates, or by examining the degree of associated lung inflammation (Hoffmann et al. 2005). Visualization of bacterial biofilms present in the lung has been performed with light microscopy (LM) on fixed tissue following different staining protocols such as hematoxylin and eosin (H&E) combined with alcian blue-periodic acid-Schiff (PAS) stain for exopolysaccharides (Hoffmann et al. 2005).

16.2.3.2 Endocarditis Models

Endocarditis, or inflammation of the endocardium, typically follows an infection of the heart valves during which a vegetation, or a mass of platelets, fibrin, microorganisms, and/or immune cells has formed (Hall-Stoodley et al. 2004). In general, infectious endocarditis is preceded by congenital valve defect or by damage to the heart valves initiated by bypass surgery, stent placement, or autoimmune injury following rheumatic fever. Several model organisms have been studied as causes of infectious endocarditis, including *Staphylococcus, Streptococcus*, and *Entercoccus spp.* (Table 16.1). The rat infectious endocarditis model is commonly used (Heraief et al. 1982), whereby a sterile catheter is placed near the aortic valve to induce sterile vegetations. The damaged valve is then infected by an inoculation of bacteria intravenously through the tail vein (Xiong et al. 2005). Aortic vegetations can be harvested at different time points to determine the extent of infection by determining CFU, or other analyses for biofilm formation.

16.2.3.3 Keratitis Models

Biofilm-implicated ocular infections include contact-lens-associated keratitis and postoperative endophthalmitis (Zegans et al. 2002), both of which have been modeled in mice (Fleiszig et al. 1994, Leid et al. 2002). In the experimental keratitis model, full-thickness epithelial abrasions were produced on the corneas of mice and inoculated with *P. aeruginosa* (Fleiszig et al. 1994). After sacrificing the mice and harvesting the corneas, CFU determination to examine bacterial load and transmission electron microscopy (TEM) to visualize bacteria within the infected tissue were performed (Fleiszig et al. 1994).

Endophthalmitis is a bacterial infection of the posterior of the eye that affects over 2,000 people a year and can result following introduction of bacteria from ocular surgery or trauma (Leid et al. 2002). A murine model of endophthalmitis was used to examine *S. aureus*ocular biofilms (Leid et al. 2002). In this study *S. aureus* was injected into the vitreous cavity and after 72 h *S. aureus* was visualized in a biofilm along the lens capsule by TEM (Leid et al. 2002).

16.2.3.4 In Vivo Models of Otitis Media and Sinusitis

Otorhinolaryngological-associated biofilm infections, or those pertaining to the ears, larynx, and upper respiratory tract are extremely prevalent in humans. Otitis media, chronic tonsillitis, and sinusitis combined affect nearly half a million people per year in US (Bondy et al. 2000, Murphy et al. 2002, Chole and Faddis 2003, Sanclement et al. 2005, Hall-Stoodley et al. 2006, Coticchia et al. 2007, Psaltis et al. 2007). With the exception of tonsillitis, where animal models appear to be scarce if not totally non-existent, there are a wide array of animal models available to study these infections. The animals currently in use for studying otitis media range from rodents to chinchillas and primates (Table 16.1). Although the animal species are diverse, the infection procedures are quite similar. Typically the bacterial species of choice are injected into the middle ear, with or without concurrent placement of pressure equalization (PE) tubes. The species of choice usually represents common causes of human ear infection including *P. aeruginosa*, *S. aureus*, and *H. influenzae* (Table 16.1). During different points in the infection, middle ear fluid and/or mucosa of the middle ear itself can be harvested and analyzed. The bacterial load involved in the infection can be monitored by determining CFU from aspirates of the middle ear, as demonstrated by Swords et al. (Swords et al. 2004). Biofilms formed along the middle-ear mucosa or on PE tubes in vivo have been imaged by fluorescence microscopy (FM), scanning electron microscopy (SEM), and CLSM [see Saidi et al. (1999). Ehrlich et al. (2002) and Dohar et al. (2005) for examples]. In vivo biofilm models for studying sinusitis have also been described for several animal species including sheep and rabbits (Table 16.1). These models resemble those for otitis media except that the bacteria are instilled into the sinuses rather than the middle ear. The collection processes, detection strategies, and even organisms studied are very similar.

16.2.3.5 Wound Models

The diseased or damaged dermis is also highly susceptible to biofilm infections. Dermal biofilms in diabetic foot and decubitus ulcers, surgical site infections, and burns affect over a million people a year in US (Boyko et al. 1996, Beckrich and Aronovitch 1999, Ramsey et al. 1999, Perencevich et al. 2003, Medina et al. 2005, Church et al. 2006, Armour et al. 2007, Klevens et al. 2007, Schaber et al. 2007, James et al. 2008). Importantly, this area of medical biofilm research is on the rise as both wound infections due to military conflicts, and chronic wounds in the growing diabetic population are increasing (DeLeon et al. 2009). Despite the prevalence of this emergent problem, few animal models have been described to study wound biofilms. Davies et al. utilized a pig surgical wound model to investigate biofilms formed by *S. aureus* in chronic wounds (Davis et al. 2008). Partial thickness wounds measuring 0.3 mm deep$\times$10$\times$7 mm were created on the paravertebral area of pigs using an electrokeratome. *S. aureus* was inoculated into the wounds, and the wounds were subsequently covered by a polyurethane dressing. After sacrifice, biopsies of the wounds were assayed for bacterial colonization by determining CFU and for biofilm formation by SEM, LM (following H&E and Gram's staining), and FM, using calcofluor white and ethidium bromide (Davis et al. 2008).

Schaber et al., used a murine burn wound model to demonstrate that acute wounds were also subject to biofilm infection (Schaber et al. 2007). In this study, mice were administered full-thickness scald burns, which were inoculated with *P. aeruginosa*. Wounds were evaluated for bacterial infection by determining CFU, and biopsies of the infected hypodermis were imaged to observe biofilm formation. Biofilm structures were imaged within the hypodermis utilizing SEM, TEM, LM, following H&E staining, FM, following immunohistochemistry with an anti-alginate antibody, and CLSM of a green-fluorescent protein (GFP)-expressing *P. aeruginosa* strain. Interestingly, biofilms were observed as early as 8 h post-infection within the burned tissue (Schaber et al. 2007).

16.3 In Vivo Biofilm Detection Methods

In order to fulfill Koch's postulates, and identify causative agents of infection, one must obtain a pure culture of a microbe and then replicate the infection in a suitable experimental host. Accordingly, hundreds of animal models of disease have been established and enabled thousands of researchers to investigate disease processes. Similarly, Parsek and Singh established four criteria to determine the role of a microbial biofilm in an infection. The first two criteria are that the bacteria are attached or associated with a surface and that there are viable cells within an extracellular polymeric matrix (Parsek and Singh 2003). However, whether by inserting an artificial surface into a subject and allowing the host microflora to form a biofilm, as in the case of oral biofilms (Zaura-Arite et al. 2001), or by inoculating an animal model with a known bacterial species to analyze biofilm formation (Schaber et al. 2007), examination of the biofilm itself can prove challenging. As with choosing an appropriate animal model, the method(s) used to detect in vivo biofilms must be thoroughly considered to ensure they are appropriate for the microorganism(s), model, and type of infection being examined. Several of the methods that have been utilized to detect in vivo biofilms are listed in Table 16.2 and described below. They are generally categorized as either "microscopic" methods or non-microscopic methods.

Table 16.2 In vivo biofilm detection methods

Detection method	Target	References
Microscopy		
Stains used with light microscopy		
Hematoxylin & Eosin	Nucleic acids and cytoplasm	Kurosaka et al. (2001), Davis et al. (2008), and Li et al. (2008)
Gram's stain	Bacteria	Monzon et al. (2001) and Davis et al. (2008)
Toluidine blue	connective tissue	Leknes et al. (2005)
Gomori's trichrome	cytoplasm & collagen	Nablo et al. (2005)
Silver	Bacteria	Nablo et al. (2005)

Table 16.2 (continued)

Detection method	Target	References
SEM	Bacteria and bacterial/host cell interfaces	Saidi et al. (1999), Leung et al. (2000), Yanagihara et al. (2000), Cho et al. (2001), Kurosaka et al. (2001), Ehrlich et al. (2002), Andes et al. (2004), Dohar et al. (2005), Kadurugamuwa et al. (2005), Nablo et al. (2005), Saygun et al. (2006), Smith et al. (2006), Chiu et al. (2007), Schaber et al. (2007), and Davis et al. (2008)
TEM	Bacteria and bacterial/host cell interfaces	Fleiszig et al. (1994), Leid et al. (2002), and Schaber et al. (2007)
Ruthenium red	Polysaccharides	Schaber et al. (2007)
FM/CLSM detection agents		
Cells		
BacLight^{TM}	Live and dead cells	Saidi et al. (1999), Ehrlich et al. (2002), and van der Borden et al. (2007)
Propidium iodide	Nucleic acid, dead cells	
Syto9^{TM}	Nucleic acid, all cells	
DAPI	Nucleic acid	Chalmers et al. (2007)
Ethidium bromide	Nucleic acid, dead cells	Zaura-Arite et al. (2001) and Davis et al. (2008)
Fluorescein diacetate	Live cells esterases	Zaura-Arite et al. (2001)
FUN® 1	Cells (live: red-orange, Dead: green)	Andes et al. (2004)
FISH		
P. aeruginosa 16S probe	Bacteria	Schaber et al. (2007)
GFP	Bacteria	Christensen et al. (2007) and Schaber et al. (2007)
EPS		
Wheat germ agglutinin	*N*-acetylglucosamine	
Calcofluor white	*N*-acetylglucosamine	Davis et al. (2008)
Concanavalin A	Glucose and mannose residues of polysaccharides	Andes et al. (2004)
Antibodies	*P. aeruginosa* alginate, and *S. gordonii*, *S. oralis*, and *Veillonella* sp.	Chalmers et al. (2007) and Schaber et al. (2007)
Non-microscopic		
X-ray	Bone infection/inflammation	Monzon et al. (2001) and van der Borden et al. (2007)
ATP-bioluminescence	Bacteria	Monzon et al. (2001)
Whole body imaging/BLI	Tagged microbes	Kadurugamuwa et al. (2003), Kadurugamuwa et al. (2005), Xiong et al. (2005), and Li et al. (2008)
Antibody detection	Biofilm-specific epitopes	Kofonow (2007) and Brady et al. (2008)

16.3.1 Microscopic Methods for Imaging Biofilms Formed In Vivo

16.3.1.1 Light Microscopy of Stained Specimens

Microscopy allows one to visualize the aggregation of microorganisms either on an implanted surface (Yanagihara et al. 2000) or on the host tissue itself (Schaber et al. 2007). The use of LM in conjunction with different stains is one of the simplest methods in which to view the bacteria involved in an infection. Stains such as H&E have been used to visualize bacteria in animal models (Fig. 16.1a, Table 16.1). Typically, stains are applied to tissue samples that have been treated with a fixative, such as formalin, embedded in paraffin, thin-sectioned, and mounted on slides. The disadvantage of H&E staining is that hematoxylin stains all nucleic acids, potentially making it difficult to distinguish microbial cells from host tissue. To specifically detect bacterial cells, some investigators have used the well-known microbial Gram's stain (Monzon 2001 #879) on tissue sections. In fact Davis et al. (2008) applied both H&E and the Brown and Brenn modified version of the Gram's stain separately to their porcine wound samples to better differentiate the host tissue from the *S. aureus* biofilm cells (Davis et al. 2008). In addition, the toluidine blue and Gomori's trichrome staining techniques utilized by Leknes et al. (2005) and Nablo et al. (2005), respectively, helped distinguish host tissue from biofilm cells. Nablo et al. (2005) also used periodic acid-Schiff and sliver to visualize aggregates of *S. aureus* in situ.

Fig. 16.1 Demonstration of biofilm detection techniques on tissue from a *P. aeruginosa*-infected mouse wound. *P. aeruginosa* forms biofilms around blood vessels in the wounds of thermally injured mice (Schaber et al. 2007), and these vascular biofilms have been detected by a number of different techniques. Some *P. aeruginosa*-infected mouse wound tissue was fixed in formalin, embedded in paraffin, thin-sectioned, then stained with H&E (**a**), or de-parafinized and analyzed by FISH with a *P. aeruginosa*-specific probe (**b**). In both images (**a** and **b**) *P. aeruginosa* cells are seen surrounding the cross section of the blood vessel, and stain purple with H&E (**a**) and red with FISH (**b**). Red blood cells are visible, and seen as pink, within the vessel lumens in both images that were captured by LM (**a**) and FM (**b**) using a 40X objective. Vascular biofilms can also be visualized by ruthenium red staining and TEM (**c**), in which a cross section of a vein, surrounded by *P. aeruginosa* cells, was imaged at 5,500X. In this image large red blood cells can be seen inside the lumen (*upper right-hand corner*) and *P. aeruginosa* cells are layered outside the vessel wall (*lower left-hand corner*). SEM was used with these same tissues to visualize aggregates of *P. aeruginosa* adhered to adipocytes, which are also coated with biofilm matrix or "bacterial flocs" (**d**). The wound tissue from mice infected with a GFP-expressing strain of *P. aeruginosa* was used to detect biofilms by CLSM (**e**). This image was reconstructed from image stacks using MetaMorph 6.1 (Universal Imaging Corporation, PA) and shows *P. aeruginosa* surrounding a longitudinal section of a blood vessel. Inset is a bright field image of the same tissue section showing the position of blood vessels within the tissue. FM was accomplished using a primary antibody raised against *P. aeruginosa* alignate, and a secondary Alexa fluor 488 antibody (Molecular Probes) (**f**). Fluorescence immunohistochemistry for alginate was performed on formalin-fixed de-parafinized tissues. In all panels arrows point to individual *P. aeruginosa* cells. Images previously published in Schaber et al. (2007) and used with permission from K.P. Rumbaugh and the American Society for Microbiology

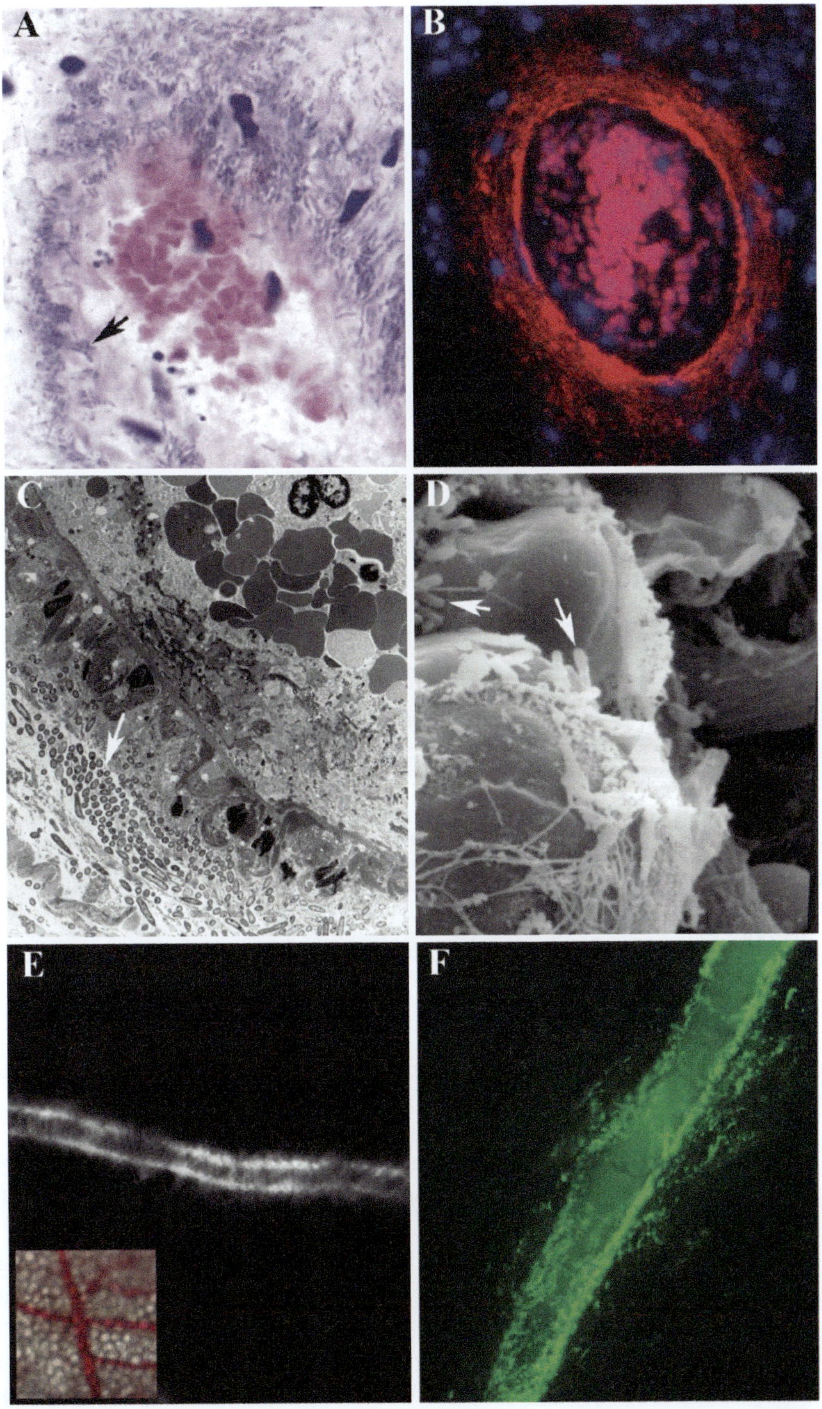

Fig. 16.1 (continued)

16.3.1.2 High Resolution of Biofilm/Host Interactions with Electron Microscopy

For higher magnification and better visualization of biofilm cells, researchers have used both TEM and SEM (Table 16.1). TEM has been used in conjunction with ruthenium red, which binds to polysaccharides, to observe thin sections of bacterial aggregates surrounded by matrix in relation to host tissue (Fig. 16.1c). Schaber et al. (2007) also used TEM in conjugation with immunogold labeling of alignate to visualize the position of EPS between individual *P. aeruginosa* cells (Schaber et al. 2007). SEM is used more frequently (see Table 16.1) than TEM and provides highly detailed images of the surface of the aggregated microbial cells as well as the host tissue (Fig. 16.1d). One disadvantage of SEM is that during preparation, samples are fixed in a series of steps which dehydrates the specimens. Due to the fact that biofilms are estimated to be 85% extrapolymeric substance (EPS) which is composed of water, polysaccharides, DNA, and proteins, this process results in the loss of the EPS structure (Donlan et al. 2002). This dehydration process causes the EPS to appear stringy, resembling cobwebs (Fig. 16.1d), in SEM micrographs; consequently the native architecture of the biofilm cannot be visualized.

16.3.1.3 Using Fluorescent Signals to Image Bacteria and Hydrated EPS with Epifluorescence and Confocal Laser Scanning Microscopy

CLSM is typically used to examine biofilm structures, including both the EPS and microbial cells. CLSM provides researchers the ability to visualize thick biofilm sections and observe the location of the microbial populations within the hydrated EPS, thus providing a more realistic image of the biofilm in situ. Furthermore, the three-dimensional architecture of the biofilm, including thickness, volume and surface area can be analyzed using CLSM in combination with analysis software such as MetaMorph (Molecular Devices, Downington, PA) and COMSTAT (Heydorn et al. 2000, Schaber et al. 2007).

Several fluorescent stains can be used in conjunction with CLSM to help identify the bacterial cells. Ethidium bromide, propidium iodide, and 4′,6-diamidino-2-phenyl indole (DAPI) stain the nucleic acids within the cells. However, ethidium bromide and propidium iodide can only enter cells with damaged membranes; therefore, they will preferentially stain the nucleic acids of dead cells. On the other hand, DAPI can permeate intact membranes, thus staining both live and dead cells. Invitrogen has capitalized on the use of these stains in order to determine the viability of bacterial cells in a biofilm. Their product LIVE/DEAD® *Bac*Light™ utilizes two different fluorescent stains; Syto®9 which stains the nucleic acids of both live and dead cells, and propidium iodide which stains only the dead cells (Invitrogen Corp., Carlsbad, CA). Consequently the live cells fluoresce in the "green: spectrum and the dead cells fluoresce in the "red" spectrum. The visualization of living versus dead cells within the characteristic mushrooms and towers of a biofilm is an important aspect of biofilm research and this product has been used extensively in vitro (see Chapter 16).

Many researchers have also used LIVE/DEAD®*Bac*Light™ to study in vivo biofilms including Ehrlich et al. (2002), who used it to analyze biofilm formation on the middle ear mucosa of *H. influenza*-infected chinchillas (Ehrlich et al. 2002). Zara-Arite et al. (2001) used ethidium bromide and fluorescein diacetate, in combination with CLSM, to examine the viability of dental biofilms treated with chlorhexidine (Zaura-Arite et al. 2001). Fluorescein diacetate is a membrane permeable dye which is converted to fluorescein by active enzymes resulting in live cells appearing green, while the dead cells appear red, due to the ethidium bromide. In addition to identifying the viability of bacterial cells, Invitrogen has developed a product, FUN®1 cell stain, to detect live versus dead fungal cells. FUN®1 permeates all cells; however, only metabolically active cells can convert FUN®1 from green to orange-red fluorescence. FUN®1 was used by Andes et al. (2004) to visualize *Candida albicans* biofilms on central venous catheters (Andes et al. 2004).

16.3.1.4 Additional Fluorescence-Based Detection Methods

Several fluorescent applications have been used to detect EPS in a biofilm. In the recently published paper by Romero et al. (2008) wheat germ agglutinin, which binds to *N*-acetylglucosamine, was used to detect EPS in a sample obtained from an intraamniotic infection (Romero et al. 2008). Both calcofluor white and concanavalin A have also been used to identify EPS components in biofilms. Concanavalin A, which attaches to glucose and mannose residues of polysaccharides, was used by Andes et al. (2004) to visualize the EPS component of the *Candida albicans* biofilm formed on a central venous catheter (Andes et al. 2004). Calcofluor white is often used to stain fungal samples as it binds to chitin, a polysaccharide composed of *N*-acetylglucosamine (Robert et al. 2008). Similar to wheat germ agglutinin, calcofluor white has also been used to detect components of the EPS (Chalmers et al. 2007, Davis et al. 2008).

Rather than staining samples with the fluorescent dyes described above, alternative methods have been used in combination with CLSM to visualize bacteria within biofilms formed in vivo. GFP-expressing *P. aeruginosa* has been used to visualize bacterial cells within a biofilm grown in vivo (Fig 16.1e) (Christensen et al. 2007, Schaber et al. 2007). Immunohistochemistry, with fluorescently tagged antibodies, has also been applied to in vivo biofilm studies. Schaber et al. (2007) used an Alexa Fluor 488-tagged secondary antibody and epifluorescence microscopy to visualize the localization of a monoclonal alginate-antibody in the biofilm matrix surrounding a blood vessel (Schaber et al. 2007). Chalmers et al. (2007) compared the resolution of streptococcal cell-surface antibodies conjugated with fluorophores (Alexa Fluor) to quantum dot-conjugated antibodies. Consequentially, they found that quantum dot-labeled antibodies successfully penetrated the EPS matrix and provided a more effective method to investigate multispecies biofilms (Chalmers et al. 2007).

Lastly, fluorescent in situ hybridization (FISH) has been used effectively to localize bacteria within the biofilms of clinical samples. FISH allows for the direct localization of bacterial DNA or RNA within a sample. DNA, RNA, or peptide nucleic acid (PNA) FISH probes are labeled with a fluorophore, typically Cy3, Cy5, or FITC, which is visualized with epifluorescence microscopy. Romero et al. (2008) used EUB338-Cy3 (Integrated DNA Technologies, Coralville, Iowa, USA), which is considered a universal probe as it binds to the 16S rRNA of the majority of bacterial species, to reveal the presence of bacteria within an amniotic biofilm sample (Romero et al. 2008). Similarly, both EUB338 and other species-specific FISH probes were utilized to examine and identify bacterial components from mucosal biofilm samples of children with chronic otitis media (Hall-Stoodley et al. 2006). In the context of in vivo biofilm models, Schaber et al. (2007) used a DNA FISH probe specific for the 16S of *P. aeruginosa* to demonstrate the perivascular biofilms formed by *P. aeruginosa* in burn wounds (Fig 16.1b). PNA FISH probes have also successfully been used to visualize biofilms in human wound biopsies (Bjarnsholt et al. 2008, Kirketerp-Moller et al. 2008, Malic et al. 2009).

16.3.2 Non-microscopic Methods of Detection

16.3.2.1 Enumeration of Bacteria by Determining Colony Forming Units

To establish the presence of live bacterial cells within a biofilm produced in vivo, the majority of papers listed in Table 16.1 have used standard microbiological methods, in other words, counting CFUs. Biofilm samples have either been scraped or sonicated from tissue or implanted devices, serially diluted and then plated on various types of agar to calculate CFUs. This method is relatively easy and straight forward, but some controversy exists concerning the interpretation of CFU data. It has been proposed that biofilm cells are difficult to culture in vitro (Costerton 2007) and require extensive technical measures, such as treatment with ultrasound to disperse them before plating (Assere et al. 2008). Therefore, care must be taken to insure the CFU obtained is not a gross underestimation of the true number of bacteria present in the biofilm.

Another disadvantage of this technique is that it may neglect the potential anaerobic or other non-culturable bacteria when attempting to ascertain the presence of viable cells within a biofilm produced in an animal model. For example, Dowd et al. (2008) revealed the presence of several anaerobic bacterial species when they identified bacteria from clinical chronic wound samples through molecular analysis, rather than traditional culturing methods (Dowd et al. 2008). A different method used by Monzon et al. (2001) to assess bacterial viability was the measurement of ATP-bioluminescence. In this assay the number of viable bacteria was correlated to the amount of ATP hydrolyzed, resulting in bioluminescence, which was measured in relative light units (Amorena et al. 1999, Monzon et al. 2001).

16.3.2.2 Whole-Animal Imaging

Several studies utilizing in vivo models of osteomyelitis have utilized X-rays to confirm the presence of infection within the bone adjacent to the implanted device (Monzon et al. 2001, van der Borden et al. 2007, Li et al. 2008). In these studies the X-rays were used as a preliminary detection technique which was then followed by determination of CFU and visualization of biofilm by LM and/or CLSM.

Within the last decade, several in vivo imaging systems have been developed with which to study biofilms. These systems detect the presence of bioluminescent bacteria within the whole animal and can be used to monitor the progression of infection without sacrificing the animal. Bacteria such as, *S. aureus*, *P. aeruginosa* and *P. mirabilis*, have been engineered to produce a bioluminescent signal, which can be detected by the imaging systems (Xiong 2005, Li 2008, Kadurugamuwa 2003, 2005). These systems have been used to examine in vivo biofilms in endocarditis, osteomyelitis, catheter, and urinary tract infection animal models (Xiong 2005, Li 2008, Kadurugamuwa 2003, 2005). However, while it is certainly beneficial to monitor the progression of the infection without sacrificing animals, these methods do not provide high resolution imaging of biofilm components, such as EPS.

16.3.2.3 Detection of Biofilms by Biofilm-Specific Antibodies

An exciting new strategy, utilizing serology to detect biofilm-specific antibodies produced by the host, has recently been described (Kofonow 2007, Brady et al. 2008). Kofonow et al. (2007), used an osteomyelitis model to show that rabbits infected with methicillin-resistant *Staphylococcus aureus* biofilms produced biofilm-specific antibodies. This technique may prove to be a good noninvasive, non-culture based diagnostic to reliably detect biofilm infections in humans, with which a simple blood test could determine presence of a biofilm.

16.4 Concluding Remarks

The last century ended with a new emphasis being placed on the way in which we grow and study bacteria involved in infectious disease. Instead of investigating microbial pathogenesis with bacteria grown at 37°C, in nutrient-rich broth, shaking in a flask, medical microbiologists, began to realize the importance of bacteria growing in biofilms to infectious disease. It is clear from the number of in vivo animal models in use (Table 16.1) that one priority for medical microbiologists in this century will be to take the knowledge gleaned about bacterial biofilms from the early in vitro studies and apply it to in vivo models. Molecular diagnostics are now making it possible to identify all the microbes involved in human biofilm disease, rather than just the predominant culturable microbes (Dowd et al. 2008a, b). However, the significance of the microorganisms identified by molecular diagnostics, and the roles they play in infection, remains to be clarified. When this hurdle is passed, this knowledge can be used to design relevant multispecies biofilm infections in animal models. Therefore, future in vivo studies should strive to mimic actual biofilm infections, as closely as possible. This includes using an animal species that displays similar disease sequelae as an infected human; using relevant microbes, attempting to establish the same types of multispecies infections seen clinically; and selecting appropriate detection methods that disrupt the biofilm communities as little as possible. Based on the methods described above and listed in Table 16.2, it is obvious that there is no set standard protocol used to detect biofilms grown in vivo. Techniques ranging from enumerating CFUs, which actually confirms the presence of living bacterial cells as opposed to bacterial biofilms, to CLSM, which can image both living and dead cells as well EPS, have been used to examine in vivo biofilms. Therefore it is up to investigators to establish a "gold standard" for detecting in vivo biofilms, with the realization that new criteria, describing what constitutes a biofilm in vivo, may diverge from those established for biofilms formed in vitro. Most importantly, investigators must take full advantage of the in vivo environment to investigate how the *host* contributes to biofilm development. Does the host cell matrix contribute to the biofilm matrix? Are the host-derived matrix components, such as heparan sulfate, collagen, fibronectin, and elastin, integrated into the bacterial EPS? If so, do they provide camouflage or further shield the bacteria from host defenses? How does the immune status of the host change biofilm development and the progression of disease? How do antimicrobial agents access and affect biofilm cells in vivo? There are several important questions that can be investigated using in vivo models to study biofilm infections, and the data generated will undoubtedly contribute to our overall understanding of bacterial communities and aid in the development of new therapeutics to treat them.

References

Agalar C, Ozdogan M, Agalar F, Saygun O, Aydinuraz K, Akkus A, Ceken S, Akturk S (2006) A rat model of polypropylene graft infection caused by Staphylococcus epidermidis. ANZ J Surg 76:387–391

Amorena B, Gracia E, Monzon M, Leiva J, Oteiza C, Perez M, Alabart JL, Hernandez-Yago J (1999) Antibiotic susceptibility assay for Staphylococcus aureus in biofilms developed in vitro. J Antimicrob Chemother 44:43–55

Andes D, Nett J, Oschel P, Albrecht R, Marchillo K, Pitula A (2004) Development and characterization of an in vivo central venous catheter Candida albicans biofilm model. Infect Immun 72:6023–6031

Bizzini A, Beggah-Moller S, Moreillon P, Entenza JM (2006) Lack of in vitro biofilm formation does not attenuate the virulence of Streptococcus gordonii in experimental endocarditis. FEMS Immunol Med Microbiol 48:419–423

Bjarnsholt T, Jensen PO, Rasmussen B, Christophersen L, Calum H, Hentzer M, Hougen HP, Rygaard J, Moser C, Eberl L, Hoiby N, Givskov M (2005) Garlic blocks quorum sensing and promotes rapid clearing of pulmonary Pseudomonas aeruginosa infections. Microbiology 151:3873–3880

Brady RA, Leid JG, Calhoun JH, Costerton JW, Shirtliff ME (2008) Osteomyelitis and the role of biofilms in chronic infection. FEMS Immunol Med Microbiol 52:13–22

Chalmers NI, Palmer Jr. RJ, Du-Thumm L, Sullivan R, Shi W, Kolenbrander PE (2007) Use of quantum dot luminescent probes to achieve single-cell resolution of human oral bacteria in biofilms. Appl Environ Microbiol 73:630–636

Chiu AG, Antunes MB, Palmer JN, Cohen NA (2007) Evaluation of the in vivo efficacy of topical tobramycin against Pseudomonas sinonasal biofilms. J Antimicrob Chemother 59: 1130–1134

Cho YH, Lee SJ, Lee JY, Kim SW, Kwon IC, Chung SY, Yoon MS (2001) Prophylactic efficacy of a new gentamicin-releasing urethral catheter in short-term catheterized rabbits. BJU Int 87: 104–109

Christensen LD, Moser C, Jensen PO, Rasmussen TB, Christophersen L, Kjelleberg S, Kumar N, Hoiby N, Givskov M, Bjarnsholt T (2007) Impact of Pseudomonas aeruginosa quorum sensing on biofilm persistence in an in vivo intraperitoneal foreign-body infection model. Microbiology 153:2312–2320

Costerton JW (2007) The biofilm primer. Springer, New York

Costerton JW, Geesey GG, Cheng KJ (1978) How bacteria stick. Sci Am 238:86–95

Costerton JW, Stewart PS, Greenberg EP (1999) Bacterial biofilms: a common cause of persistent infections. Science 284:1318–1312

Davies DG, Parsek MR , Pearson JP, Iglewski BH, Costerton JW, Greenberg EP (1998) The involvement of cell-to-cell signals in the development of a bacterial biofilm [see comments]. Science 280:295–298

Davis SC, Ricotti C, Cazzaniga A, Welsh E, Eaglstein WH, Mertz PM (2008) Microscopic and physiologic evidence for biofilm-associated wound colonization in vivo. Wound Repair Regen 16:23–29

DeLeon K, Balldin F, Watters C, Hamood N, Griswold J, Sreedharan S, Rumbaugh K (2009) Gallium maltolate treatment eradicates Pseudomonas aeruginosa infection in thermally-injured mice. Antimicrobial Agents and Chemotherapy 53:1331–1337

Dohar JE, Hebda PA, Veeh R, Awad M, Costerton JW, Hayes J, Ehrlich GD (2005) Mucosal biofilm formation on middle-ear mucosa in a nonhuman primate model of chronic suppurative otitis media. Laryngoscope 115:1469–1472

Donlan RM, Costerton JW (2002) Biofilms: survival mechanisms of clinically relevant microorganisms. Clin Microbiol Rev 15:167–193

Dowd SE, Sun Y, Secor PR, Rhoads DD, Wolcott BM, James GA, Wolcott RD (2008a) Survey of bacterial diversity in chronic wounds using pyrosequencing, DGGE, and full ribosome shotgun sequencing. BMC Microbiol 8:43

Dowd SE, Wolcott RD, Sun Y, McKeehan T, Smith E, Rhoads D (2008b) Polymicrobial nature of chronic diabetic foot ulcer biofilm infections determined using bacterial tag encoded FLX amplicon pyrosequencing (bTEFAP). PLoS ONE 3:e3326

Duarte S, Rosalen PL, Hayacibara MF, Cury JA, Bowen WH, Marquis RE, Rehder VL, Sartoratto A, Ikegaki M, Koo H (2006) The influence of a novel propolis on mutans streptococci biofilms and caries development in rats. Arch Oral Biol 51:15–22

Ehrlich GD, Veeh R, Wang X, Costerton JW, Hayes JD, Hu FZ, Daigle BJ, Ehrlich MD, Post JC (2002) Mucosal biofilm formation on middle-ear mucosa in the chinchilla model of otitis media. JAMA 287:1710–1715

Fleiszig SM, Zaidi TS, Fletcher EL, Preston MJ, Pier GB (1994) Pseudomonas aeruginosa invades corneal epithelial cells during experimental infection. Infect Immun 62:3485–3493

Gristina AG, Costerton JW (1984) Bacterial adherence and the glycocalyx and their role in musculoskeletal infection. Orthop Clin North Am 15:517–535

Gristina AG, Oga M, Webb LX, Hobgood CD (1985) Adherent bacterial colonization in the pathogenesis of osteomyelitis. Science 228:990–993

Ha KR, Psaltis AJ, Tan L, Wormald PJ (2007) A sheep model for the study of biofilms in rhinosinusitis. Am J Rhinol 21:339–345

Hall-Stoodley L, Costerton JW, Stoodley P (2004) Bacterial biofilms: from the natural environment to infectious diseases. Nat Rev Microbiol 2:95–108

Hall-Stoodley L, Hu FZ, Gieseke A, Nistico L, Nguyen D, Hayes J, Forbes M, Greenberg DP, Dice B, Burrows A, Wackym PA, Stoodley P, Post JC, Ehrlich GD, Kerschner JE (2006) Direct detection of bacterial biofilms on the middle-ear mucosa of children with chronic otitis media. JAMA 296:202–211

Heraief E, Glauser MP, Freedman LR (1982) Natural history of aortic valve endocarditis in rats. Infect Immun 37:127–131

Heydorn A, Nielsen AT, Hentzer M, Sternberg C Givskov M, Ersboll BK, Molin S (2000) Quantification of biofilm structures by the novel computer program COMSTAT. Microbiology 146(Pt 10):2395–2407

Hoffmann N, Rasmussen, TB, Jensen PO, Stub C, Hentzer M, Molin S, Ciofu O, Givskov M, Johansen HK, Hoiby N (2005) Novel mouse model of chronic Pseudomonas aeruginosa lung infection mimicking cystic fibrosis. Infect Immun 73:2504–2514

Howard ST, Rhoades E, Recht J, Pang X, Alsup A, Kolter R, Lyons CR, Byrd TF (2006) Spontaneous reversion of Mycobacterium abscessus from a smooth to a rough morphotype is associated with reduced expression of glycopeptidolipid and reacquisition of an invasive phenotype. Microbiology 152:1581–1590

James GA, Swogger E, Wolcott R, Pulcini E, Secor P, Sestrich J, Costerton JW, Stewart PS (2008) Biofilms in chronic wounds. Wound Repair Regen 16:37–44

Kadurugamuwa JL, Modi K, Yu J, Francis KP, Purchio TM, Contag PR (2005) Noninvasive biophotonic imaging for monitoring of catheter-associated urinary tract infections and therapy in mice. Infect Immun 73:3878–3887

Kadurugamuwa JL, Sin L, Albert E, Yu J, Francis K, DeBoer M, Rubin M, Bellinger-Kawahara C, Parr.TR Jr, Contag PR (2003) Direct continuous method for monitoring biofilm infection in a mouse model. Infect Immun 71:882–890

Kofonow JM, Brady RA, Vail TL, Shirtliff JG (2007) S10: 4 A non-culturing technique for diagnosing staphylococal biofilm. American Society for Microbiology Biofilms Conference, 2007, Quebec City, Quebec, Canada

Kukavica-Ibrulj I, Levesque RC (2008) Animal models of chronic lung infection with Pseudomonas aeruginosa: useful tools for cystic fibrosis studies. Lab Anim 42:389–412

Kurosaka Y, Ishida Y, Yamamura E, Takase H, Otani T, Kumon H (2001) A non-surgical rat model of foreign body-associated urinary tract infection with Pseudomonas aeruginosa. Microbiol Immunol 45:9–15

Leewenhoeck A (1684) An Abstract of a Letter from Mr. Anthony Leewenhoeck at Delft, Dated Sep. 17. 1683. Containing Some Microscopical Observations, about Animals in the Scurf of the Teeth, the Substance Call'd Worms in the Nose, the Cuticula Consisting of Scales. Philos Trans 14:568–574

Leid JG, Costerton JW, Shirtliff ME, Gilmore MS Engelbert M (2002) Immunology of Staphylococcal biofilm infections in the eye: new tools to study biofilm endophthalmitis. DNA Cell Biol 21:405–413

Leknes KN, Selvig KA, Boe OE, Wikesjo UM (2005) Tissue reactions to sutures in the presence and absence of anti-infective therapy. J Clin Periodontol 32:130–138

Leung JW, Libby ED, Morck DW, McKay SG Liu Y, Lam K, Olson ME (2000) Is prophylactic ciprofloxacin effective in delaying biliary stent blockage? Gastrointest Endosc 52: 175–182

Li D, Gromov K, Soballe K, Puzas JE, O'Keefe RJ, Awad H, Drissi H, Schwarz EM (2008) Quantitative mouse model of implant-associated osteomyelitis and the kinetics of microbial growth, osteolysis, and humoral immunity. J Orthop Res 26:96–105

Marrie TJ, Costerton JW (1984) Scanning and transmission electron microscopy of in situ bacterial colonization of intravenous and intraarterial catheters. J Clin Microbiol 19:687–693

Marrie TJ, Nelligan J, Costerton JW (1982) A scanning and transmission electron microscopic study of an infected endocardial pacemaker lead. Circulation 66:1339–1341

Marsh PD (1995) The role of microbiology in models of dental caries. Adv Dent Res 9:244–254; discussion 255–269

McCoy WF, Bryers, JD, Robbins J, Costerton JW (1981) Observations of fouling biofilm formation. Can J Microbiol 27:910–917

Minardi D, Ghiselli R, Cirioni O Giacometti A, Kamysz W, Orlando, F. Silvestri C, Parri G, Kamysz E, Scalise G, Saba V, Giovanni M (2007) The antimicrobial peptide tachyplesin III coated alone and in combination with intraperitoneal piperacillin-tazobactam prevents ureteral stent Pseudomonas infection in a rat subcutaneous pouch model. Peptides 28:2293–2298

Monzon M, Garcia-Alvarez F, Lacleriga A, Gracia E, Leiva J, Oteiza C, Amorena B (2001) A simple infection model using pre-colonized implants to reproduce rat chronic Staphylococcus aureus osteomyelitis and study antibiotic treatment. J Orthop Res 19:820–826

Murillo O, Domenech A, Garcia A, Tubau F, Cabellos C, Gudiol F, Ariza J (2006) Efficacy of high doses of levofloxacin in experimental foreign-body infection by methicillin-susceptible Staphylococcus aureus. Antimicrob Agents Chemother 50:4011–4017

Nablo BJ, Prichard HL, Butler RD, Klitzman B, Schoenfisch MH (2005) Inhibition of implant-associated infections via nitric oxide release. Biomaterials 26:6984–6990

Nallapareddy SR, Singh KV, Sillanpaa J, Garsin DA, Hook M. Erlandsen SL, Murray BE (2006) Endocarditis and biofilm-associated pili of Enterococcus faecalis. J Clin Invest 116:2799–2807

Nishioka GJ, Jones JK, Triplett RG, Aufdemorte TB (1988) The role of bacterial-laden biofilms in infections of maxillofacial biomaterials. J Oral Maxillofac Surg 46:19–25

Orlando F, Ghiselli R, Cirioni O, Minardi D, Tomasinsig L, Mocchegiani F, Silvestri C, Skerlavaj B, Riva A, Muzzonigro G, Saba V, Scalise G, Zanetti M, Giacometti A (2008) BMAP-28 improves the efficacy of vancomycin in rat models of gram-positive cocci ureteral stent infection. Peptides 29:1118–1123

Rittman BE (1982) The effect of shear stress on biofilm loss rate. Biotechnol Bioeng 24:501–506

Rittmann BE, McCarty PL (1982) Model of steady-state-biofilm kinetics. Biotechnol Bioeng 24:2291

Robert R, Pihet M (2008) Conventional methods for the diagnosis of dermatophytosis. Mycopathologia 166:295–306

Romero R, Schaudinn C, Kusanovic JP, Gorur A, Gotsch F, Webster P, Nhan-Chang, CL, Erez O, Kim CJ, Espinoza J, Goncalves LF Vaisbuch E, Mazaki-Tovi S, Hassan SS, Costerton JW (2008) Detection of a microbial biofilm in intraamniotic infection. Am J Obstet Gynecol 198:135.e1–135.e5

Rosenthal MB (2007) Nonpayment for performance? Medicare's new reimbursement rule. N Engl J Med 357:1573–1575

Saidi IS, Biedlingmaier JF, Whelan P (1999) In vivo resistance to bacterial biofilm formation on tympanostomy tubes as a function of tube material. Otolaryngol Head Neck Surg 120:621–627

Saygun O, Agalar C, Aydinuraz K, Agalar F, Daphan C, Saygun M, Ceken S, Akkus A, Denkbas EB (2006) Gold and gold-palladium coated polypropylene grafts in a S. epidermidis wound infection model. J Surg Res 131:73–79

Schaber JA, Triffo, WJ. Suh SJ, Oliver JW, Hastert MC, Griswold JA, Auer M, Hamood AN, Rumbaugh KP (2007) Pseudomonas aeruginosa forms biofilms in acute infection independent of cell-to-cell signaling. Infect Immun 75:3715–3721

Smith TJ, Galm A, Chatterjee S, Wells R, Pedersen S, Parizi AM, Goodship AE Blunn GW (2006) Modulation of the soft tissue reactions to percutaneous orthopaedic implants. J Orthop Res 24:1377–1383

Swords WE, Moore ML, Godzicki L, Bukofzer G, Mitten MJ, VonCannon J (2004) Sialylation of lipooligosaccharides promotes biofilm formation by nontypeable Haemophilus influenzae. Infect Immun 72:106–113

Tanzer JM, Thompson AM, Grant LP, Vickerman MM, Scannapieco FA (2008) Streptococcus gordonii's sequenced strain CH1 glucosyltransferase determines persistent but not initial colonization of teeth of rats. Arch Oral Biol 53:133–140

van der Borden AJ, Maathuis PG, Engels E, Rakhorst G, van der Mei HC, Busscher HJ, Sharma PK (2007) Prevention of pin tract infection in external stainless steel fixator frames using electric current in a goat model. Biomaterials 28:2122–2126

Wolcott RD, Ehrlich GD (2008) Biofilms and chronic infections. JAMA 299:2682–2684
Xiong YQ, Willard J, Kadurugamuwa JL, Yu J, Francis KP Bayer AS (2005) Real-time in vivo bioluminescent imaging for evaluating the efficacy of antibiotics in a rat Staphylococcus aureus endocarditis model. Antimicrob Agents Chemother 49:380–387
Yanagihara K, Tomono K, Sawai T, Kuroki M, Kaneko Ohno H, Higashiyama Y, Miyazaki Y, Hirakata Y, Maesaki S, Kadota J, Tashiro T, Kohno S (2000) Combination therapy for chronic Pseudomonas aeruginosa respiratory infection associated with biofilm formation. J Antimicrob Chemother 46:69–72
Zaura-Arite E, van Marle J, ten Cate JM (2001) Conofocal microscopy study of undisturbed and chlorhexidine-treated dental biofilm. J Dent Res 80:1436–1440
Zegans ME, Becker HI, Budzik J, O'Toole G (2002) The role of bacterial biofilms in ocular infections. DNA Cell Biol 21:415–420

Chapter 17
Summary and Perspectives

Thomas Bjarnsholt, Peter Østrup Jensen, Claus Moser, and Niels Høiby

17.1 Summary

The individual chapters of this book have reviewed the current knowledge, perception, and implications of biofilm infections, how to study this bacterial phenomenon experimentally, and the development of new treatment strategies.

It is evident from reading the individual clinical chapters that the biofilm definition "*A coherent cluster of bacterial cells imbedded in a matrix – which are more tolerant to most antimicrobials and the host defense, than planktonic bacterial cells*" summarizes the core elements. All the visual evidences brought forward in the chapters demonstrate areas with dense distribution of multiple bacteria within the host. Some of the pictures visualize the matrix components, but even in the pictures with no visible matrix it is evident that the bacteria are close together. The clinical chapters all emphasize the extreme problems the clinician face when treating these infections. The two immunological chapters explain and discuss the significance of the innate and adaptive host response for the outcome of the biofilm infection. These chronic infections will eventually kill the patients if left untreated; however it is not a fast and rapid death caused by the direct action of the bacteria, but a slow degradation of the surrounding tissue, which mainly results from an inappropriate immune response and may even proceed when the patients are treated.

The purpose of this chapter is to summarize the information brought forward in this book as of the problems biofilm infections causes for diagnostics, treatment, prevention and experimental research.

T. Bjarnsholt (✉)
University of Copenhagen, Faculty of Health Sciences, Department of International Health, Immunology and Microbiology, Blegdamsvej 3B, DK-2200 Copenhagen N, Denmark; H:S Rigshospitalet, Department for Clinical Microbiology, afsnit 9301, Juliane Maries Vej 22, DK-2100 Copenhagen Ø, Denmark
e-mail: tbjarnsholt@sund.ku.dk

T. Bjarnsholt et al. (eds.), *Biofilm Infections*, DOI 10.1007/978-1-4419-6084-9_17,

17.1.1 Diagnostic Problems

The initial problem or challenge with all infections is to identify the infecting organisms and the focus of the infection. This is usually not a problem for acute infections since the bacteria are readily obtained by swapping or sampling the infected area or sampling body fluid or pus from the infected part or organ of the body. For chronic infections it is usually more problematic. An exception is cystic fibrosis (CF) (see Chapter 10) in which the easy accessible purulent sputum coughed up by the patients on a regular basis harbor the bacteria. In the other infections which are reviewed in this book, it is more problematic. When it comes to bacteria in chronic wounds, the clinician faces a real problem. Since the wound bed is attached to the patient it cannot readily be removed and homogenized for routine diagnostics. The routine sampling has been using either a swab or a biopsy; however both might fail to sample the bacteria as described in Chapter 2. The swab will only collect bacteria on the surface, not the bacteria embedded in the wound bed. On the other hand, since the bacteria are very heterogeneously distributed chances are that a biopsy fails to contain any bacteria. Also for implant and catheter-related infection (Chapters 5 and 6) diagnosing the bacteria prove difficult. Five to 10 years ago the bacteria on these surfaces and biofilm in general were considered unculturable. Here the problem is surface adherence, the bacteria simply attach extremely well to the surface of the foreign bodies. It is not that they are unculturable, they have to be released from the surface, and vigorously vortexing or even mechanical scraping is not enough. The implant or catheter has to be treated with ultra sound (sonication) to release the bacteria. In the light of this it can never be ruled out that some bacteria within the persistent biofilm may be so dormant that they prove impossible to grow on any media; however, the chance that such an almost dead cell cause pathogenesis is extremely low. Such a dormant bacterium would not be able to produce any virulence factors and would only be able to resist the antimicrobial action of the host defense if protected by growing in a biofilm.

The problems of diagnosing the bacteria in these chronic infections are far from solved. Today bacteria can be detected by culturing, PCR, microscopy, or diagnostic imaging. Each method has its advantages and limitations. For culturing, the problem is to collect the bacteria, either next to the surface, which is sampled, or from the catheter or implant. On the other hand if bacteria are cultured resistance against antibiotics can easily be investigated.

PCR will detect even tiny amounts of DNA or RNA available in the sample. The first problem is to avoid eliminating the minor fraction of prokaryotic DNA/RNA by the eukaryotic DNA/RNA, which might either interfere with the PCR reaction (fairly easy to solve) or outcompete the capacity for DNA/RNA extraction. Another problem with PCR is that every DNA/RNA fragment homologous to the primers will be detected, including defragmented DNA/RNA, which may generate false positives. Additionally, just because a bacterium is present does not necessarily indicate that it contributes to the pathogenesis of the infection it might not need to be treated.

Microscopy enables direct visualization of the infecting bacteria and the surrounding tissue and inflammatory cells. Again the bacteria need to be present in the collected sample, which means many biopsies need to be analyzed for a correct diagnostics of, e.g., a wound. On the other hand if only a few bacteria are present they might be very hard and statistically impossible to observe using traditional staining such as Gram. This can be facilitated and diagnosed to the genus and even to the species level using fluorescent probes for FISH.

The final problem which may be the biggest challenge for biofilm infections in the coming years is to reveal the significance of the numerous identified microorganisms by molecular techniques. As a tool the modified Koch's criteria has been suggested and we hope the use of specific immune responses and relevant animal models will be implemented to a greater extend in biofilm research.

17.1.2 Treatment Problems

When the initial problems of diagnosing the infecting organism and revealing the significance of a chronic infection has been solved, the next problem arise; how to treat the infection. It is evident from all the clinical chapters in this book that biofilm infections cannot be treated similar to acute infections. The most efficient way to treat a biofilm infection is to mechanically remove the infection. This is sometimes possible if the focus is a catheter, an implant or an infected organ eligible for transplantation. However, this is not always possible and without risk of complications for the patients. Two main strategies to fight bacterial biofilm infections is either (i) early aggressive antibiotic treatment before the biofilm is formed, (ii) chronic suppressing antibiotic treatment when the biofilm is established, if it cannot be removed physically. As implicated in many of the chapters administering antibiotics against bacterial biofilms necessitates combinations of several different antibiotics, in high doses and for an extended period of time as described in Chapter 13.

What must be avoided is to treat a biofilm infection with an acute infection regime of single drugs in minimal doses for short periods. This may induce even further resistance and tolerance and even create opportunities for new bacteria to chronically infect the focus.

As described in Chapter 15, novel intervention and treatment strategies are on their way such as compounds blocking the inter-bacterial signaling, quorum sensing (QS), and biofilm matrix degradable drugs.

17.1.3 Prevention

The most efficient tool so far is to prevent the chronic infection in the first place. Normally the skin of the human body is acting as a barrier preventing bacterial invasion of the body. During surgery and insertion of implants this natural barrier is compromised. Due to this the surgeons have to take extreme caution not to introduce bacteria when inserting implants, injecting dermal fillers, etc. Ultra clean operation theaters, tools, cloth implants, etc. are necessary.

This is in combination with prophylactic administered antibiotics or maybe QS inhibitors in the future might prevent several chronic infections.

This can be combined with antibacterial coating of the inserted surfaces.

However in some instances, patient groups such as cystic fibrosis patients (see Chapter 10) and here early aggressive prophylactic treatment is necessary. When a chronic infection is unavoidably established chronic suppressive therapy is needed.

The best known example of prevention is against the daily-formed dental biofilms (Chapter 4). Here caries is avoidable if the teeth are mechanically cleaned for biofilm using a toothbrush.

17.1.4 Experimental

It is quite evident reading the clinical chapters of this book that more research is needed. The two last chapters of the book describe various experimental setups for studying biofilms and their related complications (Chapter 15 for in vitro methods and Chapter 16 for in vivo methods).

These two chapters describe the most used and verified methods and setups for biofilm investigation. We would like to emphasize that a model system should be chosen specifically for chronic infections and components such as antibiotic tolerance or biofilm formation, to be investigated. In addition, it is important to support the development of model systems with increased resemblance to the infection being investigated in order to narrow the gap between experimental results and clinical relevance.

Index

T. Bjarnsholt et al. (eds.), *Biofilm Infections*, DOI 10.1007/978-1-4419-6084-9,

D

GPSR Compliance

The European Union's (EU) General Product Safety Regulation (GPSR) is a set of rules that requires consumer products to be safe and our obligations to ensure this.

If you have any concerns about our products, you can contact us on ProductSafety@springernature.com

In case Publisher is established outside the EU, the EU authorized representative is:

Springer Nature Customer Service Center GmbH
Europaplatz 3
69115 Heidelberg, Germany

Batch number: 10755466

Printed by Printforce, the Netherlands